Introduction to Refractories for Iron- and Steelmaking

Subir Biswas • Debasish Sarkar

Introduction to Refractories for Iron- and Steelmaking

Springer

Subir Biswas
Refractory Technology Group,
R&D and Scientific Services
Tata Steel (India)
Jamshedpur, Jharkhand, India

Debasish Sarkar
Department of Ceramic Engineering
National Institute of Technology Rourkela
Rourkela, Odisha, India

ISBN 978-3-030-43806-7 ISBN 978-3-030-43807-4 (eBook)
https://doi.org/10.1007/978-3-030-43807-4

This Springer imprint is published by the registered company Springer Nature Switzerland AG
The registered company address is: Gewerbestrasse 11, 6330 Cham, Switzerland

Subir Biswas dedicates this book to the loving memory of his Parents & Parents in Law.

&

Debasish Sarkar dedicates this book with a great sense of Devotion to "Lord Krishna".

Preface

Despite several classic literatures including books, journal papers and internet resources, extensive demand motivated to fulfil the gap in between bookish knowledge and shop-floor experience, and thus we decided to write this *textbook* for students and professionals. In-depth thought process instigated to make an excellent link in between fundamental concepts of refractories and their judicious use of iron and steel making, prime focus of this book. So, it provides a bridge between ceramist and metallurgist in the perspective of manufacturing of iron and steel for both small- and large-scale industries. It covers basic understanding on refractory selection and operational process to mitigate the production effort. In this context, the entire book has been divided into 12 chapters, starting from "Refractories for Iron and Steel Plant" (Chap. 1) followed by "Iron and Steel Making Process" (Chap. 2); "Blast Furnace Refractory" (Chap. 3); "Hot Stove and Hot Air Carrying System" (Chap. 4); "Refractory Practice in EAF" (Chap. 5); "Refractory for Hot Metal Transport and Desulfurization" (Chap. 6); "BOF Refractory" (Chap. 7); "Refractory for Secondary Refining of Steel" (Chap. 8); "Refractory in Ladle Flow Control and Purging System" (Chap. 9); "Refractory for Casting" (Chap. 10); "Modern Refractory Practice for Clean Steel" (Chap. 11); and "Advance Material Design and Installation Practices" (Chap. 12). We believe that such concise literature is effectively helpful for a wide range of community including students, academia, researchers, novice graduate trainees, senior managers, raw material processing, refractory manufacturers, refractory procurement personalities and last but not least iron and steel manufacturers.

Chapter 1

Refractories are the essential lining materials for working interfaces and backup zone of furnaces throughout the manufacturing of iron and steel, in specific sequential and consecutive operation of forming, holding, mixing and transporting hot metal, liquid steel and slag. Despite refractory–metal direct interactions, refractory has to successively experience high temperature and corrosive environment through flues, stack or shaft and ducts. Prior to consider the refractories, recent market trend, designing parameters including predominant mechanical and thermal behaviour are explicitly discussed to provide a better insight of the subject. In this consequence, this chapter deals with the classification and description of acidic, neutral and basic refractories for different environment and application zone for iron and steel-making processes. Modern class of both shaped and unshaped refractories are highlighted starting from raw materials to installation practice. Refractory corrosion mechanism influenced by blast furnace slag, primary and secondary steel-making slag are analysed in order to understand and develop next-generation refractories.

Chapter 2

With extensive changing in the operating practice due to stringent control in product quality, introduction of new product mix and demand for high productivity, it is indeed to upgrade the refractory quality to cope up with the changed environment and refractory life improvement. Hence, starting with the master plan, it was decided to provide a brief introduction on modern iron and steel-making practice along with background of refractory choice to amalgamate the relation of refractory performance with changed escalating demand on safe performance. As small blast furnaces have been closed and replaced by large size furnaces, open-hearth steel making has been replaced by high productive basic oxygen furnaces (BOF), RH (Rurhstahl Heraeus)—degasser is the most popular device for making ultra-low carbon steel. Additional composition adjustment by sealed argon bubbling with oxygen blowing (CAS–OB) has started to operate in many integrated and large size steel plant to produce high quality alloy steel, the operating processes of those equipment have been discussed in detail.

Chapter 3

Blast furnace is likely to continue as the most efficient route to produce pig iron for its high productivity and cost optimization, for many years to come. Although a small part of iron making has been supplemented by alternate iron-making process like direct reduced iron (DRI) and smelting reduction; however, the replacement of blast furnace for steel making is a distant dream. Thus, the refractory for modern blast furnace including cast house refractories is a critical issue and is discussed. Different refractory maintenance practices prolong campaign of blast furnace life with improved productivity, and thus their key features are enrolled. Starting from the reduction of coke rate to effective hot metal through runner is a challenging task to the operator in the mind of production cost reduction and in turn, efficiency of blast furnace is analyzed.

Chapter 4

Hot stove is a thermal heat regenerator to produce and supply constant hot air to blast furnace. The present demand of increased hot metal productivity through blast furnace route requires high hot blast temperature more than 1200 °C, and it requires to optimize and upgrade the quality and design of refractory. Thus, the design of stove and their checkers have undergone major changes and demands in installation of superior quality refractories. Advantages in changed design and upgrade refractory quality from alumina dome to silica, installation of ceramic burners are extravagantly explained in this chapter. Latest development in stove design is top combustion stove that can deliver hot air more than 1300 °C; it is obvious that this operational feature has economic benefit over the conventional one. Detailed discussion has been done on critical refractory application in those type of stoves that eventually help to generate knowledge on how one can use effectively hot blast more than 1300 °C for as large as 5000 m^3 blast furnace.

Chapter 5

Owing to several advantages including flexibility to produce several grades steel, precise control, cleaner environment and minimum installation space, many mini steel plants are producing steel through electric arc furnace (EAF) route. This protocol is benefited to use sponge iron and higher share of scrap utilization. EAF supports 35% high quality alloy steel production around the globe. Different constructional and operational features, side wall and bottom/hearth refractory properties and refractory design for bottom tapping are discussed in the perspective of effective steel processing. Despite the use of three carbon electrodes in EAF, only one electrode-based direct current (DC) arc furnace is highlighted. Refractory corrosion mechanism in the presence of different impurities and subsequent slag interaction are discussed through relevant phase diagrams in order to select the refractory depending on steel compositions. Eventually, the state-of-the-art operating practice and refractory performances are summarized.

Chapter 6

Continuous steel production demands uninterrupted iron supply for steel vessels. Limited 60–80-ton hot metal ladle has several disadvantages, and thus high capacity up to 300-ton torpedo ladle is being introduced to overcome the hot metal ladle limitations. Herein, explicitly focused on the design aspects, competitive early days and recent refractory lining, plausible refractory deterioration factors and operational influence on the refractory performance to operate the torpedo ladle effectively.

Torpedo ladle refractory development and lining modules for higher campaign, effect of insulation refractories, refractory maintenance protocols, and future challenges are encountered to improve the working environment. Despite torpedo ladle, desulphurization in hot metal ladle and in situ process reactions with refractories, and wear mechanisms are discussed to fulfil the knowledge gap.

Chapter 7

Is it BOF a "heart" of any steel sector? If yes, it is mandate to take extra care to keep it healthy through refractory management, lining practice and effective operational protocols. In this backdrop, this chapter concentrates on the refractory designing for the vessel and tap hole sleeve, and analyse probable wear mechanism through ternary phase diagram. Influence of gas purging and slag splashing on the refractory life of different zones, followed by zonal lining concept and role of antioxidant are also being discussed systematically. Topics of this chapter concerns on state-of-the-art refractory maintenance practice to prolong campaign life and optimization of cost. Importance of protective slag coating on refractory performances, breakthrough in achieving vessel life >20,000 heats by introducing slag splashing and its influence in refractory performance have been explained in detail. MgO-C refractory lining wear is an inevitable circumstance and is discussed in consideration of refractory–slag interaction, reduction of MgO, metal infiltration with respect to critical pore diameter, thermal stress and spalling phenomena, impact caused by charging and mechanical erosion and abrasion. Basic philosophy has been highlighted to improve the MgO-C refractory for BOF lining. Herringbone vessel relining and maintenance practice by gunning and patching are emphasized to improve the BOF lining life.

Chapter 8

Modern steel ladle is used not only to merely transport liquid steel from BOF to caster, but also to act as a reactor vessel where refining of steel takes place to reduce impurities, carbon, alloy addition and killing (Al, Si, bi-metal) of steel to reduce oxygen and other gasses; this comprehends the necessity and critical role of refractories. Substantial modification and development have been brought into the refractory quality to eliminate carbon (C) and oxygen pick-up into steel from refractory lining for ladle furnace (LF) and CAS–OB processes. Thus, refractory design in steel ladles, energy saving issues, variation of thermal conductivity with respect to conventional graphite and nanocarbon, thermal stress-assisted cracks and refractory failure are discussed. RH–Degasser is in limelight to produce ultra-low C and high alloyed steel with an efficient productivity. Most suitable refractory practice has been suggested in consideration of slag–refractory interaction that eventually facilitate prolong ladle life and high quality steel.

Chapter 9

Continuous strive towards new technology introduced reliable flow control systems that enable safe and prolong tapping liquid steel from ladle to tundish with consistent and high casting rate. Thus, synchronization of operational parameters, appropriate refractory and analysis of probable failure are essential and encountered. Erosion of bore and corrosion of sliding surface of the slide plates, radial cracking of plates and metal leakage are the most critical issues as addressed with mention of optimize refractory quality of plates and nozzles to prolong casting time. In order to homogenous the molten steel, gas purging is an important aspect and thus different types of purging plugs, their design, refractory quality and the effect of casting sequences had also been discussed. Thermal-stressed failure and premature wear that may reduce the operational safety and promote shop-floor safety hazards are also highlighted.

Chapter 10

Continuous casting (con-cast) demands uninterrupted product length whereas in-got process is popular for batch process. In modern production protocol, in-got casting process is limited and more than 90% of global steel production follows continuous casting of slab, bloom, billet, thin slab, etc. A brief understanding including mould, pouring, choice of refractory and their failure, advantages and disadvantages of in-got casting are discussed. In spite of batch process, exhaustive con-cast in specific operational features, and refractory design and failures are delivered in the perspective of minimizing casting defects and higher yield of finished steel. Tundish is the indispensable component in the con-cast process, and it is the refractory lined last vessel before solidification of molten steel in the mould, in which different refractories in hot face and in backup for easy de-sculling are described. Causes of clogging, usage of anti-clogging refractory materials in sub-entry nozzle and other "black refractories" with latest design of inert gas purging in SEN had been mentioned.

Chapter 11

Ductility facilities reforming but elastic modulus provides the strength of steel. It all depends on constituents and phases in steel. The previous discussion emphasizes steel production is a multi-step process that has to pass through the exhaustive environment and thus possibilities of inclusions despite targeted compositions. This drawback reduces the steel performance, and therefore modern amenities demand clean steel for an extensive range of applications. An excellent choice of new class of modern refractory is required to avoid contamination and maintain desired properties of steel. Detailed discussion is concentrated on probable inclusions by refractories, the stability of refractory oxides to produce clean steel and low carbon-containing refractories through the adoption of nanotechnology like using nanocarbon and graphene. Magnesia–alumina–graphite refractories and spinel refractory are highlighted in view of the carbon reduction in a ladle and continuous casting refractories to make clean steel. Several operational features and refractory choice for large tundish including hydrogen pick-up minimization in steel through non-aqueous resin-bonded dry-vibrating mass (DVM) is encountered and the benefits of using dry vibratable mass replacing basic spray material for easy de-sculling and achieving longer sequence life are discussed.

Chapter 12

Design terminology is not only confined in a particular shape rather performance of applied components depend on the geometry and material design, together. In early days, industrial furnaces and vessels were made out of brickwork, but clean steel asks for total package including refractory composition, shape and installation practice for continuous and long period operation without interruptions for repairs. In order to accomplish steel sector demand, latest jointless monolithic especially colloidal silica-bonded cement-free castable, alumina-oxi-carbide-bonded castable, magnesia containing castable, chrome-free castable are encountered and analysed. Despite different classic endless lining protocols, installation and repair methodologies are systematically discussed. Uniform microwave-assisted heating and drying opens up a new installation practice for endless working lining that is capable of replacing partial damage zone without major replacement of original lining. Advantages of flame gunning over conventional wet gunning differentiate the performance of refractory lining and also reduction of down time of the equipment for repair and maintenance. Ceramic welding process is emphasized in which base refractory hot surface is being repaired through metal powder at elevated temperature, and exothermic reaction facilitates permanent bond and repairs the defective refractory zones.

In brief, this book has been written after gathering 25 years of hard-earned knowledge and is full of amalgamation of introductory knowledge to advance refractory practice in the outlook of iron to low carbon content steel manufacturing. So, we hope that it can serve the purpose as a textbook and will also provide a quick answer to repeatedly posed question in real-life application.

Jamshedpur, Jharkhand, India Subir Biswas
Rourkela, Odisha, India Debasish Sarkar

Acknowledgements

We would like to convey our heartfelt thanks to our parents and family members for their constant endorsement and motivation throughout the journey of writing the book. We would like to thank our students and friends for their uninterrupted co-operative actions in manuscript preparation. Thanks to PhD scholar Sarath Chandra Katakam, Laboratory of Materials Processing and Engineering, Department of Ceramic Engineering, National Institute of Technology, Rourkela, Odisha, India, for his sincere help to complete the manuscript.

We would like to acknowledge Tata Steel Ltd, Jamshedpur; NIT Rourkela, Odisha, India; Department of Science and Technology (DST/TSG/Ceramic/2011/142-G, EEQ/2017/000028), India; and Board of Research in Nuclear Sciences (BRNS, 2012/34/46/BRNS), India, for their support.

We would like to thank all the researchers for their contributions to the scientific society, which are the building blocks of the conceptual knowledge of this textbook.

Eventually, our sincere apology to those whose names are inadvertently not mentioned.

Why This Book?

This book provides unique and exhaustive topics compared to the existence of refractory books and discusses elaborately to accomplish the demand of students, shop-floor professionals, researchers and academia. Understanding the raw material selection, refractory design, tailor-made refractory development, refractory properties and their mode of applications are encountered in compliance with the objectives of iron and steel making. Cumulative information under one umbrella resolves a bridge-gap; main focus of this book. Thus, the book is considered as:

- A textbook for UG/PG students to understand the modern refractory practices of iron and steel making
- Amalgamation of interdisciplinary knowledge to boost up the refractory research
- Exhaustive modern iron and steel-making information enables refractory selection protocols
- Refractory installation and performance analysis for blast furnace to continuous casting refractories
- Excellent knowledge resource for R&D and shop floor to solve the refractory failure problems

Contents

About the Authors

Subir Biswas is Head of Refractory Technology Group at Tata Steel Ltd., Jamshedpur, India. He has held this position since 2009. Subir Biswas is graduated from the University of Calcutta, India, in 1985 with B.Tech (Ceramic Technology) and joined Carborundum Universal Ltd., India, as graduate engineer in R&D center of super refractory division. He joined Tata Steel Ltd. in 1996 and has moved through a number of positions of refractory maintenance, research and development in iron making and steel making. Subir was responsible for new refractory product development and technology, selection of refractories, application of modern refractory in steel plants for over 30 years. He has vast experience in development and manufacturing of spinel, silicon carbide, dolomite refractories and low-cement castables. He worked on advanced characterization of refractories, quality assurance and modern installation techniques of refractories in steel plant, and he was involved in various expansion projects on refractory installation in commissioning of blast furnaces and re-heating furnaces at Tata Steel. Subir is a life member of Indian Ceramic Society and Indian Institute of Ceramics and he has published appreciable number of journal papers on refractories development and installation of his credit.

Debasish Sarkar Professor and Head, Department of Ceramic Engineering, National Institute of Technology, Rourkela, Odisha, India, has 25 years of academic, research and industrial experience and published 75 peer-reviewed international journal papers and 3 Korean patents. While national patent concerns, a patent on graphene fortified MgO-C refractory for low carbon steel in collaboration with Tata Steel Limited, Jamshedpur, India, and a patent on zirconia-toughened alumina femoral head and acetabular socket for hip replacement in collaboration with IISc. Bangalore, India, are in pipeline. Professor Sarkar has written two popular books *Nanostructured Ceramics: Characterization and Analysis* by CRC Press, USA, 2018, and *Ceramic Processing: Industrial Practices* by CRC Press, USA, 2019; both grasp attention of global students, material research community and ceramic industries. Debasish is professionally involved in *Modelling, Design and Failure Analysis* of *Structural Ceramics* through association with empowered National and International experts.

Chapter 1
Refractories for Iron and Steel Plant

1.1 Introduction

Incompatible marriage results in divorce. Thus, competency and understanding expedite long life. Competitive marriage in "Refractory" and "Steel" can reduce the refractory consumption per ton of steel is a dream of all steel manufacturers. In this perspective, basic understanding comprising shaped and unshaped refractories with respect to market demand and practices, classifications, refractory designing parameters, high-temperature phase transformation behaviour and properties, effect of impurities, and corrosion phenomena are discussed. Furthermore, a brief global steel market scenario is emphasized.

1.1.1 Scenario of World Steel Production and Refractory Demand

World Steel Association published Global 2018 crude steel production data. The retrieved data for top ten steel-producing countries is given in Table 1.1. India is placed in second among high output ten countries like China, India, Japan, United States, South Korea, Russia, Germany and Turkey, with a wide gap of steel production by China [1]. Growing demand in Asian countries provokes to increase the production capacity; however, downward experience is observed in Japan steel sector economy during 2018. In order to fulfill such enormous steel production, billion tons of refractories are produced around the globe. Top players like RHI AG (Austria), VESUVIUS (UK), Magnesita (UK), KROSAKI (Japan), SHINAGAWA (Japan), Imerys (North France), HWI (USA), MORGAN CRUCIBLE (UK), SAINT-GOBAIN (France) and INTOCAST (Germany) primely monitor major share of the global refractory market [2].

S. Biswas, D. Sarkar, *Introduction to Refractories for Iron- and Steelmaking*,
https://doi.org/10.1007/978-3-030-43807-4_1

Table 1.1 Recent steel production scenario for top ten countries

Top 10 steel-producing countries				
Rank	Country	2018(Mt)	2017(Mt)	% 2018/2017
1	China	928.3	870.9	6.6
2	India	106.5	101.5	4.9
3	Japan	104.3	104.7	−0.3
4	United States	86.7	81.6	6.2
5	South Korea	72.5	71.0	2.0
6	Russia(e)	71.7	71.5	0.3
7	Germany(e)	42.4	43.3	−2.0
8	Turkey	37.3	37.5	−0.6
9	Brazil	34.7	34.4	1.1
10	Iran(e)	25.0	21.2	17.7

Source: World Steel Association

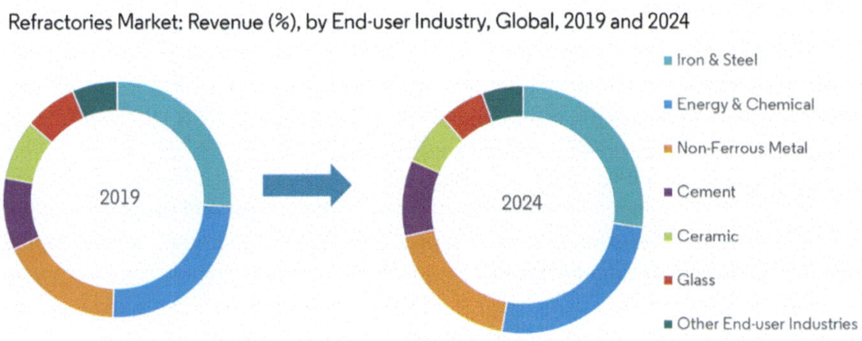

Fig. 1.1 A projected global refractories market revenue sharing by the different end-user industries in the duration of 2019–2024 [2]

Moreover, the million refractory manufacturers are actively supplying different classes of refractories around the globe despite these giant participants. In current report, the anticipated refractory market growth volume is around 5% registered by compound annual growth rate (CAGR) for the forecast duration of 2019–2024, as shown in Fig. 1.1. The prime philosophy is described as because of growing production of non-ferrous materials, increasing large infrastructure projects in emerging markets, and upcoming demand from the glass industries. However, the environmental awareness and laws restricts the disposal of refractories that eventually may hinder the actual market growth.

In spite of other end-user industries, iron and steel sector consumes lion-share market of refractories that can withstand a wide range of temperatures 260–1850 °C without major alteration of their physical properties. The prime applications of refractories in the iron and steel industry include usage in internal linings of furnaces to make iron and steel, in furnaces for heating steel before further processing, in vessels for holding and transporting metal and slag, in the flues or stacks conducted through hot gases, and others are listed. Change in global political scenario may improve the confidence and investment to fulfill the forecasted CAGR.

When we are addressing the refractory market in India as a second steel manufacturer in world, steel sector consumes near to 70% refractories, and further it is expected to increase to fulfill the cumulative steel production demand near to 300 mt in 2025. Indian refractory industry has potential to make promising product for steelmakers; however, couple of integral factors are indeed to synchronize to boost up the use of domestic raw materials and refractory market [3].

Steps to be taken to facilitate this include:

- Correction in the trade duty structure of raw material and finished products.
- Domestic procurement of products required in turnkey package.
- Focus on making domestic products cost competitive.
- Increase in export customer base for products made with domestic raw materials.

A win–win industrial environment is mandatory to establish continuous business and effective outcome for refractory manufacturer and steel-making organization, as supplier like to try supply more, and end-user try to optimize the refractory consumption per ton of steel. In general, the long-term total refractory consumption for steel making is expected to be in the range of 5–10 kg/ton, which essentially follows the trend in Japan. However, current regional rates of specific refractory consumption (kg refractories/ton steel) are China 20 kg/t; Europe and Americas 10 kg/t; Japan 8 kg/t and India 10 kg/t [2]. From this above discussion, one can envisage why refractory study is important in the perspective of iron and steel manufacturing around the globe. This refractory practice is not confined in only shaped or unshaped refractories, rather it can be classified in broader sense as acidic, basic and neutral refractories that are extensively used in different regions, and discussed systematically.

1.1.2 Modern Refractory Practices

Silica refractory is in forefront for coke oven, considered as the first region where refractory used in a steel sector to make coke from coal, as coke is an essential feed for blast furnace, but such refractory has limited interaction with iron or steel. Predominately, silica brick is the master component for coke oven lining; brief of such refractory is discussed in Sect. 1.4.1. Despite selective information on silica refractory, all other refractories are critically considered and analysed in the perspective of application zone, interaction with solid, liquid or gas during processing of iron and steel, probable installation, failure, etc. Application zone, subzone and type of all probable refractories are listed in Table 1.2 that provide an overall idea about essential demand of refractories in modern installation practice for iron and steel industries.

Table 1.2 Modern refractory practices starting from iron manufacturing to steel casting

Vessels/zones/components		Subzone	Type of refractories: Shaped	Type of refractories: Unshaped
Blast furnace		Stack	Al_2O_3, SiC, Si_3N_4-bonded SiC, Graphite	–
Blast furnace		Bosh	Al_2O_3, SiC, Si_3N_4-bonded SiC, Graphite	–
Blast furnace		Belly	SiC, Si_3N_4-bonded SiC, Sailon-bonded SiC, Graphite	–
Blast furnace		T J Area	Carbon blocks, Sialon-Bonded SiC	–
Blast furnace		Hearth	Carbon blocks with ceramic cup bricks	–
Blast furnace		Tap hole	–	Tap hole clay
Blast furnace		Trough	High Al_2O_3 bricks	Trough castable, ramming mass
Hot stove		Inner lining/checkers	Silica, HA bricks, Insulation bricks	Ramming mass, castables
Hot stove		Burner	Mullite, Mullite-andalusite bricks	Castables
Hot stove		Hot blast transport	HA bricks, Insulation bricks	Castables
EAF		Roof	Basic (Magnesite, Mag-Chrome), HA bricks, Precast blocks	Gunning castables
EAF		Side wall	MgO-C bricks, Carbon (C) blocks	Ramming masses, castables (back up)
EAF		Hearth	MgO-C bricks, C blocks	
EAF		Tapping spout	Precast blocks	Ramming masses
Torpedo ladle		Vessel lining	ASC, AC, HA bricks, Insulation bricks, insulation boards	Castable in mouth, Ramming masses and castable in intermediate lining
BOF		Vessel	MgO-C Mag-Chrome, Magnesite	Carbon ramming mass
BOF		Tap hole sleeve	Basic refractory precast	Ramming masses
Secondary refining	Si killed steel	Slag zone	MgO-C, Dolomite	
Secondary refining	Si killed steel	Metal zone	Dolomite	Castable in back up
Secondary refining	Si killed steel	Bottom	Dolomite, Al_2O_3-MgO-C (AMC)	Ramming mass, castable in back up
Secondary refining	Al killed steel	Slag zone	MgO-C	
Secondary refining	Al killed steel	Metal zone	MgO-C, AMC, Spinel, Mag- chrome in back up.	Castable

(continued)

Table 1.2 (continued)

Vessels/zones/ components		Subzone	Type of refractories Shaped	Unshaped
		Bottom	AMC,	Ramming mass
	Ca killed steel	Slag zone	MgO-C	Castable (back up)
		Metal zone	Spinel, MgO–C	
		Bottom	AMC	Basic ramming mass
RH-Degasser		Upper vessel	Mag-Chrome	
		Lower Vessel	Mag-Chrome (direct bonded, rebonded)	Basic ramming mass
		Snorkels	Direct-bonded and rebonded Mag-Chrome	Castable and Ramming mass.
Flow control and purging system		Slide plate, Well blocks, Ladle and collector nozzle	Al_2O_3-C, HA Al_2O_3-ZrO_2-C	–
		Insert	ZrO_2	–
Casting	Ingot		HA bricks and shapes	Ramming mass
	Continuous	Tundish	Dam and wear pre-cast Precast flow control devise.	Backup castable lining, hot face spray, dry vibratable mass
		Black refractory	Mono-block stopper, shroud, sub entry nozzle	–

1.2 Definition and Classification of Refractories

Classification of refractories are prerequisite to understand why alumina-silicate refractory is a suitable practice for blast furnace, but MgO-C brick for BOF. However, such classification depends on the refractory composition, slag chemistry, physical properties and mode of applications.

In consideration of different class of refractories used in iron making to steel casting, a cumulative representation of different class of refractories has been given in Fig. 1.2. Base composition of pyramid represents oxide refractories, but an upward direction signifies non-oxide refractories such as SiC and carbon-based refractories, whereas peak point is for carbon block only [4]. However, a basic idea on both shaped and unshaped refractories and their characteristics with respect to composition, temperature and application can provide in-depth understanding of future discussion.

In shaped refractories, prime products are unfired, fired, fusion cast and insulating (porous) bricks, in which first set of dense refractory is using for working lining.

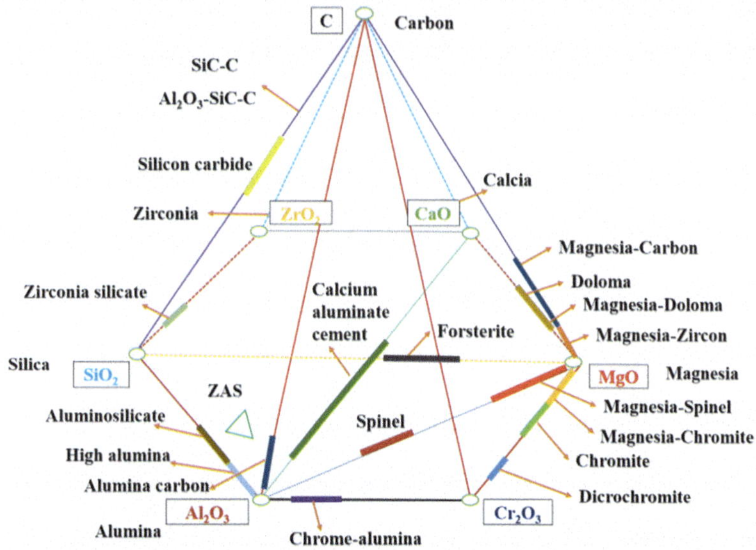

Fig. 1.2 Cumulative representation of probable refractories w.r.t. chemical compositions

However, insulating brick possesses low thermal conductivity and provides thermal barrier in the backup lining. Predominate unshaped refractories are castables, gunning mixes, ramming mixes, patching and coating materials. Castables are made of wide size range of refractory aggregates and mixed with cement as bond. The aggregate is mixed with water and casting is done to form rigid mass in lining. In modern refractory practice, no-cement castable is also introduced where different sol is used to cast and form in situ high-temperature phases. Some lightweight porous castable is also used as backup lining to maintain the thermal barrier. Gunning mixes are mixer of refractory particles which consist of binders and easily stick to the applied surface after gunning. This may be either cold or hot gunning. Ramming terminology refers a pneumatic ramming is required during application of binder (organic or inorganic) added refractory aggregates that eventually become hard through formation of ceramic bond. Mortar is a binder-added relative finer refractory mass applied through trawling that helps to join within two shaped refractories. Analogous to the mortar, the patching and coating materials are employed through spraying. Despite such class of refractories, insulating alumino silicate ceramic drawn fibres produce several refractory articles including blankets, felt, rope, 2–10-mm-thick papers, etc.

Only definite design practice is not enough to select refractory, rather cumulative accepting the high-temperature thermo-mechanical properties and chemical compatibility with iron and steel aids to select and develop new class of refractories. For example, more often acidic silica refractory has high abrasion resistance at high temperature and high mechanical strength under load. It possesses very good spalling resistance >600 °C, however, experience poor thermal shock resistance <600 °C. It has excellent corrosion resistance to acid slag and to iron oxide.

Depending on the alumina content, high-alumina (neutral) brick experiences volume stability at high temperature, thermal expansion is low up to below 80% Al_2O_3 (major phase mullite), but high above 80% Al_2O_3 (corundum phase), good load-bearing capacity and resistance to acid slag, but relatively less resistance to basic slag. Basic refractories, more popular magnesite-based refractories, exhibit high refractoriness, high basic slag corrosion resistance, high strength and refractoriness under load. Furthermore, non-oxide SiC is an excellent choice for high-temperature application, but in the presence of oxygen it oxidizes rapidly beyond 800 °C. It has also excellent abrasion resistance, thermal shock resistance and corrosion resistance towards hot metal and acid slag, as well. In consideration of different physical, thermal, mechanical, thermo-mechanical properties, are elaborated to understand how one can select the designing parameters for particular choice of application.

1.3 Refractory Design Parameters and Testing

Refractories are non-metallic and inorganic ceramic materials, composed of oxides, nitrides and carbides or combination of them and which can withstand very high temperature (1400 to as high as >2500 °C), can prevent corrosion to liquid metal and slag and are resistant to hot abrasion and erosion. Other essential properties of refractory materials are:

- Resistance to thermal fluctuation.
- Resistance to high-temperature deformation (creep).
- Volume stability at elevated temperature.
- High load-bearing capacity.

While considering the critical definition of refractory materials, background of above features is considered and discussed that eventually support to comprehend the refractory selection for particular zone of interest. In brief, the resistance to thermal fluctuation of refractory depends on the elastic modulus, work of facture, thermal expansion coefficient and thermal conductivity, and resistance to high-temperature deformation (creep) is controlled by the fundamental properties like thermo-elastic behaviour, deformation under load at high temperature (RUL) and on-site operating temperature analogous to PCE. The volume stability at elevated temperature associates with permanent linear change (PLC), reversible thermal expansion (RTE), as well as the operating temperature, and finally high load-bearing capacity depends on apparent porosity, bulk density, crushing strength, linear change, refractory composition (presence of glassy phase), etc. However, these properties are further controlled by the composition, microstructure, porosity content and their distribution. Basic philosophy staring from the physical properties to critical thermal and thermo-mechanical properties is encountered in order to emphasize as a backbone of refractory properties and enable "refractory design" for particular zone of interest during iron and steel making.

1.3.1 Density and Porosity

Rapid inspection of bulk density (BD) and apparent porosity (AP) assures the first-stage quality control of shaped and unshaped refractories. Standard testing protocol specifies a method for the determination of the bulk density, apparent (open) porosity, and true (open + closed) porosity of dense shaped refractory products. Closed porosity comes across as particular voids entrapped within the matrix and difficult to penetrate conventional liquid like water or kerosene in normal atmospheric pressure and temperature, whereas Archimedes process easily estimates the open porosity content, only. In brief, following equations are used to determine these parameters:

The bulk density,

$$\rho_b = \frac{m_1}{(m_3 - m_2)} \times \rho_{liq}\ g/cc \quad (1.1)$$

The apparent porosity,

$$\rho_a = \frac{(m_3 - m_1)}{(m_3 - m_2)} \times 100\% \quad (1.2)$$

The true porosity,

$$\rho_t = \frac{(\rho_{theoretical} - \rho_b)}{\rho_{theoretical}} \times 100\% \quad (1.3)$$

where m_1 is the mass of the dry test piece, m_2 is the apparent mass of the suspended test piece, m_3 is the mass of the soaked test piece, ρ_{liq} is the theoretical density and $\rho_{theoretical}$ is the theoretical density.

High-dense compact enhances the elastic modulus and load-bearing capacity and avoids early-stage corrosion aggravated by molten metal. However, the porous body reduces the thermal conductivity that facilitates insulation properties for backup lining. In this context, the influence of density and porosity on different mechanical and thermo-mechanical properties is summarized in consecutive sections. Several fabrication techniques fulfilled the desired density or porosity; nevertheless, the starting raw material size grading and composition are critical parameters for sintered density. Appropriate coarse-to-fine ratio provides highest tapping density that accomplishes appreciable room temperature pressing or casting green density. In order to achieve desired density, modification on particle grading or pressing module or workmanship is desired. However, composition plays a serious role during high-temperature processing or in situ application temperature. For example, high alkali content produces low eutectic phase in alumina-silicate refractories as well as different degree of grain orientations alter the resultant microstructure, resulting in the formation of weak region and early stage of failure through metal penetration in weak or porous zone. It is worthy to remember that the resultant refractory density or porosity (size, shape and distribution) amends at service temperature and strictly depends on the composition. In cumulative, the green density depends on initial stage of particle grading; however, sintered density regulates by grading, composition and temperature, where true porosity is always higher than apparent porosity. Despite refractory density and porosity, the operating condition and metal chemistry are responsible parameters to achieve high metal throughput.

1.3.2 Permanent Linear Change

Continuous pore removal provokes shrinkage, and in situ phase transformation results in either expansion or shrinkage, depending on the coefficient of the thermal expansion (CTE) behaviour of new phase. Permanent linear change (PLC) sounds irreversible dimensional change and in obvious it is critical at definite temperature for a particular set of composition. Usually, the PLC is measured by Eq. (1.4), where

$$\mathrm{PLC}\ (\%) = \frac{L_\mathrm{f} - L_\mathrm{i}}{L_\mathrm{i}} \times 100(\%) \tag{1.4}$$

where L_i = initial sample length before heating, L_f = final sample length after heating.

In order to describe the importance of such physical properties on refractory lining, let us consider the utility of high-alumina brick in hot metal ladle at 1550 °C. The high-alumina bricks are widely used in backup of steel ladle where the bricks are experiencing the temperature as high as 1400 °C. Expansion of the bricks in each cycle is preferable to keep the ring tight and bricks would not allow falling from the refractory lining. The PLC value is the utmost important factor to adjust the mortar joints also and the value to be maintained at 0.1% expansion. The stable mullite ($3Al_2O_3.2SiO_2$) phase is formed in high-alumina bricks, which involves 3–5 × 10^{-6}/K thermal expansion. Recent development is the use of spinel ($MgAl_2O_4$) forming bricks in steel ladle, which provides +ve PLC during repeated heating.

1.3.3 Crushing Strength

Terminology indicates the capability to withstand load-bearing strength; in other words, the refractory is subjected to survive an optimum compressive strength before crushing or failure. However, the importance of such data is started from green brick fabrication where an essential strength is required to handle prior to different stage of transportation and firing to obtain high degree of crushing strength under compressive (load/area) mode of loading. It is generally measured as the fracture load at room temperature and known as cold crushing strength (CCS), but time-dependent deformation at elevated temperature and constant stress (preferentially tensile) refers as creep. Green body to useable refractories experiences wide range of crushing strength because of bonding systems, either early-stage organic bonding, hydraulic bonding or high-temperature ceramic bonding. For example, molasses-bonded green and dried alumina refractory has less strength than phosphate-bonded alumina refractories; however, both have appreciable high strength at their application temperature. Low-cement castable exhibits more crushing strength at 200 °C (hydraulic bond) compared to 1000 °C, and further increase is noticed due to ceramic bonding at elevated temperature. Resin-bonded black refractory exhibits crushing strength near to ~500 kg/cm^2 at 300 °C, which is much higher compared to conventional oxide-based refractory bricks at this temperature. The strength behaviour predominately depends on the bonding, in preference grain size and pore content. Some important relations among different variables are given below:

Strength vs. grain size

$$\sigma = \sigma_0 + k_\sigma D^{-1/2} \quad (1.5)$$

where σ_0 is the intrinsic stress, k_σ is the Hall-Petch coefficient and D is the grain size.

Strength vs. pohre volume

$$\sigma = \sigma_0 \exp(-bP) \quad (1.6)$$

where σ_0 is the intrinsic stress at zero porosity, b empirical constant, P is the porosity and D is the grain size.

1.3.4 High-Temperature Deformation Under Compressive Load

Standard testing procedure defines a method for determining the deformation of dense and insulating shaped refractory products subjected to a constant load under conditions of progressively rising temperature (or refractoriness-under-load) by a differential method, with rising temperature. The test may be carried out up to a maximum temperature of 1700 °C. Refractoriness under load (RUL) is the material ability to resist the specific conditions of time, temperature and load. This actually depends on the point of softening and the quantity of glass or melt phase inside the refractory system. It is a foremost property of refractory in the view of practical applications and the property quantifies the refractoriness under a specific load. In this process, deformation initiation or composition of refractory softening is evaluated under a steady load against a raising temperature. The load and temperature both are involved in RUL; it may concern the broad class of thermomechanical properties. RUL explains the use of safe temperature for a refractory under the combined effect of a fixed load and heat; 0.2 MPa specific load is maintained during RUL testing.

The refractories are always subjected to some load from the furnace dead weight, lining of refractory, from the product materials and charge, from the flue gas, flame, air, etc. Furthermore, this load is unequally distributed through the lining of refractory and changes with time; cyclic loading is being predominate sometimes. Grains prefer to slide over one another within the refractory subjected to the external load, whenever a minute amount of liquid phase is present (or developed) in between them at elevated temperatures. Thus, application of refractory beyond the RUL temperature is restricted. Generally, the low RUL values of refractories are noticed in comparison to its refractoriness (Pyrometric Cone Equivalent, PCE). Deformation occurs due to grains sliding under load at elevated temperatures on the liquid phase formation in the refractory matrix and is responsible for such a large difference. This implies RUL temperature more reliable design and application temperature compared to PCE temperature, only.

1.3.5 Deformation in Bending

This parameter is different from elastic modulus. It describes the failure of refractory at peak load when refractory material is subjected to experience bending mode of loading at elevated temperature. For example, the estimation of 3-pt point bending or flexural strength of MgO-C at 1400 °C (commonly known as hot modulus of rupture) is a common practice in order to understand the suitability for working lining slag zone application in steel ladle. Is it really required when crushing strength data is already available? Yes, it is. In actual, it is a true indicator to estimate the performance and adaptability of a refractory at high temperature, analogous to application environment, whereas the crushing strength predominate indication of the suitability for the use of any particular refractory construction. In convention, Eq. (1.7) is adopted to measure the high-temperature flexural strength, generally known as hot modulus of rupture (HMOR),

$$\sigma_F = \frac{3P_b l}{2bd^2}\ (\text{MPa}) \quad (1.7)$$

where P_b is the bending load in N, l is the span length, b is the width of the refractory specimen in mm and d is the height of the refractory specimen in cm.

Refractory experiences thermal stress in combination with alteration of physical–chemical environment because of either formation of additional phase or metal infiltration, and results in deformation, crack initiation and rupture of the refractory. While considering the MgO-C refractory, the impurity content like SiO_2 in dead burnt MgO grain for low eutectic phase results in low HMOR at 1400 °C compared to low silica content in fused MgO; thus, composition plays a significant role to control over the refractory performance.

1.3.6 Elastic Modulus

Steel has high elastic modulus (E) than rubber, as metal alloy experiences very low strain under high stress compared to polymer. It describes the rigidity or stiffness of material or, in other words, the resistance to being deformed elastically under stress. Standard elastic modulus data library for homogenous substances are available in literature, and we quite often calculate the "E" value for heterogeneous system through rule of mixture. In obvious, porosity has to be encountered during such estimation, as any additional pore content reduces the elastic modulus of dense refractory. Some classic relations within elastic modulus and porosity are given in following [5]:

$$E = E_0 \exp(-bP) \tag{1.8}$$

$$E = E_0(1 - b_1 P) \tag{1.9}$$

$$E = E_0\left(1 - \exp^{[-b_2(1-P)]}\right) \tag{1.10}$$

where E_0 is the Young's modulus at zero porosity, b is the empirical constant and P is the porosity volume percentage. The empirical constant $b = 4$, $b_1 = 2$–4 and $b_2 = 0.5$ for 0–40 vol%, <6 vol% and ≥50 vol% porous matrix, respectively.

Refractory experiences heterogeneous composition and multivariant phases including both crystalline and glassy phase either during sintering or in situ during contact with iron and steel. Thus, it is a complicated phenomenon and difficult to predict any theoretical value until unless enough information on the type and content of phases, cracks and percentage porosity are available. Refractories have to be exposed into thermo-mechanical environment and abuse thermal shock, mechanical impact, abrasion and corrosion. Such incidence facilitates microstructural change in the heterogeneous refractory matrix, thereby affecting resultant properties and unavoidable change during operation. Despite corrosion related to the chemical interaction, thermal shock is one of the predominate properties and extensively depends on elastic modulus, E. Maximum thermal gradient (ΔT_{max}) can be represented as,

$$\Delta T_{max} = \frac{\sigma_f(1 - v)}{E\alpha} \tag{1.11}$$

where σ_f = Fracture strength, v = Poisson's ratio, α = Coefficient of thermal expansion.

In this consequence, the elastic modulus data provide the better insight of material that eventually assist to understand and analyse the failure behaviour of used refractories.

1.3.7 Mechanical Stress Assisted Crack Propagation

The crack propagation behaviour prediction and analysis of single-phase ceramics or composites during mechanical stress are well established compared to crack-assisted failure of heterogeneous refractories under thermo-mechanical stress and corrosion phenomena. Under mechanical stress, the Griffith energy criteria describe that the crack is like to propagate when energy release rate is greater than consumed energy rate, and mathematically expressed as:

$$\sigma_{\mathrm{f}}\sqrt{\pi c_{\mathrm{critical}}} \geq \sqrt{2\gamma E} \tag{1.12}$$

where σ_{f} = applied fracture stress, c_{critical} = critical crack length, γ and E are the surface energy and elastic modulus, respectively. The energy-absorbing phenomenon is well accepted when crack is subject to propagate in refractories and it occurs through "process zone" analogous to advancing crack and "wake region" follows the linking up with preformed advancing crack front. In practical, the propagation of large cracks is much difficult to extend than short cracks; in other way, when the cracks grow or extend further as there is extensive probability on the interactions in the larger wake region of the cracks, they result in R-curve, known as crack growth resistance curve. Experiencing non-linear R-curve behaviour in obvious has more reliability compared to flat toughness versus pre-crack length data. In spite of crack front initiation, heterogeneous refractory matrix experiences aggregate bridging and ligament bonding by in situ formation of low eutectic phase (for example, silicate phase). Continuous thermal shock, infiltration through pore or weak pocket, and expansion/contraction due to in situ phase formation exaggerate cracks in the refractories. Furthermore, the operational condition and metal chemistry simulate the viscosity and their fluid dynamics that eventually synchronize the refractory-metal interaction facilitate to penetration and corrosion.

1.3.8 Thermal Stress Assisted Crack Propagation

Refractory is an example of high-tech structural ceramics that is eventually used in large volume around the globe. In obvious, it experiences high magnitude of thermal stress apart from mechanical stress during application. Thus, without consideration of thermal stress one cannot judge the failure of refractories either during initial heating or cyclic ordered day-to-day operational practices. Thus, thermal shock is coming into the picture that facilitates predominate quasi-static crack propagation and less effective kinetic crack growth. While considering the kinetic growth, the kinetic thermal shock damage resistance (K_{T}) depends on the elastic modulus (E), work of fracture (γ_{wof}) and thermoelastic fracture stress (σ_{f}), and simply it can be represented as:

$$K_T = \frac{E\,\gamma_{wof}}{\sigma_f^2} \tag{1.13}$$

The work of fracture can be obtained from load displacement curve,

$$\gamma_{wof} = \int \frac{Pdu}{2A} \tag{1.14}$$

where $\int$Pdu represents the obtained area under the load-displacement curve, A is the cross-sectional area, and factor 2 is included as two new surfaces generated during fracture. Thermoelastic defines the effect of temperature on the stretching or contracting of an elastic material that eventually depends on the system temperature, coefficient of thermal expansion (CTE) and heat capacity. Analogous to Eq. (1.13), an additional parameter coefficient of thermal expansion (α) is further required to introduce and explain the quasi static growth damage resistance (S_T) phenomenon and represented as:

$$S_T = \left(\frac{\gamma_{wof}}{E\,\alpha^2}\right)^{1/2} \tag{1.15}$$

In consideration of one-dimensional expansion behaviour, the strength of refractory system can be expressed as the product of CTE and elastic modulus as a function of temperature T:

$$\sigma = \alpha\,E\,(T) \tag{1.16}$$

From the above Eqs. (1.13–1.15), both the damage resistance capability is inversely related to the (σ^2/E), and this parameter (σ^2/E) refers as strain energy at the stress level of "σ" [6]. In actual, this stored elastic strain energy is the prime driving force for the crack propagation during thermal shock damage of refractories. Thus, very dense and high strength refractory is much more inclined to thermal shock damage. Hall Petch relation (Eq. 1.5) predicts that strength increases inversely with reduction of the square root of grain size. Despite pore filling and accomplishing highest density, proposition of smaller grain size is not at all effective for thermal shock resistance.

Thus, common fact envisages that refractory fabrication demands combination of certain percentage of larger and hard grains or aggregates in order to act as thermal shock stoppers. This exercise reduces the cumulative strength of brick and stored elastic strain energy that eventually play a critical role to create a rising R-curve and decrement of crack propagation. In brief, utility of high strength brick for thermal shock damage resistance is not a good practice, and hence the raw material selection and refractory design is a critical concern in the outlook of load-bearing and thermal shock resistance applications, together.

1.3.9 Thermal Conductivity

Heat loss from working interface to backup lining depends on how fast heat dissipates through the refractory and joining mortars. Rapid heat conduction facilitates the heat transfer and reduces the thermal gradient, resulting in enhancement of thermal shock resistance behaviour of refractories. Thus, high thermal conductivity is responsible for rapid heat loss as well as increment of thermal shock resistance. In order to design refractory composition and shape, one has to come across both the factors along with the unavoidable metal-refractory interaction. This heat conduction phenomenon is the exploration of the phonon coherence and their synchronization through heterogeneous matrix is a challenging task. Under steady-state condition, the thermal conductivity (k) defines the amount of heat passing in unit time through unit surface in a direction normal to the surface through a temperature gradient. This quantity can be represented in Eq. (1.17) and unit is $W.m^{-1}.K^{-1}$.

$$\frac{\Delta Q}{\Delta t} = k\,A\,\frac{\Delta T}{\Delta x} \tag{1.17}$$

where Q is the heat passing through the surface area A, normal to the heat transfer, $\Delta T/\Delta x$ is the temperature gradient along the "x" direction of the heat transfer. In contrast to metal (cloud of free electrons), ceramic possess low thermal conductivity because of non-availability of free electrons in ionic and covalent bond. Thermal conductivity (k) depends on how thermal energy transport is taking place and this phenomenon depends on the cumulative effect of migration of free electrons, related to the electronic thermal conductivity (k_e) and lattice vibrational waves (phonons), and referred as k_{ph}. In pure metal, k_e is predominant, whereas phonon-insisted transport mechanism facilitates heat conduction in alloy and ceramics. Thus, thermal conductivity of refractory is influenced by phonon mean free path that may shorten during scattering by defects, boundary and other phonons (a harmonic vibration). The phonon-assisted thermal conductivity, k_{ph}, can be equated as Eq. (1.18) [7].

$$k_{ph} = \frac{1}{3}Cvl \tag{1.18}$$

where C is the specific heat, v is the velocity and l is the mean free path of the phonon. The atom vibrations are dependent on each other, rather strongly coupled with neighbouring atoms. So, it is more systematic in well-ordered crystalline atoms and exhibit high thermal conductivity of quartz than amorphous silica-based glass. In consideration of temperature effect, the influence of phonon mean free path is minimum; moderately predominant boundary effect decreases the conductivity near to zero at 0 °K. However, at some low temperature, the conductivity increases to maximum limit and further decreases above Debye temperature (highest normal mode of crystal vibration). At sufficiently high temperature, the phonon interaction may shorten the resultant mean free path "l" near to the lattice spacing, and thus the conductivity is expected to be independent of temperature. In situ phase formation, microstructure alteration and cavity formation may vary the thermal conductivity at high temperatures. Despite crystallinity, the heat flow characteristic also differs due to the crystallographic (a, b, c) anisotropy. For example, non-cubic sapphire having thermal conductivity along the *C*-axis is 35 W/mK, and 32 W/mK along the *A*-axis. In order to achieve specific thermal conductivity and design refractory, one has to consider the following factors:

(a) Chemical composition, crystallinity and crystal structure of raw materials.
(b) Previous heat treatment or refractory processing temperature.
(c) Microstructure and physical properties of the material.
(d) Pore content, size and distribution phenomena.
(e) Temperature for the interest zone of application and probable in situ phase formation.

1.3.10 Thermal Expansion Behaviour

Temperature increases the atomic vibration and mean position where repulsion between atoms predominate than attraction that critically leads to atomic separation distance corresponding to the lattice expansion. Crystal tends to expand and become more symmetrical at elevated temperature and results in increment of specific volume. Lattice vibration-assisted volume change increases the energy and results in the expansion of the system.

The coefficient of thermal expansion (α) is a relative increase in length per unit temperature increment, and represented as Eq. (1.19)

$$\alpha = \frac{dL}{L_o}\frac{1}{dT} \tag{1.19}$$

where dL is increment length, L_o is the initial length, dT is temperature rise. In obvious, the metal has relatively high CTE than ceramics. At relatively low temperature, the CTE increases rapidly and follows nearly constant plateau above the Debye characteristic temperature (θ_D). For example, the rate of expansion of MgO is higher than Al_2O_3, where MgO experiences $D_{temperature}$ ~750 K compared to ~1050 K for Al_2O_3 [8]. In the same time, it is essential to encounter that the "reversible thermal expansion" behaviours that describe the contraction during cooling as well. These are very critical aspects when additional phase occurs during high-temperature treatment.

Like most of the other materials, refractory materials expand when heated and contract when cooled. This effect of heating on linear change of material is called "Reversible Thermal Expansion (RTE)". The thermal expansion graphs are available on most of refractory products, as shown in Fig. 1.3 [9]. These graphs demonstrate the expansion rate at specific temperature range. The rate of thermal expansion, expressed as "coefficient of thermal expansion", is different for different types of refractory oxides.

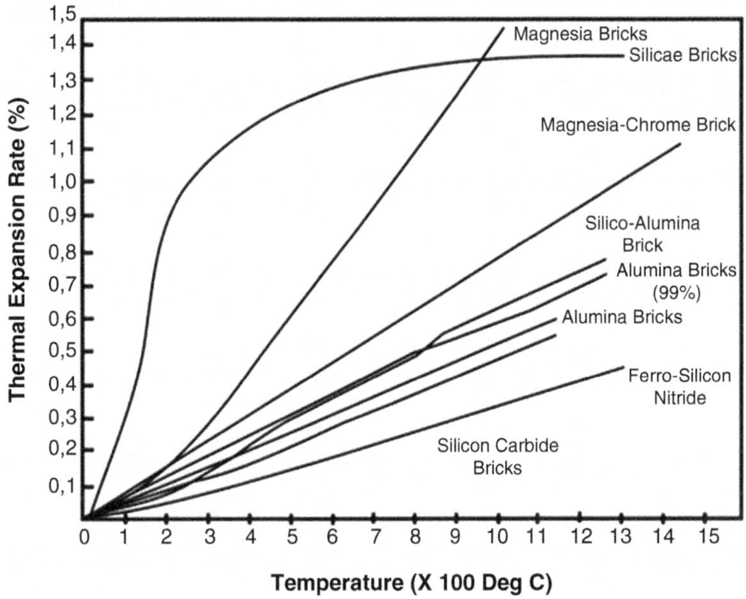

Fig. 1.3 Thermal expansion of different refractory oxides [9]

Refractory materials can undergo mineral formation or phase transformations when heated; few of them are reversible. These phase changes involve volume expansion or contraction. Upon cooling of those materials to room temperature, the volume of the material would possibly be larger or smaller to the original size of the product, which is referred as permanent volumetric or linear change. Change in volume which occurs on heating is one of the most important physical properties of refractory materials for designing the refractory lining in integrated shapes and special attention is to be given to determine provision for keeping gaps for the thermal expansion at operating temperature. For lining silica bricks in coke oven and blast furnace stoves, adequate expansion allowance is to be provided for the irregular expansion of silica bricks during heating. Severe failures have been experienced due to inadequate expansion allowance in silica bricks lining in hot stove dome. However, no irregularities in expansion are observed in magnesia, corundum or alumino-silicate bricks.

1.3.11 Thermal Stress and Shock

Differential thermal expansion coefficient of multiphase materials or anisotropy thermal expansion induces thermal stress. In order to understand the magnitude of thermal stress, let us consider Eq. (1.19) and $E = \sigma/\varepsilon$, where E is elastic modulus and σ and ε ($=dL/L_o$) are stress and strain, respectively. Combining CTE and elastic modulus, the thermal stress can be represented as Eq. (1.20), which may be destructive as is well above than the commonly used refractories.

$$\sigma_t = \alpha E(T - T_o) \tag{1.20}$$

This incidence affects the elastic instability and aggravates crack propagation. For example, alumina experiences thermal stress 3040 MPa at T–T_o = 1000 K in consideration of CTE, $\alpha = 8.1 \times 10^{-6}$, and $E = 380$ GPa, T and T_o are final and initial temperatures, respectively. This thermal stress is near to nine times compared to flexural strength $\sigma_F = 350$ MPa. The thermal stress significantly varies with temperature difference; however, beyond critical temperature difference (ΔT_c) the material experiences extensive thermal shock that extends the cracks kinetically to a new length and results in lowering the strength. If the crack propagation is not stopped in refractory matrix, it becomes the prime failure cause of entire structure.

In brief, the thermal shock resistance of refractory is an ability to withstand sharp changes in temperature during high-temperature operation. Theoretically, the maximum temperature jumps sustainable by the solid ΔT can be represented by Eq. (1.21) [10].

$$\Delta T = \frac{k\,\sigma_F}{E\,\alpha} \tag{1.21}$$

where σ_F = flexural strength, E = elastic modulus and α = coefficient of thermal expansion, and k = thermal conductivity, respectively. In consideration of Eq. (1.21), one has to synchronize the following factors to avoid failure of refractories due to thermal shock:

(a) Reducing the thermal gradient to avoid formation of high thermal stress through slow temperature change.
(b) Increasing the strength and thermal conductivity.
(c) Reducing the elastic modulus and coefficient of thermal expansion.
(d) Increasing toughness through crack bridging or phase transformation.

Despite thermal shock behaviour of refractory matrix, the high local thermal stresses in metals merely lead to a slight local plastic deformation; that penetration can lead to the propagation of cracks in ceramic materials as well.

The grain size and grain boundary volume have also influence on the strength that influence thermal shock resistance. In a recent article, the competitive ΔT for both single crystal and polycrystalline alumina has been evaluated and plotted in Fig. 1.4 [11].

The retained strength of single crystal dramatically decreases compared to polycrystalline alumina beyond the critical temperature range (ΔT_c). Thus, selection of adequate grain size grading and composition formation are important criteria for refractory manufacturing.

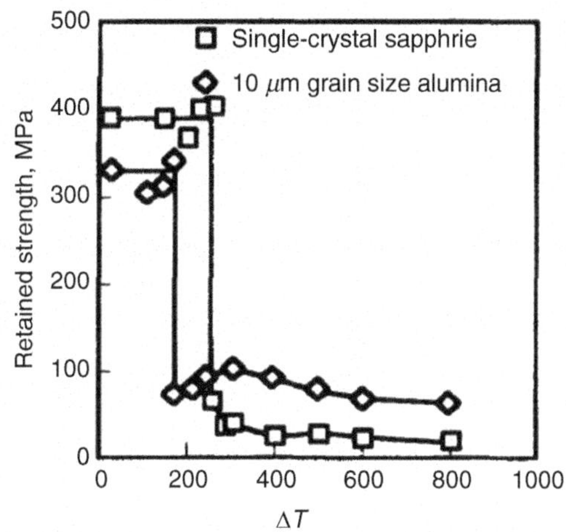

Fig. 1.4 Thermal shock measurement data for both single crystal and polycrystal alumina

1.3.12 Wear Behaviour

Wear is the removal of the surface material or damage of surface from one or both of the two contact surfaces in a relative motion to each other; most of the cases through surface interactions preferentially surface asperities facilitate wear. The damage due to wear precedes the loss of actual material, and it may also have occurred independently. Under service, several wear mechanisms including chemical wear, mechanical wear, and thermal shock are the predominant wear mechanisms encountered by refractories in iron- and steel-making vessels. Typical wear mechanisms attacked on the different zones of blast furnace stack such as chemical attack (all shaft), erosion (upper stacks), thermal shock (lower stacks) and heat load (bosh) are shown in Fig. 1.5.

1.3.13 Difference Between Corrosion, Erosion and Abrasion

Refractories are preferentially used at very high temperatures and they experience both load and corrosive domains. These domains actually have direct contact with melted (liquid metal or slag) phases, facilitate refractory corrosion or wear out or consumption at elevated temperature. Wear of refractory is measured by the thickness and mass loss with respect to the exposed refractory face; results of a chemical attack by a fluid (molten metal slag) that eventually corroding and in situ refractory reaction approaching chemical equilibrium in the contact zone between the fluid and refractory.

Erosive wear occurs due to the impact of solid or liquid particles against the refractory surface, which is eventually involved in several wear mechanisms that are predominantly controlled by the material particle density, impingement of angle, velocity impact and the size of the particles. The impingement angle is the angle

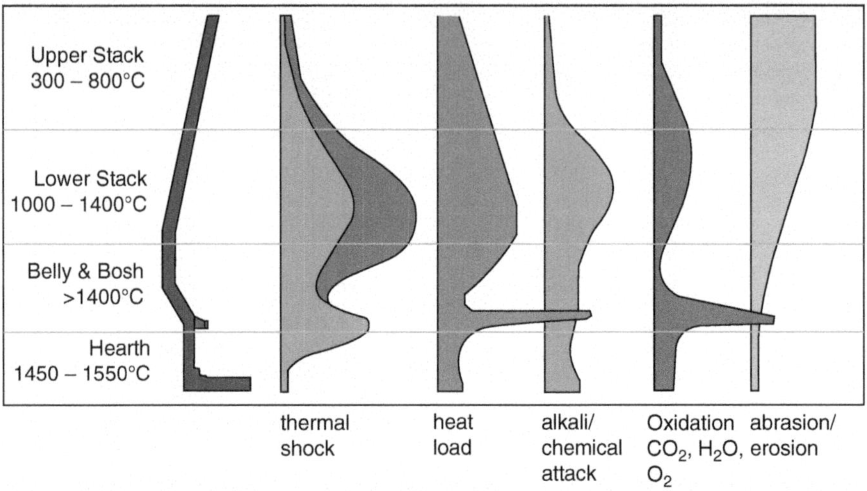

Fig. 1.5 Basic BF refractory wear mechanisms [12]

between the trajectory of the particle and surface erode before impact. A small impingement angle recommends the processes of wear similar to abrasive wear. If the particles are very hard, similar kind of abrasive wear may happen; however, liquid erodent particles are not involved in abrasive wear mechanism, rather influenced by repetitive impact stress, only.

Abrasion was related to the destruction or damage of material during friction. It is the interaction between the surfaces and due to the mechanical action of the opposite surfaces in relative motion; the surfaces undergo the deformation and material removal from the surface. The initial mechanical contact between tribosurfaces and the degree of relative motion are crucial and predominant factors. Abrasion wear is also defined as a phenomenon of dimensional loss of one solid, loss of material and with or without any actual decoupling due to the interaction between the faces of solids actually bounding within the working domain. Abrasion actually depends on the present working domain, like the sliding direction, type of load (means reciprocating, rotary or impact), temperatures and speed. It also depends on the various types of opposite bodies such as liquid, gas or solid, and type of ranging contact between single and multiphase is important.

Refractories are subjected to experience continuous friction or rubbing action due to the charge movement or materials of product in use, and also the processing conditions at high temperature under mechanical thrust. The action of rubbing of the charge and particles of the product on the refractory materials of porous surfaces leads to rapid material removal and wear. The solid charge material moving downwards in the blast furnace is an example of refractory wear and sliding abrasion [13].

1.4 Shaped Refractories

1.4.1 Silica Refractories

Several classic literature has already mentioned details of silica brick; however, our motive is to emphasize concise text that is mandate to resolve the refractory quality and cause of failure during application in critical zone like coke oven batteries and hot blast stoves. The important considerations including phase change phenomenon, impurities, addition of mineralizer, consolidation protocol and high-temperature processing are discussed.

1.4.1.1 Raw Materials and Processing

Thermodynamically stable quartz is the prime resource for SiO_2 refractory brick, and it is found in natural raw materials known as quartzite along with certain degree of impurities like Fe_2O_3, TiO_2, Al_2O_3 and CaO, essential to synchronize in view of the low eutectic liquidus temperature as well as high-temperature application. In consideration of application criteria, this class of acidic refractory is further classified as super duty refractories maximum limit of impurity $Fe_2O_3 + Al_2O_3 \leq 0.5\%$, but high duty refractories comprise more impurities in the range of 0.5–2 wt%. Different fractions of washed and cleaned low impurity (<0.5 wt%) quartzite, 3 wt% milk of lime as binder as well as mineralizer to facilitate formation of tridymite or cristobalite, green binder like cellulose and certain amount of moisture as plasticizer are required constituents to make super duty silica refractories. Despite such essential raw materials, several optional mineralizers, binders and abrasion resistance materials are added to achieve tailor-made properties. Conventional pressing followed by sintering provides the desired phase, but sintering schedule including heating–cooling and retention at particular temperature are critical issues during silica brick manufacturing. What happens during silica brick sintering?

Before discussing such incident, the polymorphic phase transformation and their corresponding volume change in the particular temperature region is indeed, and listed in Table 1.3. These phase transformations are slow process and thus require a substantial long preheating and cooling zones in a continuous kiln to accommodate prolong heating and cooling schedule (Fig. 1.6).

Consistent slow heating and cooling is mandatory to accommodate the volume change during firing of silica bricks. Commercial silica brick consisted chiefly of cristobalite, tridymite and un-converted quartz and showed a large expansion up to 300 °C, followed by a considerably smaller but regular expansion to 550 °C. From 550 to 600 °C the rate of expansion was considerably increased, but above 600 °C the change in dimensions was small. Silica bricks made of quartzite with lime bond has a large expansion up to 550 °C and also an expansion observed from 550 to 600 °C due to the inversion of quartz. The cooling curve nearly coincides with the heating curve so that the abrupt changes in length at 150–250 °C and 550–600 °C

Table 1.3 Polymorphic transformation and volume change of quartz with respect to temperature

Transformation temperature (°C)	Polymorph change (low pressure)	Volume change (%)
117–163	$\alpha \Leftrightarrow \beta_1 \Leftrightarrow \beta_2$ tridymite	+0.5
220–280	$\alpha \Leftrightarrow \beta_2$ cristobalite	2.0 ⇔ 2.8
573	$\alpha \Leftrightarrow \beta$ quartz	0.86 ⇔ 1.30
870	β quartz ⇔ β_2 tridymite	14.4
1250	β quartz ⇔ β cristobalite	≈17.40
1470	β_2 tridymite ⇔ β cristobalite	≈3.0
1713	β cristobalite ⇔ liquid	–

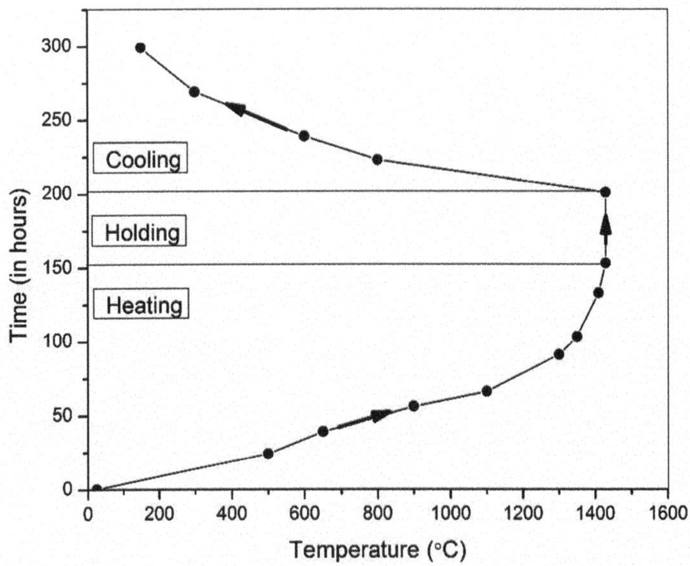

Fig. 1.6 A typical firing profile of silica brick

were reversible and were evidently due to the inversion of *a* to *β* cristobalite, and *a* to *β* quartz, respectively. It was apparent that the firing of green bricks during manufacturing in the kiln causes the major transformation of large part of quartz into other forms of silica. The large expansion at low temperature, less than 550 °C, is due to inversion of cristobalite present in the silica bricks. It would demand very slow and careful heating with no rigid ties when the brickwork was in first time preheating. It would be source of danger from cracking and falling of bricks, when the brickwork is made to cool down below 550 °C and below 300 °C in particular. It requires very slow cooling to absorb the expansion–contraction of the bricks. For high-temperature applications, desired tridymite and cristobalite phase can be achieved in between 1420–1430 °C and 1480–1500 °C, respectively. However, the firing results in relatively less expansion than expected 4% as pore accommodates certain expansion within the matrix. Targeted phase stabilization can be done through optimum addition of CaO and Fe_2O_3. For example, up to 2–3 wt% hydrated lime expedites the formation of stable cristobalite phase (Dale theory) while 2 wt% lime along with 0.5 wt% Fe_2O_3 are effective mineralizer to obtain tridymite phase.

1.4.1.2 Effect of Impurities on Eutectic Temperature

In this context, the major impurities like CaO, Fe_2O_3, TiO_2 and Al_2O_3 are summarized. Impurity CaO content in the range of 7 mol% reduces the eutectic temperature to 1650 °C; however, beyond this critical content, the system transforms to immiscible liquid phase of tridymite and dramatically lowering the melting temperature to 1430 °C, eventually much lower than the application temperature. Analogous to CaO, 4 mol% FeO impurity reduces the eutectic temperature up to 1685 °C; beyond this limit, the system temperature drastically reduces the liquidus temperature to 1189 °C. The presence of TiO_2 impurity lowers the liquidus temperature to 1550 °C. Alumina near to 5 mol% in silica system assists to sharp fall in eutectic liquidus temperature from 1723 to 1587 °C. Figure 1.7a–d, a typical schematic, represents the formation of binary phases in the presence of impurities; furthermore, several low-melting tertiary compounds are also expected in this process and this information is pointed out in several chapters. Thus, control over starting raw material and impurities are important criteria to avoid early failure in application temperature.

1.4.1.3 RUL vis-a-vis PCE

Common incidence deployed refractory application under load can withstand relatively lower temperature compared to without load, pretty obvious, and thus most of the refractory exhibit high PCE than RUL. However, pure silica refractory exhibits almost similar RUL to fusion point PCE without formation of low eutectic phases, in other words impurity facilitates formation of low eutectic phases and enhance the difference between RUL and PCE. Preferentially, high silica content makes high

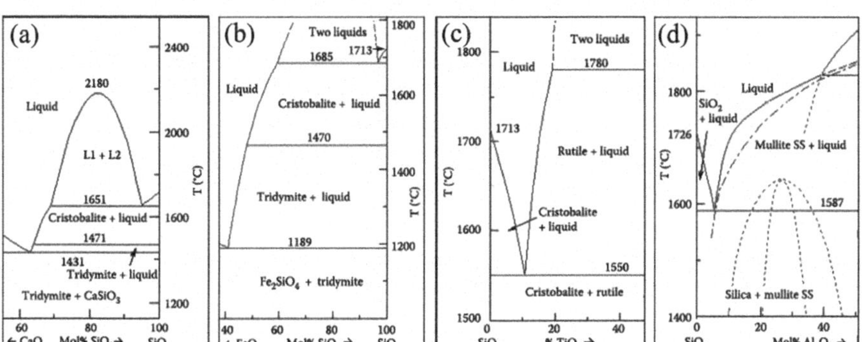

Fig. 1.7 Effect of impurities on the silica system; (**a**) CaO, (**b**) FeO, (**c**) TiO_2, (**d**) Al_2O_3

viscous liquid phase that results in negligible deformation under load and thus near to equal PCE.

1.4.2 Alumina-Silicate Refractories

Early days but sustainable alumina-silicate refractories are predominately classified into fireclay and high-alumina refractories. The high-alumina refractories can further be classified as:

- 42–50% Al_2O_3, containing calcined clay grog and other high-alumina materials.
- 55–62% Al_2O_3, containing kyanite, sillimanite and andalusite
- 65–80% Al_2O_3 containing andalusite, mullite and bauxite
- More than 80% Al_2O_3, containing bauxite, fused alumina, sintered alumina and corundum.

1.4.2.1 Fireclay Refractories

Fireclay is the oldest traditional-type refractory made of clay minerals, which has its essential constituents of hydrous alumino-silicates and presence of free silica, iron oxide and alkalies as impurities. Fireclay deposits are seldom pure hydrous alumino-silicates; the presence of impurities determines the resultant properties and quality of refractory made from them. Fireclay bricks are mainly consisting of oxides component of Al_2O_3 and SiO_2, the SiO_2 content is in the range of 50–80%, while Al_2O_3 content varies from 15 to 42%. Fe_2O_3 content is maximum 3.0% and of alkalies, about 3.5%. Depending on Al_2O_3 content and respective application temperature (differing refractoriness), fireclay refractory is classified as low heat duty, medium heat duty (MHD) and high heat duty (HHD).

Fireclay products are manufactured from natural raw refractory-grade hydrated plastic clay as bond, calcined clay, pyrophelitic clay and high-temperature fired refractory clay (chamotte) as aggregate. Naturally occurring clay occurs as sedimentary rock consisting of fine grained (<2 μm) crystallized water containing alumino silicate with a laminated layer structure. The commonly used clay minerals are kaolinite, fireclay and halloyside, having the basic formula $Al_2O_3.2SiO_2.2H_2O$ and $Al_2O_3.\ 2SiO_2.\ 4H_2O$, in the case of halloyside. This group of clay is known as *kaolin group of clay* and has a two-layer crystal lattice. The other types of clays occur in *Montmorilonite* and *illite* group of clay and have three-layer crystal lattice, and both groups have the formula $Al_2O_3.4SiO_2.nH_2O$, differing with the amount of varying amounts of foreign ions present in the lattice structure and plasticity when mixed with water. Plasticity occurs when mixed with water as the layer structured clay minerals are surrounded by a thin film of water within the particles and reduces the cohesion forces between the particles.

During heating up to 550 °C, kaolin loses the chemical water of crystallization and an intermediate phase known as *metakaolin* ($Al_2O_3.2SiO_2$) is formed. At above 925 °C, the kaolinite lattice disintegrates and metakaolin phase becomes unstable. Initially, there is no reaction with Al_2O_3 and SiO_2 observed up to 950 °C, when *mullite* ($3Al_2O_3.SiO_2$) starts forming as a stable phase and this phase is acting as a high-temperature bond in alumino-silicate refractory system. Above 1100 °C, mullite is the major phase along with cristobalite and glassy phase as shown in the following reactions:

$$Al_2O_3.2SiO_2.2H_2O \rightarrow Al_2O_3.2SiO_2 + 2H_2O \left(\text{at } 900 - 925^{\circ}C\right) \quad (1.22)$$

$$3\left(Al_2O_3.2SiO_2\right) \rightarrow 3Al_2O_3.2SiO_2 + 4SiO_2 \left(\text{at } 950 - 1100^{\circ}C\right) \quad (1.23)$$

During the process of drying and heating, clay shrinks significantly and densification occurs, and that is why raw clay cannot be used in brick making. Thus, it requires high-temperature calcination at above 1250 °C, and above the phase transformation, densification takes place through 8–10% volume shrinkage. The manufacturing of fireclay bricks is done by crushing and grinding of the calcined clay and fireclay grog in proper grading and proportioning to attend adequate packing density. Graded grog mixed with clay (as binding substance) and water in a mixer machine to attend adequate plasticity for brick shaping. The amount of bond clay and water varies and depends on the mode of shaping as follows:

- Plastic pressed and hand moulding: 25–35% plastic clay and 10–15% water.
- Sami-dry pressed: 20–25% plastic clay and 4–6% water.
- Dry pressing in high capacity press: 5–10% bond clay and 1.5–3.0% water.

Slow drying process is required to drive out the total moisture content at 110–150 °C. Faster drying would cause for cracking of refractory bricks. Firing is usually done in chamber kiln or tunnel kiln with a temperature range between 1250 and 1500 °C, depending on brick composition (Al_2O_3 content and presence of fluxes). Depending on the presence of alkalies and other fluxes, at high temperature, the fireclay bricks may be composed of mullite and low-melting liquid phases. The amount of glassy phase is about 35–50% with raw materials containing low Al_2O_3 and high in fluxes (>3.5%) when special raw material with low amount of fluxes (<1.5%) is firing at 1400 °C, glassy phase can be reduced to 20% for a fireclay brick of 40% Al_2O_3 with content of mullite as high as 55% and cristobalite of 20–25%.

The presence of glassy phase has a direct influence on high-temperature properties of fireclay bricks, because of chemical composition and softening behaviour. The glassy phase softens as a result of its alkali content and formation of low-melting phases with the reaction of alkalies with Al_2O_3 and SiO_2. Further, silica-rich alkaline glassy phase starts softening at around 1000 °C, but duration of complete softening is significantly high due to high viscosity of siliceous glass. As a result, the presence of impurities, softening behaviour and presence of glassy phase influence the thermo-mechanical load-bearing response, such as refractoriness under load (RUL), creep and high-temperature crushing strength. Lower thermo-mechanical

Table 1.4 The properties of fireclay refractories

Properties	Unit	Low duty fireclay	Medium duty fireclay	High duty fireclay	Super duty fireclay
Maximum service temperature	°C	1000	1200	1350	1550
Chemical analysis					
Al_2O_3	%	24–26	30–32	36–39	40–42
SiO_2	%	66–70	62–65	55–57	52–54
Fe_2O_3	%	2–2.5	2–2.5	1.3–1.5	1.0–1.2
Alkali	%	2.5–3.0	1.5–2.0	1.2–1.5	1.0–1.2
Physical properties					
Porosity	%	19–21	18–20	16–18	14–16
Bulk density	g/cc	2	2.1	2.12	2.35
Mod. of rupture at RT	Mpa	5–5.5	5.5–5.8	7.5–8.5	8.5–9.0
Cold crushing strength	Mpa	25–28	30–33	34–36	35–38
Permanent linear change @ 1400 °C @ 1550 °C	%			(±) 0.3	(±) 1.2
Refractoryness under load under 2 kg/cm^2 load	(ta) °C			1400	1600

strength is the limitation of such fireclay bricks in critical high-temperature load-bearing areas and limiting application temperature. The properties of fireclay refractories are shown in Table 1.4.

1.4.2.2 Sillimanite Group of Raw Materials

The sillimanite group of materials consists of three types of materials, sillimanite, kyanite and andalusite, and this group of materials are used to produce 50–60% Al_2O_3 class of bricks, all exhibit characteristic permanent expansion on heating due to decomposition and formation of mullite. The chemical composition of all the three members of sillimanite group is the same as $Al_2O_3.SiO_2$, but their crystal structure differs widely according to their formation condition and origin, depending on the pressure and temperature and the time of formation, as shown in Fig. 1.8. When the pressure and temperature are both high, the mineral formed with a chemical composition $Al_2O_3.SiO_2$ is *Sillimanite*. If the pressure is high, and in moderate temperature, *Kyanite* is formed and in the condition of high temperature under moderate pressure, *Andalusite* is formed.

The three sillimanite group minerals are not found in Al_2O_3–SiO_2 phase diagram at a pressure of 1 atm. This phase diagram shows only the single alumino-silicate compound, *Mullite* ($3Al_2O_3.SiO_2$). Absence of the three minerals in the binary diagram is because they are geologically formed at high pressure and elevated temperature within the earth. None of the three minerals of the sillimanite group

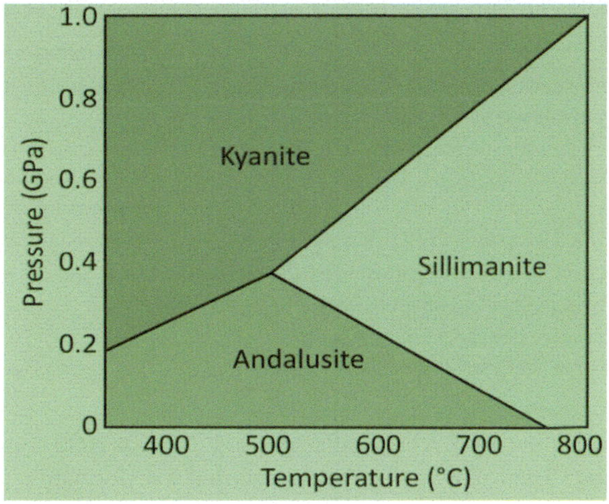

Fig. 1.8 Schematic representation for the phase change of andalusite—kyanite—sillimanite with respect to temperature and Pressure

Table 1.5 Variation of sillimanite group of minerals and mullite

Properties	Kyanite	Andalusite	Sillimanite	Mullite
Chemical formula	Al_2O_3.SiO_2	$Al_2O_3.SiO_2$	$Al_2O_3.SiO_2$	$3Al_2O_3.2SiO_2$
Crystal structure	Triclinic	Orthorombic	Orthorombic	Orthorombic
Crystal habit	Lathlike	Prismatic	Fibrous	Acicular
Type of silica after decomposition	Cristobalite	Amorphous	Amorphous	N/A
Specific gravity	3.6	3.15	3.2	3.05
Transformation temp	1310	1410	1540	–
Vol. change (%)	14	Negligible	4–5	–

are in equilibrium phase at 1 atm pressure. The properties of three sillimanite group minerals including mullite is shown in Table 1.5.

Andalusite

Andalusite is a natural ore, discovered in Spain in 1789. It belongs to sillimanite group of minerals and it composed of around 59% Al_2O_3 and 38% SiO_2. Iron oxide and alkalies are common impurities of this mineral. However, this mineral is easily form mullite when fired at high temperature because of Al_2O_3 and SiO_2. Andalusite is one of the few minerals which can be used in refractories without pre-firing. It transforms to mullite and cristobalite in the temperature range from 1200–1500 °C, accompanied by volume increase of 3–5% (theoretically 3.7%). Andalusite has a low impurity, but silica content is as high as 38–40%, preferentially the grains are andalusite crystals consist of very low open porosity. The chemical composition of andalusite is bioxides of aluminium and silicon: $Al_2O_3.SiO_2$. Andalusite transforms

Table 1.6 Advantages of using andalusite in high-alumina bricks and monolithics

Unfired refractory	Fired refractory
(Castable, gunning mass, ramming material. Chemical-bonded and resin-bonded bricks)	(Fired bricks and shapes)
The andalusite transforms to mullite progressively during in use.	During firing of bricks, andalusite transforms to mullite partially or fully depending on firing temperature and high temperature soaking time.
Good resistance to CO attack and to alkali	Very good resistance to CO and to alkali
Good resistance to corrosion by the blast furnace slag and hot metal up to 1700 °C	
Low thermal conductivity (close to chammotte) Positive permanent linear change Good resistance to thermal shock	

to mullite (87.6%) and silica glass (12.4%) at 1000–1400 °C depending on the grain size. The following shows the thermal transformation process:

$$3(Al_2O_3.SiO_2) = 3Al_2O_3.SiO_2 + SiO_2\ (\text{glass}) \quad (1.24)$$

During mullitization of andalusite, the excess of silica is transformed to silica glass. The glass phase on the interface of andalusite and mullite forms a bond within andalusite and newly formed mullite. From a temperature range of 1000–1450 °C, the glass phase softens within the mullite that eventually allows stress absorption due to volume change with increase of temperature. Thus, such transformed mullite from andalusite has more thermal stability than other types of mullite. For further improvement of high-temperature strength (HMOR and Creep resistance), calcined alumina or reactive alumina is added to form secondary mullite through reaction of remaining glassy SiO_2. Advantages of using andalusite in monolithic refractory as well as making bricks are summarized in Table 1.6.

Kyanite

Kyanite, the name is derived from the Greek word, *Kuanos*, sometime called as *Kyanos*, meaning deep blue is a typically blue silicate mineral, rich in alumina. It crystalizes in triclinic system with a specific gravity of 3.6. When heated for a long time, it is slowly converted into mullite and silica at a temperature of 1310 °C; however, at higher temperature the conversion is rapid with a significant volume expansion and consequent reduce in specific gravity. In order to have a volume stability to use as refractory raw material, kyanite must be calcined above 1420 °C till it achieves a constant specific gravity. The effect of heat on kyanite is shown in Fig. 1.9.

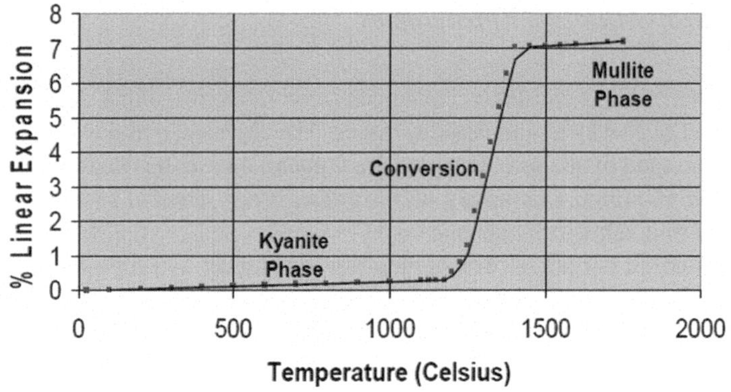

Fig. 1.9 Effect of temperature on the linear expansion of kyanite

1.4.2.3 Mullite and Andalusite

Mullite is a mineral existing under the form of $3Al_2O_3.2SiO_2$ or $2Al_2O_3.SiO_2$ and provides attractive properties of refractory. Unfortunately, it rarely occurs as natural raw material. Consequently, the development of mullite in commercial refractory has to be achieved by firing various alumino-silicates.

Among the alumino-silicate raw materials, andalusite is the best natural source for mullite. Andalusite transforms to mullite at considerably low temperature, with a minor volume change of 4.0% volume expansion. Unlike most of the other silicates, mullite formation from andalusite does not involve any dehydration stage and does not generate intrinsic porosity.

Thermal shock resistance and other thermo-mechanical properties are considered as one of the most important features of mullite and mullite-based refractories. Mullite has very low thermal expansion coefficient compared to bauxite, corundum, fused and other sintered aluminas. Expansion coefficient of mullite is $4.5–6.5 \times 10^{-6}\ K^{-1}$, whereas other refractory materials have the value of above $8.10^{-6}\ K^{-1}$.

Mullite is one of the most important ceramic materials, due to high thermal stability in air, high corrosion resistance even in harsh environments, low thermal expansion associated with excellent thermal shock resistance, high creep resistance and low thermal and electrical conductivity. Together with acceptable mechanical properties such as strength and fracture toughness, and the relatively low materials costs, these outstanding properties make mullite ceramics suitable for a wide field of technical applications. This includes traditional technical products such as porcelain, tableware, and refractories, but also advanced materials such as heat exchangers, filters, catalytic convertors, substrates, electronic packaging devices, components for optical systems and turbine engines and heat shields. In recent years, mullite has also been used to produce dental ceramics and substrates for antibacterial applications.

1.4.3 High-Alumina Refractories

Increased thermal and physico-chemical stress on refractory lining demands higher quality materials for many industrial furnaces. The chemical composition of fireclay products is limited by the use of clay as a raw material base. It is therefore necessary to produce high-alumina refractory with a different raw material base. Properties of the high-alumina refractories depend on the two major phases, mullite and corundum and associated glassy phase. The technology of fireclay and high-alumina refractories is traditionally explained in Al_2O_3–SiO_2 phase diagram as shown in Fig. 1.10.

The phase diagram demonstrates that mullite ($3Al_2O_3.SiO_2$) is the only congruently melting stable compound among all alumino-silicate systems, with having composition of 72% Al_2O_3 and 28% SiO_2, and the softening temperature is as high as 1850 °C. The refractoriness of alumino-silicate refractory is primarily dependent on the amount of mullite phase present in the refractory composition. Various thermo-mechanical properties of entire group of alumino-silicate refractory can be explained by the Al_2O_3-SiO_2 phase diagram. The essential facts of the phase diagram are summarized as follows:

1. Mullite ($3Al_2O_3.SiO_2$) is the only stable compound that forms in the Al_2O_3-SiO_2 system and is highly refractory in nature. The theoretical composition of mullite is 71.6% Al_2O_3 and 28.4% SiO_2 on weight basis and it is congruently melting at 1850 °C. Near to the theoretical composition of Mullite, a solid solution of mullite is formed.
2. From 0% Al_2O_3 (100% SiO_2) to 72% Al_2O_3, the complete solid phase exists below the eutectic temperature of 1595 °C, hence, with any composition within this alumina range, can withstand a temperature about 1500 °C without showing any sign of liquid formation. Above 1595 °C, a refractory in this composition

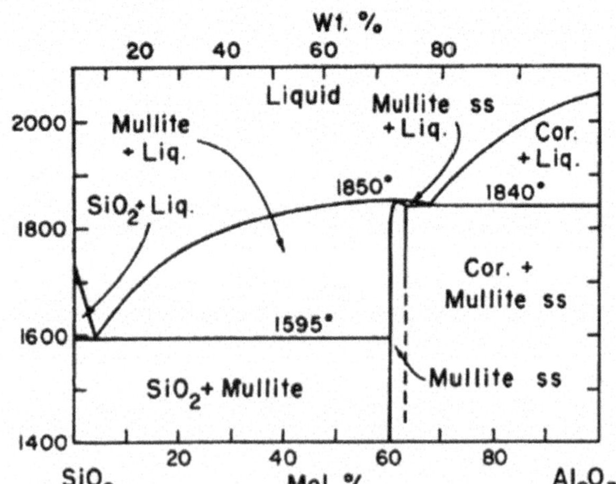

Fig. 1.10 A typical Al_2O_3-SiO_2 binary phase diagram

Table 1.7 Mechanical and thermal properties of mullite and corundum

Properties	Mullite	Corundum
Crystal system	Orthorombic	Trigonal
Density (g/cc)	3.15	3.96
Fracture toughness ($MPa.m^{1/2}$)	1.5–3.0	4.5
Bending strength (Mpa)	150–400	500
Linear thermal expansion coeff	4.2×10^{-6}/°C	8×10^{-6}/°C
Thermal conductivity (W/Mk) at 1400 °C	2.5–2.8	3.45
Melting point (°C)	1850	2050

Table 1.8 Raw material resource for different grades of alumina

Natural	Synthetic
Naturally occurring hydroxides	Fused corundum
Sillimanite group of materials	Mullite
Bauxite	Calcined Alumina
High-grade clay (>45% Al_2O_3)	Sintered Corundum

range is not suitable to use as partial liquid phase of cristobalite appears with the stable solid phase of mullite.

3. With the composition of 72% Al_2O_3 and above, no sign of melting occurs up to 1840 °C, and with increase in Al_2O_3 content, refractory properties improved. Pure Al_2O_3 has a melting point of 2050 °C. The properties of alumino-silicate refractories largely depend on the presence of two crystalline phases, mullite and corundum and the siliceous glassy phase. Mullite is having various advantages in its properties such as high thermal stability, low thermal expansion, high bending strength and creep resistance. Resistance to thermal shock is excellent due to its low thermal expansion and high work of facture. In the absence of glassy phase formation in grain boundary, mullite can retain >90% of its room temperature strength at more than 1600 °C. The comparative properties of mullite and corundum are given in Table 1.7.

For the production of high-alumina bricks, natural raw materials and synthetic materials are used; in consideration of availability and targeted properties several raw materials are used and summarized in Table 1.8.

1.4.3.1 Forms of Aluminium Hydroxides

Aluminium hydroxide occurs in its most naturally abundant α-trihydrate form (gibbsite) or as a mono hydrate (diaspore or boehmite). The other important crystalline forms of aluminium hydrates are bayerite, diaspore, boehmite and dawsonite. Synthetic bayerite or β-aluminium hydroxide is commercially produced by reacting CO_2 with sodium aluminate liquor at ambient temperature or bayerite precipitation from a super saturated aluminate liquor. Diaspore is another crystalline form of naturally occurring hydrate. The largest concentration of diaspore is found in

Table 1.9 Comparison of different aluminium hydroxides

Aluminium trihydroxide	Gibbsite	Alpha alumina trihydrates	Comparison of different aluminium hydroxides
	Bayarite	Beta-alumina trihydrates	Chemical formula: Al(OH)3, crystal system: monoclinic, specific gravity: 2.53
Aluminium oxide hydrates	Boehmite	Alpha alumina monohydrate	Chemical formula: AlO(OH), crystal system: orthorhombic, specific gravity: 3.02–3.07
	Diaspore	Alpha–beta alumina monohydrate	Chemical formula: AlO(OH), crystal system: monoclinic, specific gravity: 3.1–3.4
	Corundum	Alpha alumina	Chemical formula: Al_2O_3, crystal system: triclinic, specific gravity: 3.95–4.1

Chinese and Russian bauxites. The major demand of diaspore is in order to manufacture high-alumina refractories either by itself or by bonding with flint or plastic clay as per content of alumina needed in the finished product. The domestic refractory manufacturers are using diaspore, analysing Al_2O_3 56–62%, Fe_2O_3 1–4%, TiO_2 0.8–1.5%, Pyrometric Cone Equivalent (PCE) 36 (min.). In 2013–2014, about 20,400 tones diaspore was consumed in refractory industry. A competitive crystal structure and properties of different aluminium hydroxides are summarized in Table 1.9.

1.4.3.2 Bauxite Natural Source of Alumina

Bauxite is mainly originated from weathered rocks containing alumina. It is a naturally occurring raw material composed of one or more hydroxide of alumina minerals, together with silica, iron oxides, titania, alumino-silicate clays and other impurities in minor amount. The principle alumino oxide minerals are:

- Gibbsite (Al_2O_3. $3H_2O$).
- Diaspore (Al_2O_3.H_2O) [also known as α–AlO (OH)].
- Boehmite (Al_2O_3.H_2O) [also known as γ–AlO (OH)].

Bauxite is usually composed of gibbsite and boehmite in different proportions. At 900 °C, gibbsite loses the chemically bonded water and is converted to γ–Al_2O_3, and on heating to 1250 °C, it is converted to corundum (α–Al_2O_3). Similarly, at 800 °C, boehmite is converted to δ–Al_2O_3, and on further heating to 1100 °C, it is converted to corundum. Diaspore is converted to corundum on heating at 550 °C. In the presence of silica, at high temperature, mullite is formed and the quantity of mullite formation depends on the available silica content and this conversion involves volume expansion.

Refractory grade bauxite is calcined bauxite and only 1–2% of the globally mined bauxite is usable to fulfill the stringent requirement of maximum 2.0% Fe_2O_3. Corundum is estimated as the major phase after calcination of bauxite. The main

Table 1.10 Properties of Chinese, Guyanese and Indian bauxite

Properties	Chinese	Guyanese	Indian
Al_2O_3 (%)	86–88	88–89.5	85–86
SiO_2 (%)	6.5	6.5	8.0–10.0
Fe_2O_3	1.5–1.75	1.25–1.50	3.5
TiO_2	3.0–4.5	3.0	3.5–4.5
CaO + MgO	0.8–1.0	1.0	4.5–5.5
Alkalies	0.3	0.4	0.8
BD (g/cc) min	3.4	3.25	3.2
Corundum phase (%)	70–80	60–70	60–65
Mullite Phase (%)	10	15	15–20

resources are located in China, having total Chinese ownership of global refractory grade bauxite resources of more than 90% Al_2O_3.

Refractory-grade bauxites potentially have high refractoriness (1840 °C) as it can be deducted from Al_2O_3-SiO_2 system, where these materials are represented on the basis of their two main constituents: Al_2O_3 and SiO_2. Interesting to note is that such system exhibits low RUL (1450–1550 °C) compared to high refractoriness (PCE). This behaviour is attributed to the presence of liquid phases at lower temperature due to the presence of impurities in the bauxite, mainly TiO_2, Fe_2O_3 and alkali oxides. Such a low RUL value not only depends on the amount of liquid phases, but also on the distribution of liquid phases around the grain boundaries.

Presence of large amount of CaO causes disruption of direct bonding between corundum and mullite phases due to the formation of increasing amount of low-melting phases of anorthite and glass of silica and alkalies. Presence of TiO_2 forms $Al_2O_3.TiO_2$ as intermediate compound with high refractoriness (more than 1800 °C); however, part of this dissolves in glassy matrix, tending to lower the viscosity of glass, which are detrimental to high-temperature properties.

Presence of Fe_2O_3 increases the glassy phase by the formation of $Fe_2O_3.Al_2O_3$ which resulted in reducing the strength at elevated temperature. With TiO_2 it reacts to form $TiO_2.Fe_2O_3$ (Pseudo-brookite) as low-melting phase and significantly reduces the value of refractoriness under load. So, presence of both the oxides, TiO_2 and Fe_2O_3, in higher level would result in lowering the strength at elevated temperature. Presence of alkalies, MgO and CaO further deteriorates the high-temperature properties. To satisfy the high-temperature operating demand of refractory products, a high alumina, low iron content bauxite, generally with Fe_2O_3 content of 2.0% (max) after calcination, is favourable. The requirement of alumina content on calcined basis is 86% (min). TiO_2 content is 4.0% (max) and trace of alkali are the requirement of refractory grade bauxite. The properties of calcined bauxite of different sources are shown in Table 1.10.

1.4.3.3 Synthetic Alumina

With the technological development on product quality in ferrous, non-ferrous, cement, glass and petrochemical sectors, the operating parameters for production in those units became very stringent with respect to higher operating temperature and severe corrosive atmosphere. Hence, it demands superior quality refractory with superior high-temperature strength and resistant to physico-chemical reactions. Refractories made of higher alumina content with very low level of impurities are only suitable to manufacture those refractories; naturally occurring raw materials are not suitable due to higher level of impurities. Hence, it is required to use synthetically made high–alumina-enriched raw materials to use along with naturally occurring raw materials or only synthetic raw materials as per requirement. The main synthetic alumina sourced raw materials used in refractory bricks or monolithic manufacturing are brown fused alumina (BFA), white fused alumina (WFA) and white sintered alumina, fused mullite and sintered mullite.

Synthetic raw materials are of two types, Fused and sintered with varying alumina content and level of impurities. The input material for manufacturing those synthetic raw materials, white fused alumina, white tabular alumina and mullite is *calcined alumina*. Brown fused alumina is made of bauxite by the process of electrofusion.

Calcined alumina is produced by heat treatment of aluminium hydroxide which comes from the refining of bauxite by the Bayer process. Refractory-grade calcined alumina contains 99.0–99.5% with <0.2% soda. In normal Bayer process, the aluminium hydroxide in the bauxite is being digested and form soluble sodium aluminate, according to the reaction (1.25):

$$2Al(OH)_3 + 2NaOH \rightarrow 2NaAlO_2 + 4H_2O \tag{1.25}$$

$$2NaAlO_2 + 3H_2O + CO_2 \rightarrow 2Al(OH)_3 + Na_2CO_3 \tag{1.26}$$

The output of the extraction process is separated by dilution and cooling into a clear, filtered sodium aluminates liquor and an insoluble residue called red mud. The filtrate is treated with bubbling carbon dioxide to achieve $Al(OH)_3$ precipitate and soluble Na_2CO_3 (Eq. 1.26). The obtained aluminium hydroxide is calcined at 1200–1400 °C in a rotary kiln to produce calcined alumina. This Bayer-processed calcined alumina is fused in an electric arc furnace (EAF), solidified and ground to produce white fused alumina aggregate. The product is high purity alumina of more than 99.5% Al_2O_3 with high refractoriness and hardness. It is dense with large crystal size, which gives high corrosion resistance towards hot metal and slag. The volume stability is high when it is fired at high temperature, while degree of sintering is lower than sintered aggregates.

Brown fused alumina is produced by electrofusion of high-purity bauxite with having low alkali and iron oxide. Bauxite, fire wood and scrap are charged in EAF and fused to remove impurities in the form of ferro-alloys and results in electro-fused brown coloured alumina with more than 96.0% Al_2O_3, <0.5% F_2O_3 and < 1.5% SiO_2. Coke and iron bearings are added as reducing agent to lowering the impurities

(SiO_2, TiO_2, Fe_2O_3 and alkalies). Ferro-silicon is obtained as a by-product. The total world market is to be about 1.2–1.5 million tons and this is split between abrasive and refractory application. China accounts for about half of the global BFA supply. Degree of sinterability at elevated temperature is high due to the presence of TiO_2, which forms $TiO_2.Al_2O_3$, and hence improve the toughness. The porosity of BFA is lower than that of white fused alumina, which helps to reduce water absorption, amount of water requirement in monolithic refractory application. Refractory made of brown fused alumina has high abrasion resistance, thermal shock resistance and high corrosion resistance, though the corrosion resistance towards molten metal and slag is inferior to that of white fused alumina because it has a little higher impurity content.

1.4.3.4 Sintered Alumina

White sintered alumina, often called *Tabular alumina*, has a wide application for its excellent high-temperature properties and thermal shock resistance in particular. Its name is derived from the large, typically 100 to >400 μm, flat table like corundum crystals constituting sintered α-Al_2O_3 in hexagonal structure. Tabular alumina is distinctly different from other dense crystalline alumina such as white and brown fused alumina. The dense tabular corundum was first prepared in 1934 by Alcoa and it was commercially available in 1940 by the Aluminium Ore company.

For manufacturing tabular alumina, the calcined alumina feed stock was used with controlled surface area for high reactivity during sintering at elevated temperature. The fine powder stock was agglomerated by pelletizing process to form dense balls of desired size of 25–40 mm diameter. Green balls must have adequate green strength for handling and transporting to drying oven. The sintering takes place in a convertor, called Curtis convertor, developed by Alcoa or in shaft kiln, at 1820–1950 °C, also manufactured in high-temperature rotary kiln.

Good-quality tabular alumina is characterized by its large α-Al_2O_3 crystalline structure. High amount of closed porosity and very little open porosity result in very low water absorption. Although the total porosity of white fused alumina and tabular alumina is about the same level, there are significant difference in their grain porosity. The open porosity of fused alumina is two to three times greater than that of tabular alumina. Most of the porosity in white fused alumina consists in large sized open pores in large quantity, whereas over 50% of the total pores in tabular alumina is in closed pores. The porosity in the fused alumina is large in size and inter connecting open pores. This grain structure apparently accounts for lower grain strength compared to tabular alumina. In the presence of large numbers of closed pores and of controlled pore size, tabular alumina is superior in thermal shock resistance compared to white fused alumina and other alumino-silicates. Those closed pores help to arrest the crack propagation through the grains and grain boundary in mullite-bonded tabular alumina-based refractory. The above-mentioned properties of tabular alumina make it an excellent refractory raw material when incorporated into refractory enabling to obtain the following advantages:

Table 1.11 Competitive porosity and water adsorption of tabular alumina, WFA and BFA

Properties	Tabular alumina	White fused alumina	Brown fused alumina
Porosity (%)			
Open	3.6–5.1	12–14	8–10.5
Closed	6.2–6.9	1.2–1.8	1.0–2.5
Total	10.5–13.5	14–16	10–12
Water absorption (%)	1.1–1.7	3.8	3.8
Apparent Sp. Gravity	3.7	3.94	3.92
Grain density	3.5	3.45	3.4

- High fusion point.
- Excellent hot load resistance—resistance to creep.
- Good thermal shock resistance compared to white fused alumina.
- Good hot abrasion resistance.
- Low reheat shrinkage.
- High purity.

Competitive properties of tabular alumina and white fused alumina are given in Table 1.11.

1.4.3.5 White Fused Alumina and Tabular Alumina

1. White Tabular Alumina is characterized by its large α-alumina crystalline structure, high amount of closed porosity and very little open porosity. Very low water absorption.
2. Advantages of coarse crystalline Al_2O_3, similar to tabular grain size, have also been demonstrated for excellent thermal shock resistance. High amount of closed pores exhibits thermo-mechanical shock resistance through branching of crack propagation effect.
3. Thermal shock damage resistance of dense polycrystalline α-Al_2O_3 improves with increased grain size.
4. Although the total porosity of the tabular and fused alumina aggregates are about the same level, there are marked difference in their grain porosity. The open porosity of the fused grain is two to three times greater than that of tabular alumina as shown in the following table.
5. Most of the porosity of fused alumina consists of large sized open pores, whereas most of the porosity in tabular alumina is small sized closed pores. The porosity in the fused alumina is concentrated in large, connecting, open pores. This grain structure accounts for lower grain strength of fused alumina compared to the tabular alumina, which contains closed pores evenly distributed throughout its structure.
6. The nominal 0.2% Cr_2O_3 addition to white fused alumina gives a characteristics pink fused alumina exhibiting superior thermo-mechanical properties and it also improves thermal shock resistance.

Brown fused alumina grain is made using bauxite as feed material for fusion. Unlike the fused white and pink alumina, it exhibits a very high particle density (3.91 g/cc) and $< 2.0\%$ grain porosity. The grain strength of brown fused alumina is much higher than white and pink fused alumina, but about half the strength of tabular alumina.

Bauxite is considered as the prime raw material during manufacturing of 76–85 wt% alumina-containing high-alumina refractories. To develop mullite matrix in order to improve thermal shock resistance and other high-temperature properties, sillimanite group of material with calcined alumina are added. Bauxite bricks are not CO resistant due to the presence of high level of iron oxides. This must be considered for deciding its application in iron-making areas. Chemically bonded bauxite bricks are the most suitable refractory for rotary kiln in cement industry, aluminium melting furnaces, different heat treatment furnaces and refuse incinerators.

High-fired mullite bricks are used in the critical areas in different industries due to its following properties:

- High thermal shock resistance.
- Very high creep resistance up to 1650 °C.
- Very high refractoriness under load, >1700 °C.
- Volume stable.

Sintered or fused mullite is used to make 70–76% alumina bricks. High-fired mullite bricks with zircon addition provides very good service results in glass industry, crown of glass melting furnace, regenerator and other critical application areas, above 1700 °C service temperature.

In the range of 90–98% alumina refractory range the main raw materials are sintered or fused corundum with corundum–mullite bond matrix. The bricks made of sintered alumina with corundum-mullite matrix have excellent thermal shock resistance and creep resistance, and dense structure provides very low permeability resulting in good corrosion resistance. This refractory is largely used in flow control system of steel ladle and in induction furnaces. Thermal shock resistance can be further improved by adding zirconia. Properties of high-alumina bricks grades are shown in Table 1.12.

1.4.3.6 Corrosion of High-Alumina Refractories in the Presence of FeO and Fe_2O_3

1. Under reducing condition, the matrix forms liquid at 1210 °C, where a small amount of iron oxide is adsorbed by mullite and tridymite phases. Eventually it facilitates to form iron cordierite solid solution with mullite above 1210 °C.
2. High-alumina content expedites mullite and corundum phase formation, where no liquid forms below 1380 °C, even after absorption of considerable amount of FeO.

Table 1.12 Competitive properties of different grade of alumina refractories

Base material	% Al_2O_3	% SiO_2	% Fe_2O_3	% TiO_2	% Alkalies	Bulk density (g/cc)	RUL (ta)	Rte at 1000 °C	Thermal conductivity at 1000 °C (W/mK)
White fused alumina	>99.0	0.3	0.2	0.1	<0.4	3.14	>1750	0.8	3.3
Sintered alumina	90–96	4–6	<0.4	0.2	<0.5	3.0–3.2	>1750	0.8	2.5–3.3
Bauxite bricks	80–85	10–12	<2.5	<3.5	<0.8	2.8	1560	0.68	1.8
Fused mullite	72–75	22–24	0.3	0.1	<0.3	2.65	>1700	0.6	1.75
Sintered mullite	72–75	24–26	<1	0.2	<0.6	2.55	>1700	0.6	1.75
Andalusite	59–62	35	<1.2	<1	<0.8	2.5–2.6	1680	0.55	1.75
Sillimanite, kyanite and technical alumina	55–65	33–43	<1.5	<1	<0.8	2.55–2.65	1620	0.55	1.65
Calcined clay+ Grog	50–55	35–42	<2	<1.5	<1.5	2.4–2.55	1500	0.48	1.6–1.8

3. In normal atmosphere, the liquid formation in fireclay bricks is preferentially above 1380 °C. In high-alumina bricks, in the presence of $Al_2O_3 > 72\%$, liquid appears above 1460 °C.
4. In oxidizing atmosphere, Fe_2O_3 reacts with Al_2O_3 to form a solid solution and it can absorb significant amount of Fe_2O_3 without showing any liquid formation. The Fe_2O_3–Al_2O_3 forms a solid solution with mullite at high temperature and liquid forms above 1460 °C.

1.4.4 Magnesite Refractories

The magnesia refractories are classical type of basic refractory, and they are characterized by high refractoriness, with having high melting point of MgO (2850 °C) and corrosion resistant to basic steel-making slags. But they have few disadvantages such as high thermal expansion coefficient, poor thermal shock resistance and hydration. To improve the performances, many refractories based on MgO are combined with other materials such as alumina, chrome and carbon. Presently, use of magnesite refractories are restricted in backup lining LD-vessels and steel ladles. Magnesite refractory is produced from magnesia, and referred as magnesium oxide, MgO and the term periclase is the name of the mineral and it contains less than 10% impurities. Magnesia is also produced synthetically from seawater or brines and is called seawater magnesia.

1.4.4.1 Dead Burned Magnesium Oxide

The natural source of magnesia is magnesite ($MgCO_3$); deposits in Austria and China are famous worldwide; the other sources are in North Korea and Russia. It is mined underground or by strip mining and processed. The mined natural magnesite is made into usable refractory aggregate by calcining or high-temperature firing and thus the resultant high-temperature fired product is called dead burned magnesia. Naturally occurring magnesite has the theoretical weight percentage of MgO as 47.6, so about half the weight of magnesite is lost due to CO_2 evolution during sintering or dead burning. They are classified according to the iron content and CaO/SiO_2 ratio in the brick composition. Commercial grade MgO brick demands minimum 80 wt% of MgO, in order to achieve high performance.

1.4.4.2 Seawater Magnesia

MgO can be produced from seawater or brines or salt deposits, all containing common soluble compounds, magnesium chloride ($MgCl_2$). Usually, 2.15 g/l of Mg^{++} are dissolved in brine as chlorate and sulphate salts, and they can be extracted from seawater through replacement reaction as shown below:

$$MgCl_2 + Ca(OH)_2 = Mg(OH)_2 + CaCl_2 \tag{1.27}$$

$$Mg(SO)_4 + Ca(OH)_2 = Mg(OH)_2 + Ca(SO)_4 \tag{1.28}$$

The Mg(OH)2 thus obtained is generally fired in a rotary kiln or shaft kiln to produce magnesia clinkers. Its characteristics are different depending on firing temperature. Fine powder made by low-temperature sintering is very reactive, called reactive magnesia, and the clinkers fired in high temperature are called hard burned or dead burned, having well-developed crystalline structure. The MgO is thus produced as per the following equation:

$$Mg(OH)_2 = MgO + H_2O\ (\text{at } 900^{\circ}C) \tag{1.29}$$

Herein, the obtained MgO crystal size is about 150–350 μm, and are suitable to use as pure quality refractory aggregates.

1.4.4.3 Fused Magnesia

Densely crystalline MgO aggregate is produced by melting natural magnesite ($MgCO_3$) or seawater magnesia by melting in electro-fusion process at >2800 °C, followed by controlled cooling and crushing. The crystal size of fused magnesia ranges from 800 to 1000 μm, and nowadays the use of fused magnesia is very popular to produce high-quality MgO-containing refractories.

1.4.4.4 Characteristics of Magnesite bricks

There are various raw materials used to make magnesite bricks. Based on bonding system and manufacturing protocols, the product types are classified as fired and unfired. To produce fired magnesite bricks, properly graded magnesia aggregates are mixed with binders, pressed in high-capacity press, followed by drying and firing. The properties of magnesia bricks depend on firing temperature and schedule. Magnesite has poor sinterability due to its very high refractoriness, and hence the products have low strength at high temperature. To improve the thermo-mechanical properties, alumina-enriched raw material, such as bauxite, is added. Usually, bauxite contains more than 85% Al_2O_3 and therefore it contains free alumina in addition to the aluminosilicate compound such as mullite. Both the free alumina and mullite react with MgO to form magnesia–alumina spinel ($MgAl_2O_4$) and forsterite (Mg_2SiO_4), according to the following reactions:

$$MgO + Al_2O_3 = MgAl_2O_4\ (\text{Spinel}) \tag{1.30}$$

$$MgO + SiO_2 = Mg_2SiO_4 \text{ (Forsterite)} \tag{1.31}$$

Once the free alumina present in bauxite is consumed, MgO further reacts with mullite to form additional spinel and forsterite. The reaction is described as follows:

$$3Al_2O_3.2SiO_2 + 7MgO = 3MgAl_2O_4 + 2MgSiO_4 \tag{1.32}$$

The reaction with mullite starts at 1200 °C and it is almost completed after 1550 °C.

The manufacturing process of unfired bricks is similar to that of fired magnesite bricks, but the bricks are not fired. Some of the bricks are clad in steel casing or plates to provide increased strength and thermal shock resistance. In situ sintering during application provides the analogous properties of fired magnesite bricks.

1.4.4.5 Effect of Impurities

The common impurities like SiO_2, Al_2O_3, Fe_2O_3 and B_2O_4 are present in magnesite refractory raw material. The properties of the bricks predominately depend on the CaO/SiO_2 ratio, whereas Fe_2O_3 and B_2O_4 act as fluxing agents and they must be maintained at as low as 0.06%. When the CaO/SiO_2 ratio is greater than 2.0, all the parts of Fe_2O_3 combine with CaO or MgO to form calcium-ferrite, CaO. Fe_2O_3(C2F), or magnesium ferrite spinel ($MgO.Fe_2O_3$), respectively. Along with these phases partial presence of calcium silicates exhibits high thermo-mechanical properties. Figure 1.11 represents the ternary $CaO\text{-}MgO\text{-}SiO_2$ phase diagram and their typical analysis in order to understand the influence of refractory composition. If the weight ratio of CaO/SiO_2 is greater than 2.8, tricalcium silicate, Ca_3SiO_5, and free CaO can exist as the impurity or formation of additional liquid phase at 1850 °C. An exact 2.8 ratio can yield only tricalcium silicate. If the weight ratio of CaO/SiO_2 is between 2.8 and 1.87, tricalcium silicate and dicalcium silicate, Ca_2SiO_4, can be present, while the temperature of initial liquid formation is 1790 °C; at just 1.87, only dicalcium silicate can exist. When the CaO/SiO_2 weight ratio exists in between 1.87 and 1.4, dicalcium silicate and merwinite, $Ca_3MgSi_2O_8$, can be represented and an initial liquid can appear in the magnesia aggregate at 1575 °C. Further, weight ratio CaO/SiO_2 reduction in the range of 1.4–0.93 can produce a combination of merwinite and monticellite, $CaMgSiO_4$; with this additional-phase mineral assemblage with MgO, liquid can begin to be generated if the temperature reaches 1490 °C. A CaO/SiO_2 wt. ratio of less than 0.93 can yield the accessory-phase minerals monticellite and forsterite (Mg_2SiO_4), and a temperature of 1500 °C can begin to initiate melting in the MgO grain.

For these reasons, these impurities need to be kept low. The iron oxide content of MgO grains needs to be maintained low as it forms low-melting phases with a wide range in CaO/SiO_2 ratio. Boric oxide is also a very undesirable impurity, and thus the

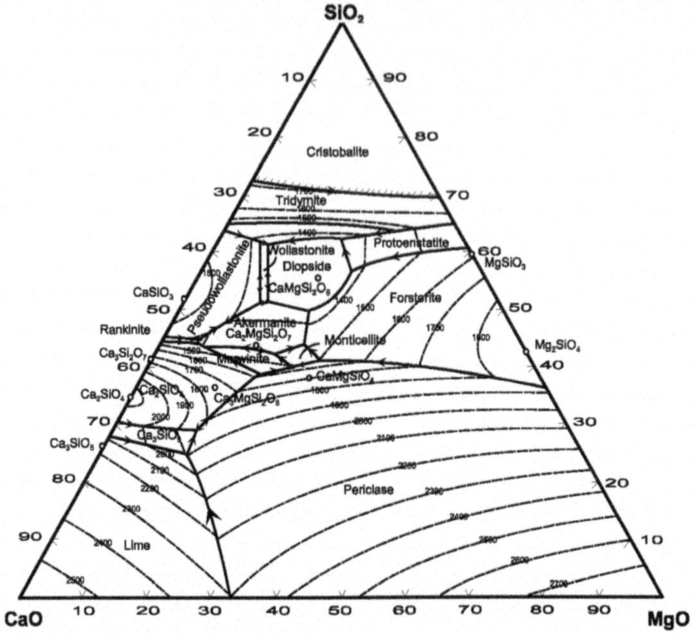

Fig. 1.11 CaO-MgO-SiO$_2$ phase diagram

boron content to be maintained within 0.02 wt% to achieve good hot strength of magnesite refractories.

1.4.4.6 Effect of MgO Crystallite Size

The crystal size of MgO plays a vital role in corrosion resistance of magnesia refractory in the presence of steel-making slag, as in refractory corrosion is a function of available surface area. Prior average crystallite size of dead-burned MgO ranged from 25 to 100 μm; current synthetic, dead-burned magnesium oxide products are available with crystallites ranging from 100 to 200 μm. Moreover, when the best corrosion resistance is needed, the industry trend is to use refractories containing at least some fused MgO whose crystallite size can be measured in the several millimetres range and individual MgO grains can be composed of fragments of single MgO crystals.

1.4.4.7 Poor Thermal Shock Resistance of MgO Refractory

The coefficient of thermal expansion of an essentially pure MgO refractory is very high; for example, at 1400 °C, the linear expansion of a fused MgO or iso-statically

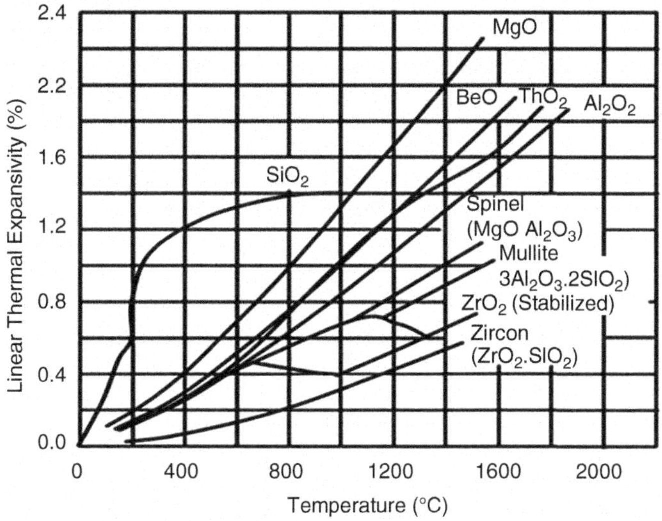

Fig. 1.12 Relative thermal expansion of refractory oxides [14]

pressed and fired MgO (99 wt.%) with having minimum purities and bulk density of 3180 kg/m^3 is about 2%.

Of course, incorporating other refractory raw materials with magnesia in a refractory body may alter this value. Care is required in using these materials to achieve adequate thermal shock resistance. The relative thermal expansion of different refractory oxides is shown in Fig. 1.12.

1.4.5 Dolomite Refractory

Dolomite refractory is mainly MgO–CaO composition, based on the natural mineral dolomite, which is a double carbonate of magnesium and calcium ($MgCO_3.CaCO_3$). The mineral dolomite, as mined, cannot be used directly as a raw material. So this must be calcined and sintered at high temperature to yield mixed oxide aggregate (MgO.CaO). Theoretically, the mineral contains 30.4% CaO (54.3% $CaCO_3$), 21.7% MgO (45.7% $MgCO_3$), and 47.9% CO_2. As other carbonate rocks, dolomite deposits are fairly common throughout the world. The impurities typically consist of silica, alumina, iron, manganese, sulphur and phosphorous. Refractory-grade dolomites are usually high enriched magnesia, which are considered to contain a minimum of 20% MgO (42% $MgCO_3$) and a maximum of 2.5% impurities.

Dolomite bricks are made of sintered dolomite and fused dolomite. They consist mainly of around 60% CaO and 40% MgO. In dolomite, CaO and MgO remain as solid solution, as shown in the phase diagram (Fig. 1.13). MgO grains are embedded

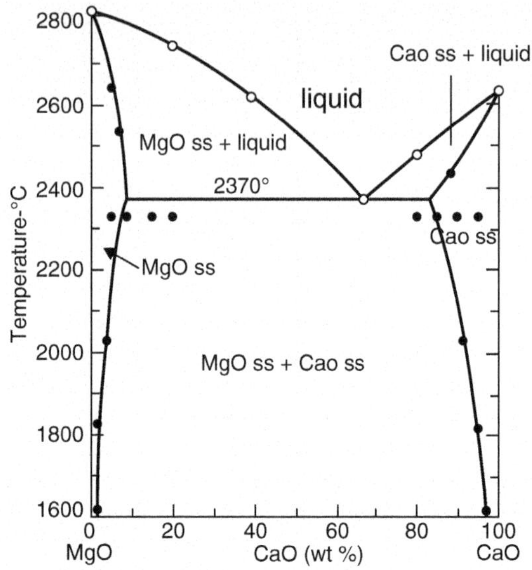

Fig. 1.13 CaO–MgO phase diagram

in fine matrix of CaO and the matrix is unsaturated. Hence, the CaO matrix can easily combine with other oxides, such as MgO, TiO_2 and ZrO_2.

1.4.5.1 Direct-Bonded Doloma Bricks (Fired)

Fired direct-bonded dolomite bricks are manufactured by mixing graded dolomite in different fractions, mixed with non-aqueous organic binders, followed by pressing and firing at high temperature. The fired bricks are impregnated with tar or pitch to improve the corrosion and hydration resistance. During high-temperature firing, ceramic bond is formed in the matrix and develops the strength. Direct bonding between the doloma grains is achieved by firing the brick through a tunnel kiln. The organic binder provides green strength to the brick for handling and stacking in the kiln, until the ceramic bond is developed.

Doloma brick shows a high degree of direct bonding. In the past, there has been some reluctance to using direct-bonded bricks for steel-making applications because of their susceptibility to thermal shock. Significant improvement in the thermal shock resistance of the brick is accomplished through the addition of small amounts of zirconia. The enhancement is attributed to microcracking caused by the formation of calcium zirconate as a result of reaction between the lime in the doloma and zirconia. Further improvements to the slag resistance of both direct-bonded and carbon-bonded doloma bricks have been made through the addition of high-purity magnesia. Doloma bricks that have been enriched with MgO are typically used in areas of ladles and AOD vessels where high slag resistance is required. High thermal conductivity and hydration are major limiting factors of dolomite refractories and several process conditions are adopted to overcome such incidences.

1.4.5.2 Chemically Bonded Dolomite Bricks

Tar-bonded tempered dolomite bricks were in use since last few decades. Due to ever-increasing environmental problems it becomes more difficult to produce and use tar/pitch-bonded products. Pitch belongs to the group of components which can cause cancer, if the content of benzo (A)-pyren exceeds 50 ppm. Pitch binders usually used in bricks contain about 10,000 ppm benzo (A)-pyren. Therefore, several precautions are decreed including additional health control of the workers in manufacturing and users' end.

Thus, many efforts had been made to substitute the tar/pitch binder. In recent, a lot of resin-bonded dolomite bricks are in use. The base of the binder is a thermo-setting phenol-formaldehyde resin. Because of the hydration problem of dolomite sinter, the binder needs to be moisture free. A chemically bonded dolomite brick with 3–4% residual carbon effectively improves the corrosion resistance due to non-wetting characteristics of carbon and enriched thermal shock resistance. Moreover, carbon also protects hydration of dolomite.

1.4.5.3 Direct-Bonded Dolomite Bricks with Improved Thermal Shock Resistance

Fired, direct-bonded dolomite bricks do not have organic compounds, which are the causes of volatiles or fumes. From that point of view, it is the only dolomitic refractory available in large quantity to use worldwide in rotary kiln for lime calcination and in AOD vessel to produce stainless steel. It does not cause any environmental hazards as it is free from any organic compounds. The direct bonding of these bricks with low porosity is not as flexible as the carbon-bonded bricks. Therefore, the recent dolomite brick development is concentrated to produce fired bricks with improved properties where thermal shock resistance is the very important factor in the perspective of application. Several methods are known from the refractory science to improve mechanical properties of oxide materials, which can be adopted to improve dolomite bricks. Few of them are:

- Zirconia and TiO_2 added refractory to improve thermal shock resistance and hot strength.
- Toughened by fibres.
- Formation of micro-cracks to arrest crack propagation by branching effect.

The zirconia addition offers a suitable possibility to improve the thermal shock resistance of burnt dolomite bricks, and compensate disadvantages compared to carbon-bonded bricks.

1.4.5.4 Dolomite Brick Selection Criteria

Higher steel temperature ranges and enlarged steel–refractory contact time are consequence in the secondary steel-making operations. High-alumina refractories had reached to their higher thermal limit and thus the necessity of highly basic slag and clean steel quality requires a basic lining. Preliminary characterization of the products allied to theoretical consideration allowed to select dolomite for the lining. The potential of dolomite had been improved by carbon addition and other additives. Enriching dolomite by adding MgO has improved its durability and became compatible to highly basic lining. The different solicited parts of the lining were optimized in order to reach a balanced wear as follows:

1. Tar-bonded dolomite is replaced by resin-bonded Dolo-carbon bricks to use in steel ladle, and due to low MOE, tar or tar/resin-bonded dolomite were used in convertor lining.
2. Impregnated dolomite is used in impact zone of steel ladle.
3. Magnesia-enriched dolomite is in use.
4. Chemically bonded dolomite to improve thermal shock and hydration resistance.
5. Direct-bonded dolomite bricks with improved thermal shock resistance.

1.4.5.5 Specification of Dolomite Bricks

The oxide component of dolomite bricks is prepared from raw dolomite with a molar ratio $CaCO_3/MgCO_3$ of approximately 1 with very low impurity content of <2.0%, mined from selected pits and calcining a high-density clinker in shaft or rotary kiln at a temperature of 1800–2000 °C. The clinker is crushed and grinded as per requirement of brick-making operation. Brick-processing protocol implies two major bonding system: fired direct bonded and unfired carbon bonded as described above.

Unfired carbon-bonded bricks are prepared from graded dolo sinter mixed with approximately 4.0% dehydrated tar/pitch or resins and other organic binders with high carbon yield. In preference the graphite and carbon black are added for carbon-enriched brick. However, in case of pitch-bonded system, the pitch is preheated at 60–80 °C; in few cases the dolomite mix is also heated during mixing to avoid hydration. After pressing, the bricks are tempered or cured at a temperature between 200 and 300 °C. The final bond is developed in preheating at service. Above 600 °C a strong carbon bond is formed as a result of coking the binder. Above 1000 °C, under oxidizing condition the carbon bond is partly oxidized and dolomite grains would sinter together to form a direct bonding in hot face. Specifications are shown in Table 1.13.

Table 1.13 A competitive dolomite brick composition and properties with respect to bonding system and base materials

Properties	Unit	Carbon-bonded dolo bricks	Carbon-bonded Mag-Dolo bricks	Burnt Mag–Dolo bricks	Pitch impregnated dolo bricks
Chemical					
MgO	%	42–45	64–70	55–64	55–64
CaO	%	53–58	35–38	42–46	42–46
Al_2O_3	%	<1.0	<0.8	<1.0	<1.0
Fe_2O_3	%	<1.0	<0.5	<1.0	<1.0
SiO_2	%	<1.0	<1	<1.0	<1.0
Physical					
Bulk density	g/cc	2.9	2.96	2.85	2.95
Porosity	%	4.5–5.5	3.6–4.0	16–18	3–3.5
MOR at 1400 °C	kg/cm^2	60–80	65–85		
Thermal expansion (800 °C)	%	1.6	1.65	1.2	1.2
Thermal conductivity (1000 °C)	W/mK	2.4–2.6	2.4–2.6	2.0–2.2	2.2–2.4
Cold crushing strength	kg/cm^2	450–500	550–600	550–700	700–800
RUL	ta (°C)			> 1700	>1700

1.4.6 MgO-C Refractories

The evaluation and use of magnesia refractories in combination with carbon started in early 1970s. In early 1950s pitch-bonded dolomite and later magnesia-enriched tar-bonded dolomite developed primary for the basic oxygen furnaces. In those days, some of these linings in the basic oxygen furnaces lasted only 100 heats. Significant improvement in vessel refractory performances is noticed after introduction of pitch-bonded magnesia bricks. About that time magnesia purity became a factor and special low boron, 96% MgO grains having a lime to silica ration of >2.0% was used extensively.

In 1980s, the development of resin-bonded magnesia graphite, first with higher carbon content and then with the addition of antioxidant to prevent oxidation, has escalated the life of BOF. Recently fused magnesia with larger crystal size sintered magnesia were introduced to further improve the corrosion resistance. The following three types of MgO-C bricks are in use in lining BOF and steel ladles:

- Sintered magnesia (97% MgO) with medium quality graphite (95% C).
- Fused magnesia (>98% MgO with large crystals) and pure graphite (> 98% C) and addition of antioxidants.

- Pure sintered magnesia (99% MgO) with pure graphite and addition of antioxidants.

The different quality MgO-C bricks used in BOF and steel ladles and role of different antioxidants in carbon-containing refractories had been described in detail in Sects. 7.1 and 8.3.1.

Depending on the production process and different mix composition, the carbon-bonded magnesia bricks are further classified as follows:

- Tar-impregnated fired magnesia bricks.
- Carbon-containing impregnated magnesia bricks.
- Resin-bonded MgO-C bricks.

1.4.6.1 Production Technology of Fired Carbon-Bonded Magnesia Bricks

Up to firing stage these bricks are manufactured as same process as the fired magnesite bricks. The fired bricks are then placed under vacuum at 150–200 °C and impregnated with molten tar or pitch and finally tempered and the impregnated tar yields carbon in the process of pyrolysis. In this process, the residual carbon obtained 2–4%. Due to impregnation of tar and consequently pyrolysis facilitates the carbon filling and deposition into the open pores and thus reduces permeability, concerning the properties, these bricks only differ from fired magnesite bricks by reduction of slag infiltration.

1.4.6.2 Production Technology of Carbon Bond Impregnated Magnesia Bricks

Pitch is used as a binder in this type of bricks. After mixing magnesia and pitch at 100–200 °C, the bricks are shaped by hot pressing and finally tempered at 250–300 °C to obtain sufficient hot strength. The addition of carbon black, use of dehydrogenation agents and pitch impregnation under vacuum are the further development in process technology to improve the carbon bond. The carbon black can be dispersed in pitch and improves the high-temperature strength. The carbon-bonded magnesia carbon bricks are characterized by residual carbon content of more than 7%.

1.4.6.3 Production Technology of Resin-Bonded MgO-C Bricks

Sintered and fused magnesia with larger crystal size are used as the source of MgO; graphite is mainly used as a source of carbon; other carbon-containing materials are carbon black, hard pitch and powder resins. Purer quality MgO and graphite have improved the physico-chemical properties and corrosion resistance at high operating

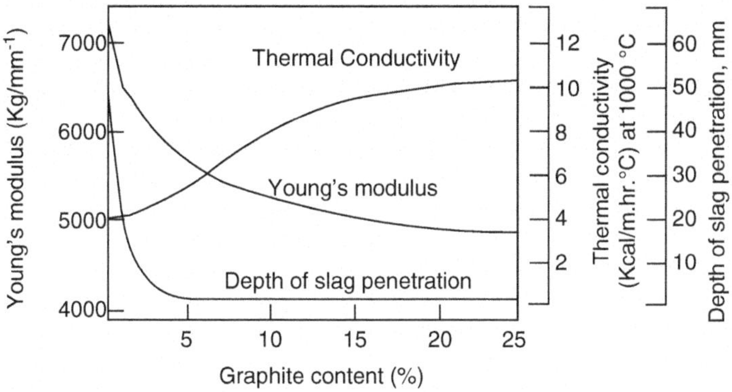

Fig. 1.14 Effect of graphite addition in properties of MgO-C bricks [15]

temperature. The carbon yield has been improved up to 20–25%. Effect of graphite addition on the properties of MgO-C bricks is shown in Fig. 1.14.

Phenolic resin is a synthetic poly-condensation product made of phenol and formaldehyde, usually used as synthetic resin. Refractory manufacturers are using two types of resins, "resol" or "novolak", either in powder or in liquid form, and are added in magnesia graphite mix. In contrast to pitch, pyrolysis of the resins occurs in solid stage. Antioxidants, such as Si, Al, metal powders and B_4C, are used in the mix to improve oxidation resistance and high-temperature strength. The resin-bonded magnesia–carbon bricks are shaped in cold condition using high capacity pneumatic or hydraulic presses, followed by curing those bricks at a temperature of 300–350 °C to develop adequate strength. The properties of different quality of MgO-C bricks are summarized in Table 1.14.

1.4.7 MgO–Cr_2O_3 Refractories

Advances in basic refractories have led to the replacement of conventionally used magnesia-chrome brick with modified varieties of bonds and aggregates. The problem with magnesia bricks was its high sensitivity to thermal shock resistance and a tendency of shrinkage when exposed to high temperature for long periods. With chrome bricks the problems of thermal shock and poor hot strength were restriction to its usage. With rapid research composite refractory of chrome and magnesite came to be manufactured with different types of bond systems and corresponding properties specific to each bond system type. Hence, categorization can be made under chemically bonded, silicate-bonded, direct-bonded and rebonded types of magnesia-chrome brick. Direct-bonded and rebonded magnesia-chrome refractories have been used in the RH vessel linings due to their good mechanical and refractory properties like high hot temperature strength, resistance to slag attack and dimensional stability.

Table 1.14 Chemical and physical properties of different quality of bricks

Properties	Unit	Fired pitch impregnated	Pitch bonded	Resin-bonded MgO-C
Chemical analysis				
MgO	%	95–97	88–97	92–98
Al_2O_3	%	0.1–0.5	0.1–0.6	0.5–3.0
Fe_2O_3	%	0.2–3.0	0.2–6.0	0.2–2.0
CaO	%	0.5–4.0	0.6–4.0	0.6–3.0
SiO_2	%	0.1–1.5	0.1–2.0	0.2–2.0
Residual carbon	%	1.5–2.0	5.0–6.0	14–22
Physical properties				
Bulk density	g/cm^3	3.1–3.15	2.96–3.0	2.75–2.85
Porosity	%	1.0–5.0	3.0–7.5	2.0–4.0
Cold crushing strength	N/mm^2	40–80	30–50	30–40
Mod. of rupture at 20 °C at 1400 °C	N/mm^2	10.0–20.0 10.0–20.0	5.0–10.0 6.0–8.0	8.0–12.0 10.0–16.0

The reaction between chrome ore and magnesite outlines the fundamental chemistry of the magnesite chrome bricks. Magnesia chrome bricks exhibit good corrosion resistance, high melting temperatures, low thermal conductivity, good thermal strength and excellent thermal shock resistance. The refractoriness (PCE) of magnesia chrome brick is over 2000 °C, and the refractory under load (RUL) is above 1550 °C. It has good steady volume in high temperature and the thermo-mechanical spalling performance is better than magnesia bricks. They are used mainly in the secondary refining furnaces, especially degassing furnaces, cement kilns and non-ferrous metal refining furnaces. Despite these advantages, magnesia-chrome-based refractory linings also suffer from a serious disadvantage. The chrome within the magnesia-chrome bricks can react with the lime and alkali to form hexavalent chrome which is a toxic material requiring special disposal measures.

Magnesite chrome bricks can be either silicate bonded or direct bonded. The balance of properties of the bricks is a function of the magnesite-to-chrome ratio. The changing properties as a function of ratio of magnesite to chrome ore makes it possible to choose products best suited for an individual application. Silicate-bonded bricks have a thin film of silicate minerals that surrounds and bonds together the magnesite and chrome ore particles. Direct-bonded bricks have the direct attachment of the magnesia to the chrome ore without intervening films of silicate. Direct bonding is obtained by combining high-purity chrome ores and magnesites and firing them at extremely high temperatures. The direct-bonded bricks are used under more severe service conditions. Chemical-bonded magnesite chrome and conventionally manufactured chrome magnesite bricks do not have the high-temperature strength, load resistance or slag resistance of burned compositions.

Rebonded magnesia chrome brick is made with synthetic fused magnesia, shaped under high pressure and fired at ultra-high temperature. It is used in tapping hole of

AOD furnace, the lower vessel of vacuum chamber and snorkel of RH furnace, and the slag line.

1.4.7.1 Chrome Ore

Chrome ore is primely composed of two constituents, a chrome grain and a gangue material of magnesium silicate composition (mainly forsterite) that surrounds the chrome grain and occupies cracks and fissures between the chrome grain. The chrome grain consists of a solid solution of spinel compounds with having a general formula RO. R_2O_3 for example (Mg,Fe)O. $(Al,Cr,Fe)_2O_3$. The chrome spinels have high refractoriness with melting point more than 1800 °C. The density of chrome ore varies from 3.8 to 4.5 g/cc depending upon the spinel composition and the proportion of gangue present. Chrome is an important refractory material that can increase the refractory bricks erosion resistance ability.

1.4.7.2 Bond Aggregate System: Synthetic and Rebonded

There are mainly two types of aggregate systems used in the manufacture of magnesia-chrome bricks, synthetic and rebonded system. The basic variation in both is the type of grain used for fabrication of brick. In the synthetic system brick an admixture of refractory grade periclase and refractory grade chrome ore is taken, pressed and fired at 1700 °C to form the sintered brick. The conventional direct-bonded and silicate-bonded bricks have been manufactured using the synthetic aggregate mechanism.

The rebonded system was developed using pre-reacted magnesia and chrome grain to fabricate the brick. In this system, the grain used to fabricate the rebonded fused grain is made by fusing a mixture of magnesia and chrome ore in an electric arc furnace at 2450 °C. The melt is then poured into moulds and allowed to cool and the material is crushed. Minor amount of binders is added and the material is pressed and fired at 1550 °C to manufacture the bricks. The brick structures are composed of large, fused aggregates embedded in a matrix of finer fused grains. Comparative chemical analysis and physical properties of sintered and rebonded brick are shown in Table 1.15.

1.4.7.3 Different Types of Magnesia-Chrome Bricks Based on Bonding System

Chemically Bonded

These are unfired bricks where a bonding chemical (sodium silicate/sodium sulphate) is added to the common raw materials, magnesite and chromite, and mixed together. The pressed shapes are dried and a chemical bond develops slowly due to

Table 1.15 Chemical and physical properties of different class of mag-chrome bricks

Properties	Sintered magnesia chrome	Rebonded magnesia chrome
Bulk density (g/cc)	2.95–3.00	3.35–3.45
Apparent Porosity (%)	17–20	12–15
Chemical composition (wt%)		
MgO	62–66	50–55
Cr_2O_3	14–17	20–27
Temperature of formation (°C)	Fired at 1550 °C	Fusion at 2450 °C, Fired at 1500 °C

reactions of the chemical with the refractory material. The bricks are cured at 60–100 °C for reactions to take place. This type of brick is not suitable for high-temperature application as the chemical bond is stable only up to 700 °C beyond which the bond disintegrates.

Silicate Bonded

These are fired bricks in which in situ forsterite bonding occurs, and considered as the prime boning mechanism in these bricks. The periclase grain reacts with the gangue (with the silicate impurities) in the chromite and magnesium silicate bonding occurs on firing at high temperature around 1200 °C that binds the periclase and chrome grains. Due to the presence of large amount of silicate impurity, there is formation of low–melting-phase liquids and glassy phase that result in low hot strength and low thermal shock resistance of these bricks, and hence these are not considered very feasible for high-end applications.

Direct Bonded

Direct-bonded bricks may be defined as those in which the silicate films, which typically surround the chrome ore and periclase grains in the conventional bricks, are replaced to a greater or lesser extend by direct periclase–spinel and periclase–periclase bonds. Such bricks are characterized by high hot strength, improved resistance to slags and dimensional stability at temperature up to 1800 °C. These bricks are fired (during manufacturing) at high temperatures of 1700 °C and use superior quality high-purity raw material to minimize the gangue and silicate phase. The mechanism of bond formation follows a series of steps that begin with temperatures around 1500 °C and the final direct bonding is completed by 1700 °C. The direct bond forms during cooling of the bricks by the precipitation of spinel from solution in liquid silicates at the height firing temperature. There is direct bond development between periclase and spinel with negligible silicate bond. With increasing the firing temperature, there is more direct bonding of periclase and spinel, and Fe_2O_3 and Cr_2O_3 from the chrome spinel diffuse into the magnesia grain, forming the secondary spinel. Thus, the microstructure of direct-bonded brick includes both the primary chromite spinel and secondary chromite spinel. These bricks have good strength at high temperature, volume stability and good corrosion resistance.

When magnesia–chrome and chrome–magnesia bricks are fired at high temperature, each particle of chrome ore is surrounded by a rim of silicate, forsterite ($2MgO.SiO_2$) composition. Similarly, the periclase crystals are surrounded by films of silicates. And the composition of silicate with ratio of MgO to SiO_2 of less than 2.0 forms the silicate gangue. During the course of firing, the gangue minerals melt at a temperature depending on their composition, but typically at around 1200 °C, low-viscous phase migrates toward chrome ore surface, where they react with the magnesite matrix to form solid silicates of higher melting points. However, no remarkable change is noticed during further heating at 1500 °C. Upon heating to 1600 °C the silicate films around the chrome particles migrate out of the matrix of the brick, augmenting the silicate films originally associated with the periclase and leaving gaps around the chromite grains.

As the firing temperature raised to above 1700 °C, the periclase rims around the chrome particles become a major feature of the microstructure. A crystal boundary precipitation gave a high degree of direct bonding after firing to temperature around 2000 °C. Saturation of the liquid silicate by spinel facilitate during firing and the direct bond develops on cooling either by precipitation of spinel from solid solution in the periclase or by the crystallization of periclase and spinel from solution of the liquid silicates at high temperature. Thus, the degree of direct bonding of a "direct-bonded brick" gradually diminishes as the temperature is raised during high-temperature operation. However, at all temperatures below the original firing (brick processing) temperature, some bonds independent of the silicate phase are common to enhance the mechanical strength.

Rebonded

The major raw materials to manufacture such bricks are magnesia and chrome ore, which are melted together in an electric arc furnace to produce the fused magnesia-chrome aggregate. This aggregate is size graded, mixed, pressed and fired to make bricks. These bricks have excellent thermal properties and corrosion resistance. The uses are same as direct bonded but these bricks are used for applications that involve severe conditions. Manufacturing of fused grains and their using to production of bricks are the prime importance of rebonded brick, different from conventional bricks. Conventional bricks, including direct-bonded bricks, are fabricated of mixture of calcined magnesia and chrome ore; none of the materials are fused, although in most cases the magnesia has been calcined at above 1500 °C, to form periclase.

The grain used to manufacture rebonded bricks is made by fusing or melting a mixture of magnesia and chrome ore, in an electric arc furnace at a temperature around 2400 °C. After the magnesia and chrome batch had been fused, the material is cooled in slow rate of cooling for proper crystallization, crushed and grinded to different fractions as per requirement of brick-making process. After pressing, the bricks are fired at a temperature above 1550 °C.

Semi-rebonded

Semi-rebonded magnesia-chrome brick is a kind of refractory product with the partial fused magnesia chrome used as the raw material to burn. The main mineral composition of semi-rebonded magnesia-chrome brick is periclase, spinel and small amount of silicate. The granule is of opacified structure, the spinel in periclase is full-grown, and the matrix part is of metasomatic structure. The direct combination of crystal is periclase–pericalse or periclase–spinel, silicate phase exists in isolated state between intergranular. The thermal shock resistance of semi-rebonded magnesia-chrome brick is better than the direct-bonded magnesia-chrome bricks. Properties of rebonded $MgO\text{-}Cr_2O_3$ bricks is shown in Table 1.16.

Advantages of using this type of brick are:

1. Compact structure.
2. Low porosity.
3. Good high-temperature strength.
4. Thermal shock resistance is better than rebounded magnesia-chrome brick.

1.4.7.4 Effect of Fused Magnesia–Chromite Quality on Rebonded Bricks

The bricks containing electro-fused magnesia-chromite of different $MgO:Cr_2O_3$ ratio has different properties with respect to thermal shock resistance and corrosion resistance. The ratio greatly determines the strength-related properties of the bricks as well. Improved sintering property is attribute to the $MgO: Cr_2O_3$ ratio reduction and subsequent increasing the direct bonding phenomenon. Use of reduced ratio of $MgO:Cr_2O_3$ produces a brick having high hot strength, but at the same time decreases its spalling resistance, resulting from increased MOE and decreased thermal conductivity as per the following equation of thermal shock index R:

Table 1.16 Properties of rebonded magnesia-chrome bricks

Properties	Unit	Type1	Type2	Type3	Type4
MgO	%	50	55	60	65
Cr_2O_3	%	35–37	28–30	15–18	18–20
Porosity	%	16	17	16	16
Bulk density	g/cc	3.35	3.25	3.05	3.1
Cold crushing strength	Mpa	45	40	40	40
RUL	ta	>1700	>1700	>1700	>1700
Thermal shock resistance at 950 °C air cooling	Cycle	40	45	40	40

$$R = W(1 - \gamma).k/E\ a \quad (1.33)$$

(W = work of facture, γ = poison's ratio, k = thermal conductivity, E = MOE and a = Thermal conductivity).

Effect of $MgO:Cr_2O_3$ ratio in Mag-Chromite fused clinker is summarized below:

1. The lower the ratio of the feed fused clinker, higher the hot strength of the bricks.
2. At a carbon level normally associated with secondary steel making, it is more advantageous to use a magnesia–chromite bricks of low $MgO:Cr_2O_3$ ratio.
3. The lower the ratio in the brick produced, it shows less resistance to thermal shock.

1.4.7.5 Competitive Features of Direct-Bonded, Rebonded and Semi-rebonded Bricks

1. Direct-bonded bricks have superior thermal shock resistance but low corrosion resistance towards oxidizing slags of vacuum degassers.
2. Rebonded bricks have superior corrosion resistance but low thermal shock resistance.
3. To optimize both the properties, semi-rebonded bricks had been developed, which has better thermal shock resistance than rebonded bricks as well as better corrosion resistance than direct-bonded bricks. Hence the use of semi-rebonded bricks have been popular in steel-making processes.
4. The comparative corrosion resistance and thermal shock resistance in the presence of FeO-enriched slag, of direct-bonded and semi-rebonded bricks are shown in Fig. 1.15.

1.4.7.6 Corrosion of Magnesia:Chrome Bricks

With an increase in quantity of Cr_2O_3 in mag-chrome bricks, the corrosion rate decreases. Moreover, higher weight percentage of electro-fused mag-chrome clinker reduce the corrosion rate in both rebonded and semi-rebonded bricks. In the case of direct-bonded bricks, sintered magnesia particles react with FeO at high temperature forming solid solution with adequate expansion. Due to the expansion, direct-bonded bricks experience cracks and subsequently wear out. In the case of semi-rebonded bricks, the corrosion resistance in the presence of FeO at high temperature is better than direct-bonded bricks. In semi-rebonded bricks, the spinel precipitated at grain boundary hinders the intrusion of Fe-oxides. This seems to be one of the reasons why the semi-rebonded bricks are superior in corrosion resistance compared to the direct-bonded bricks. It has been confirmed that the presence of electro-fused mag-chrome particles in semi-rebonded bricks exhibit a high Fe-oxide resistance at high temperature.

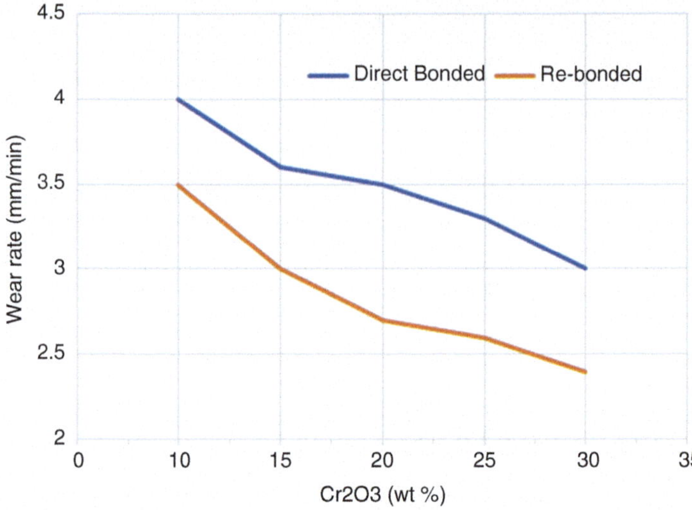

Fig. 1.15 Effect of Cr_2O_3 on the wear of direct-bonded and semi-rebounded MgO-Cr_2O_3 brick in the presence of iron oxide-enriched slag

1.4.8 Spinel Refractories

Spinel can refer to a group of minerals of the same structure. Spinel exists in two different types, normal spinel type, having crystal structure AB_2O_4, and inverse spinel of structure $BABO_4$, with reference to distribution of the cations and A represents divalent cation and B represents trivalent cation. The common trivalent cations are Aluminium (Al^{+3}), Ferric iron (Fe^{+3}), and chromium (Cr^{+3}) and divalent cations are magnesium (Mg^{+2}), ferrous iron (Fe^{+2}), etc. The spinel structure normally forms a cubic closed packing comprising 32 oxygen ions in the octahedral sites of which are located 16 numbers of B type cations and 8 numbers of A type cations are located tetrahedral sites.

Depending on the composition of the spinel, the edge length of the unit cell ranges from 8.1 Å to 8.55 Å. Aluminium-containing spinel has the closest spacing, iron-containing spinel has the widest, and chromium-containing spinel has the intermediate values of the lattice spacing. Few common types of spinel are shown in Fig. 1.16. Among those group of spinels, only magnesia–alumina spinel (MgO.Al_2O_3) and magnesia-chrome spinel are widely used as a refractory material because of their excellent high-temperature properties and thermal stability.

The most common spinel and the only compound in the binary system MgO–Al_2O_3 is magnesium–aluminate spinel with a stoichiometric ratio of 71.8% Al_2O_3/28.2% MgO. Magnesium–aluminate spinel is a highly refractory material, with a melting point of 2135 °C. One important feature of all types of spinels is the tendency to form substitutional solid solution among the spinels or with other

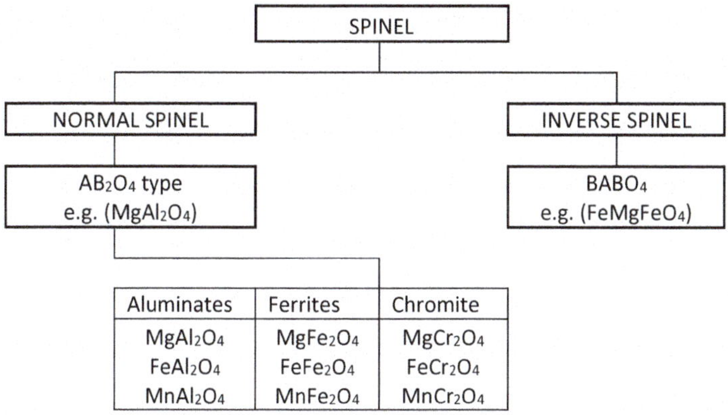

Fig. 1.16 Characterization of spinels

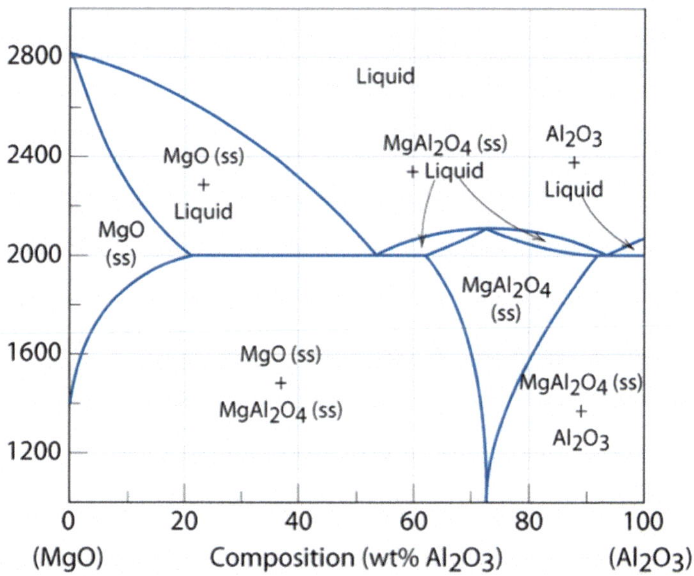

Fig. 1.17 $MgO–Al_2O_3$ phase diagram

oxides, as shown in $MgO-Al_2O_3$ phase diagram (Fig. 1.17). For $MgO.Al_2O_3$ spinel both magnesium and aluminium cations can be replaced by others with similar size.

The $MgAl_2O_4$ structure shows an increasing phase region that increases with temperature, specially towards higher alumina content. The $MgAl_2O_4$ can form solid solutions with its components, MgO and Al_2O_3 [16]. This allows to produce alumina-enriched spinel as well as magnesia-enriched spinel. Pre-reacted alumina-enriched spinel has the capability to absorb FeO and consequently retard the

Table 1.17 Comparative critical properties

Properties	Unit	Spinel	Corundum	Periclase
Chemical composition		$MgAl_2O_4$	Al_2O_3	MgO
Melting point	°C	2135	2015	2800
Density	g/cm^3	3.6	4.1	3.8
Specific heat	J/kgk	1076	1130	1202
Thermal expansion coefficient	$\times 10^{-6}$	7.6	8.8	13.5
Thermal conductivity	W/mK	5.8	6.3	7.2

infiltration of slag into spinel-containing refractory. The comparative properties of spinel, MgO and corundum are shown in Table 1.17. Low thermal expansion, favourable thermal conductivity and high thermal shock resistance of spinel advocates as one of the suitable materials in refractory use.

1.4.8.1 In Situ and Preformed Spinel

$MgAl_2O_3$ spinel can be added in refractory composition in two different methods:

1. **In situ spinel formation**: Spinel is formed in situ by adding magnesia in high-alumina refractory or adding alumina in magnesia-enriched refractory. At high temperature during firing of bricks or preheating and high-temperature application of high-alumina castable, MgO reacts with Al_2O_3 to form spinel and it involves high volume expansion. Preformed and in situ spinel-containing refractory could be very similar with respect to chemical composition and bulk density of the product, but they show distinct difference in other physical properties, such as thermal expansion, hot strength and slag corrosion resistance. In situ formed spinel in high-alumina refractory show a stoichiometric composition but no alumina reaches solid solution. Stoichiometric spinel can absorb FeO or MnO by substitution of MgO, which improves corrosion resistance in steel-making slag.
2. **Preformed spinel**: Pre-reacted spinel does neither have the expanding behaviour nor any unwanted hydration problem due to the absence of free MgO. This can be added in higher quantity in high-alumina refractory as there is no adverse effect of high expansion. Preformed spinel is pre-synthetized either by process of fusion or by sintering the in-put raw materials alumina and magnesia. The grains of higher crystal size can be produced by fusion process, which improves corrosion resistance. The processes involve batch-wise fusion of suitable MgO and Al_2O_3-rich initial raw materials in an electric arc furnace or sintering in a rotary kiln at about 1900 °C. Fused spinel has the crystal size >500 μm and that of sintered spinel, 50–70 μm.

1.4.8.2 Features of Spinel as a Superior Refractory Material

The important technical properties of spinel compared to magnesia and corundum are summarized below:

1. The difference in thermal expansion of these phases creates micro-cracks at high temperature, responsible for superior thermal shock resistance through micro-crack toughening of spinel-containing refractory.
2. The decrease in density involves volume expansion in spinel formation from MgO and Al_2O_3, needs to be considered and has advantages in lining steel ladle and formulation of castables.
3. It is important to use the high-purity raw materials to produce spinel as a preformed raw material to use in refractory bricks and castable. When bauxite is used in place of pure synthetic fused or sintered material, the resulting spinel performs worse. The low-melting phases are formed in the presence of high SiO_2, Fe_2O_3 and other impurities.
4. The optimum spinel addition in high-alumina refractory is in the range of 15–30%, preferably 20–25%. Lower quantity of spinel addition does not improve corrosion resistance, and higher amount of spinel results in high slag penetration, as spinel does not react with infiltrated slag.
5. In situ spinel formation in refractory bricks and castable is advantageous for high-temperature strength and thermal shock resistance, but it has an adverse effect of expansion during spinel formation, and results in cracking of refractory and lowering strength. Hence, addition of MgO in high-alumina refractory to form in situ spinel is optimized by about 3–5% MgO to from 12 to 18% spinel, and preformed spinel is also added to optimize spinel content for high corrosion resistance.

1.4.9 Silicon Carbide Refractories

Silicon carbide (SIC) is virtually not found as a natural raw material, but is produced by way of electric arc process using petroleum coke and silica sand as starting materials reacting them at a temperature of above 2000 °C. It is also called "Carbo-rundum" and the refractories made of SiC is called "Carborundum Refractories". The reaction of carborundum (SiC) takes place in two stages:

$$SiO_2 + 2C = Si_{(vapor)} + 2CO_{(g)} \tag{1.34}$$

$$Si + C = SiC \tag{1.35}$$

or

$$SiO_2 + C = SiO_{(g)} + CO_{(g)} \tag{1.36}$$

$$SiO_{(g)} + C = Si + CO_{(g)} \quad (1.37)$$

$$Si + C = SiC \quad (1.38)$$

The quartz and carbon react at about 1000 °C and the reaction accelerates above 1600 °C. The SiC is manufactured in a rectangular electric resistance batch furnace. Carbon electrode is used and over the electrode the SiC is formed and deposited. A core of coke or graphite serves as resistance. After spreading the charge, which consists of a mixture of ground coke and quartz sand, taken in stoichiometric ratio with 1–2% of NaCl and 5–10% of fine sawdust, the current is passed. The product thus manufactured consist of crystalline SiC, graphite and admixtures of Fe and Al oxides, which are removed by washing with water and dilute sulphuric acid. The sawdust provides the necessary porosity after burning to provide passage to escape out the CO and other gases, addition of NaCl to the charge eliminates the impurities in the form of volatile chlorides of iron and aluminium. The process of SiC formation takes 25–36 h. The furnace is cooled in 5–6 days. A 1500 kw furnace produces 6500–6800 kg of SiC in 36 h with a power consumption of 8.7 kw-h per kg of SiC produced [17]. Pure carborundum is colourless and consists of transparent crystals. The commercial products are usually green or black depending on the presence of impurities. The properties of crystalline SiC are given below:

Specific Gravity: 3.17–3.23.
High Mohr Hardness: 9.5–9.75.
High Thermal conductivity: 18–22 W/mK, at 600 °C and 16–20 W/mK at 1300 °C.
Low thermal expansion: $4.3–4.5 \times 10^{-6}$ from room temperature to 1700 °C.

The use of SiC in refractories has started in late 1960s for Blast Furnace. The main advantages of SiC refractories are:

- High strength at elevated temperature.
- Excellent thermal shock resistance and thermal stability.
- High thermal conductivity.
- Resistance to corrosion and erosion in iron-making slag and metal.

Refractory bricks made of SiC and mixes of SiC grains alone cannot sinter in true sense like alumina, mullite or clay. This is why, the refractory manufacturers endeavour to develop different binding systems, which retain the attractive properties of SiC raw materials. Different bond types are used in manufacturing silicon carbide refractories, summarized as follows:

(a) Oxidic or Silicate bond:
The oxide-bonded SiC bricks are formed by adding a refractory clay, silica and mullite to the silicon carbide grains and clay acts as a sintering aid. Mixtures of SiC, clay, Al_2O_3 and SiO_2 are pressed in high-capacity presses, dried and fired at around 1400–1550 °C in an oxidizing atmosphere. Silicate phase is formed during firing the bricks, which is a weakest link in the bonding chain. Generally, the hot strength is affected due to the presence of the liquid silicate phases at high

temperature. Amount of SiC present in final silicon carbide refractory composition is around 80%.

(b) Nitride or oxy-nitride bond.
Si_2ON_2 (silicon oxy-nitride), Si_3N_4 (Silicon Nitride) or SiAlON (silicon aluminium oxy-nitride) are the bond phases for SiC bricks. The pressed bricks made of graded SiC, silicon metal powder and Al_2O_3 are fired in the absence of oxygen and under nitrogen atmosphere at around 1450–1550 °C. During firing, the nitrogen gas reacts with silicon metal forming in situ silicon nitride (Si_3N_4) and this nitride gets accumulated at the grain boundary and filling up the pores in the brick structure. Hence, in silicon metal containing compositions, the dimensional change is very marginal. The stable in situ nitride phases are formed during prolong firing and they bring about both the excellent high-temperature strength and corrosion resistance.

In oxy-nitride-bonded materials, formed silicon nitride further reacts with silica to form Si_2ON_2 phase in the fired shapes. The compositions containing both silicon powder and alumina, the formed Si_3N_4 reacts with Al_2O_3 and SiAlON is formed in the products.

(c) Direct bonded (self-bonded).
The bricks made of graded SiC, silicon powder and carbon fired at around 1400–1500 °C, in the absence of oxygen, to form in situ SiC bond within the bricks. Both the SiC modifications (β-SiC and α-SiC) are used as high-temperature bond. The low-temperature phase, β-SiC, is formed below 1400 °C and α phase is formed at high temperature. These products have higher high-temperature properties and corrosion resistance than oxide-bonded products. In direct-bonded SiC refractories, firing conditions along with amount of free graphite or carbon and silicon metal added are required to be precisely controlled to control the formation of reaction bonding within the bricks and shapes. Those refractories contain more than 90% SiC in the compositions. Mixtures of α-SiC grains with very fine reactive α-SiC powder can be bonded at around 2400 °C in vacuum or in the presence of inert gas by re-crystallization of SiC. In the perspective of refractory applications, considered properties are excellent thermal shock resistance, hot strength and corrosion resistance. Properties of different qualities of silicon carbide bricks are shown in Table 1.18.

1.4.9.1 Oxidation Reactions of SiC

High strength and high resistance against thermal shock of SiC refractories are widely applied in metallurgical furnaces. However, they are weak in oxidation resistance and are oxidized in the presence of oxygen and specially in water vapour. Oxidation occurs in the presence of oxygen by the following reactions:

Table 1.18 Properties of SiC bricks

Properties	Unit	Oxide, silicate bond	Bonds with nitrogen		Direct bond	
			Si_3N_4	SiAlON	β—SiC	Re-crystallized a—SiC
Chemical properties						
SIC	%	50–75	70–80	70–80	90–95	>98
SiO_2	%	2.5–15	0.5–3.0	2.5–6	<3.0	<1.0
Al_2O_3	%	0–15	0–1	3.5–10	nil	<0.1
Fe_2O_3	%	0.1–0.8	0.1–0.8	0.1–0.8	<0.5	<0.1
Minarology						
SiC	%	75–90	70–80	70–80	90–95	> 98
Si_3N_4	%	nil	15–25	4.5–15	2.5–5.0	nil
Si_2ON_2	%	nil	nil	9.5–20	nil	nil
SiAlON	%	nil	nil	9.5–20	nil	nil
Mullite	%	0–10	nil	nil	nil	nil
Crystobalite/tridamite	%	1.5–10	< 5.0	nil	0–3.0	<1.0
Corundum	%	0–10	nil	<5.0	nil	nil
Glassy phase (amorphous)	%	2.5–5	< 2.5	< 5.0	0–3	nil
Physical properties						
Maximum service temperature	°C	1450	1600	1650	1450	1600
Bulk density	g/cc	2.4–2.7	2.5–2.7	2.6–2.8	2.5–2.7	2.5–2.7
Porosity	%	17–20	13–18	12.5–18	14–20	14–18
MOR at 1400 °C	MPa	15–25	35–50	40–60	30–40	40–50
Thermal expansion coefficient	$\times 10^{-6}$	5.5–5.8	4.8	4.9	4.9	4.8
Cold crushing strength	MPa	50–120	150–300	150–300	130–180	150–200
RUL (t 0.5)	°C	1550	>1700	>1700	>1650	>1700
Thermal conductivity at 1200 °C	W/mK	9.5–10	15–20	10.5–18	10.5–15	18–20

$$SiC_{(s)} + 3/2\,O_2 = SiO_{2(s)} + CO_{(g)} \tag{1.39}$$

$$SiC + O_2 = SiO_{(g)} + CO_{(g)} \tag{1.40}$$

In the case of the reaction in Eq. (1.39), SiO_2 is formed over the SiC grains and SiC shows a weight increase. A protective layer of SiO_2 forms and it is a form of passive oxidation which only occurs to the extent of SiO_2 layer formation and the

SiO_2 layer thus formed prevents further oxidation. In the case of reaction (1.40), because the SiO gas is formed, and it escapes out from the SiC surface, a weight loss of SiC occurs during the oxidation and it is referred to active oxidation as the reaction proceeds continuously.

Passive oxidation takes place on the condition of high partial pressure of oxygen; the oxide film of SiO_2, formed on the surface of SiC, has superior adhesion and prevent further oxidation as the diffusivity within SiO_2(s) is low. Conversely, active oxidation takes place in the condition of low partial pressure of oxygen. The SiO (g) and CO(g) are generated and the protective layer of SiO_2 is not formed on the surface of SiC.

1.4.9.2 Oxidation of SiC in the Presence of Water Vapor

In the presence of water vapour, SiC reacts directly with water vapour as shown below:

$$SiC(s) + 3H_2O(g) = SiO_2(s) + CO_2\ (g) + 3H_2(g) \tag{1.41}$$

Whenever water vapor co-exist with oxygen, the oxidation reaction becomes more severe, compared to oxidation by oxygen alone. Furthermore, the H_2O reacts with the SiO_2 protective layer to form SiOH, disrupting the glass network, and the protective layer become porous and it is unable to protect further oxidation. In a water vapour atmosphere, the oxidation of SiC accelerated compared to SiC oxidation by air or oxygen, only.

1.4.9.3 Oxidation of SiC in CO Atmosphere

The oxidation reaction of SiC by CO gas is described by the following reactions:

$$SiC(s) + 2CO = SiO(g) + 2C(s) \tag{1.42}$$

It generates SiO gas which further reacts with CO according to:

$$SiO(g) + CO(g) = SiO_2(s) + C(s) \tag{1.43}$$

Thus, resultant yield becomes:

$$SiC(s) + 2CO = SiO_2(s) + 3C(s) \tag{1.44}$$

By this mechanism, SiC experiences active oxidation under the condition of very low partial pressure of oxygen, $P_{(SiO\ +\ CO)} > P_{O2}$, and form protective SiO_2 layer.

1.4.9.4 Oxidation of SiC in the Presence of Alkali ($Na_2O + K_2O$)

When SiC is exposed in the presence of alkali, it is widely known to accelerate the oxidation of SiC. It is believed that the SiO_2 formed on the SiC surface further reacts with alkali to form low-melting compound and diffusivity of oxygen through the low-melting SiO_2 layer increases and therefore the oxidation process steadily progressed. However, nitride-bonded and oxy-nitride-bonded SiC show better alkali resistance.

1.4.9.5 Damage of SiC Refractories

The oxidation is encountered one of the main factors for the damaging of bricks and castable lining made of SiC refractory materials. Nitride and oxy-nitride-bonded are better oxidation resistant than oxide-bonded and direct-bonded SiC bricks. The damage due to oxidation depends on the various gas atmosphere and partial pressure of oxygen, experienced by SiC refractories, as summarized below:

- The oxidation of SiC is active and most severe in the presence of water vapor and oxygen.
- In the presence of low P_{O2}, protective layer of SiO_2 on SiC does not form and then it processes active oxidation, but in the presence of high P_{O2}, protective SiO_2 layer forms and it protects further oxidation.
- Oxidation of SiC brick begins above around 750 °C and increases with increase of temperature. It attends maximum at around 1250 °C. Significant shrinkage of SiC bricks takes place in oxidizing atmosphere within a temperature range of 1400–1500 °C.
- With regard to Fe and oxidation reaction in the presence of molten steels, oxidation by FeO and dissolution of SiC in molten pig iron are considered. The oxidation of SiC by FeO depends on the following reaction:

$$SiC + 3FeO = SiO_2 + 3Fe + CO \quad (1.45)$$

In the slag-metal interface of blast furnace trough, FeO is continuously supplied from slag film, and due to high turbulence in hot metal flow, protective layer of SiO_2 is removed from the surface of SiC, and oxidation of SiC continues.

- In the presence of high O_2 ppm in hot metal, dissolution of SiC into hot metal occurs by the following equation that eventually influence the Si metal concentration in hot metal.

$$SiC(s) = [Si] + [C] \quad (1.46)$$

In consideration of the above facts, selection of bonding system, manufacturing protocol and application zone, and installation/operational practice determine the performance of non-oxide ceramic-based SiC refractories.

1.4.10 Zircon and Zirconia Refractories

Zircon is zirconium silicate ($ZrSiO_4$), the main raw material for zircon brick is zircon sand of 100–300 μm particle size. Ground zircon, called zircon flour, from milling of zircon sand, is also used. The zircon raw materials have no plasticity, so fireclay or organic binder is added during mixing. The firing temperature of the shaped product is 1400–1500 °C, depending upon the type and amount of binders.

Zircon bricks have good wear resistance, less wettability by molten metal and slag, and hence good corrosion resistance. Alumina is added to zircon to form zircon-mullite composition. The bricks made of this composition are more resistant to thermal shock and corrosion than zircon bricks. In steel making, zircon refractory is used in tundish nozzle, due to its non-wetting property with liquid steel, and has excellent spalling and erosion resistance.

1.4.10.1 ZrO_2

In recent years, zirconia (ZrO_2) has a wide application in steel plant refractories due to its high corrosion resistance and low thermal conductivity. It has high expansion coefficient, which helps to compensate shrinkage in monolithic and prevents crack formation at high temperature.

South Africa is the most important deposit site of ZrO_2 mineral, baddeleyite found in nature. In addition, ZrO_2 is produced synthetically from $ZrSiO_4$ through fusion process. ZrO_2 is polymorphic in nature, and displays different crystal structure in different temperatures range, with no change of chemistry. It exists in three crystalline forms, monoclinic, tetragonal and cubic. Pure ZrO_2 has a monoclinic structure at room temperature and stable up to 1170 °C, then above this temperature it is converted to tetragonal with a volume contraction of about 5.0%, while tetragonal to cubic transformation takes place at about 2300 °C. The conversion process is reversible during cooling; tetragonal to monoclinic transformation occurs with a volume expansion of 4–5%, inducing high compressive stresses in the material. The transformation of phases in ZrO_2 and linear change with increase of temperature is shown in Fig. 1.18. Due to irregular volume change with temperature, there would be an intensive crack formation if a brick consists of only pure zirconia. Such volumetric change of pure ZrO_2 can be arrested by adding MgO, Y_2O_3, CaO or other rare earth materials, as called stabilizers, and the process is known as "Stabilization of ZrO_2".

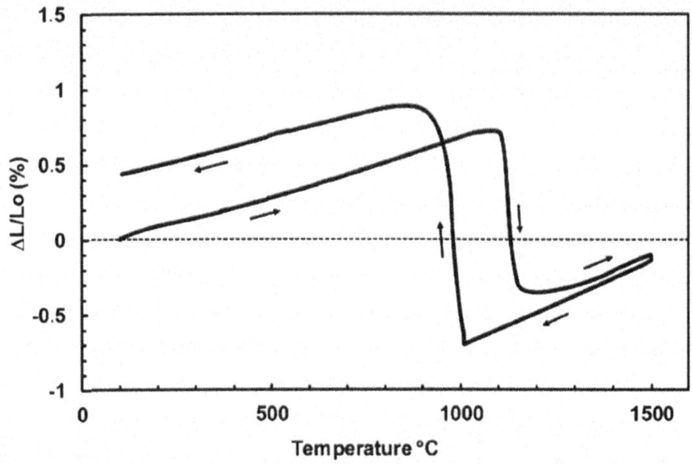

Fig. 1.18 Linear change of pure ZrO_2 with temperature [18]

1.4.10.2 Stabilized Zirconia

Stabilized zirconia is a mixture of ZrO_2 polymorphs obtained at room temperature, by the addition of stabilizers. Two types of stabilized materials are in use, fully stabilized and partially stabilized zirconia (PSZ). In fully stabilized ZrO_2, the crystal structure is stabilized in cubic form at room temperature and no further polymorphic transformation takes place during heating and cooling. It shows linear expansion and the expansion is higher than un-stabilized ZrO_2.

With the addition of stabilizing oxides in concentration less than those required for complete stabilization, zirconia can also be partially stabilized in a multi-phase form, known as PSZ. It consists of cubic ZrO_2 as a major phase, and monoclinic and tetragonal zirconia precipitates as the minor phase. To date, ZrO_2 stabilized with Y_2O_3 shows the best mechanical properties, and MgO-stabilized zirconia is widely used in refractory industries; the latter one has some cost benefit. The comparative thermal expansion behaviour of un-stabilized (monoclinic), partially stabilized and fully stabilized (Cubic) is shown in Fig. 1.19. The manufacture of zirconia products starts with stabilized or partially stabilized zirconia raw materials which is ground, mixed and graded material mixed with organic binder, shaped in high-capacity press or in isostatic press, dried and fired at more than 1700 °C.

Excellent corrosion resistance is the characteristic of ZrO_2. The main reason is that it is not easily reacting with other oxides except for SiO_2. Mixture of alumina and zirconia and zircon ($ZrSiO_4$) with Al_2O_3 addition form high corrosion-resistant refractory. Zircon with alumina forms zirconia-mullite grains, manufactured by fusion process, and graded raw material is used to form bricks, and used for good

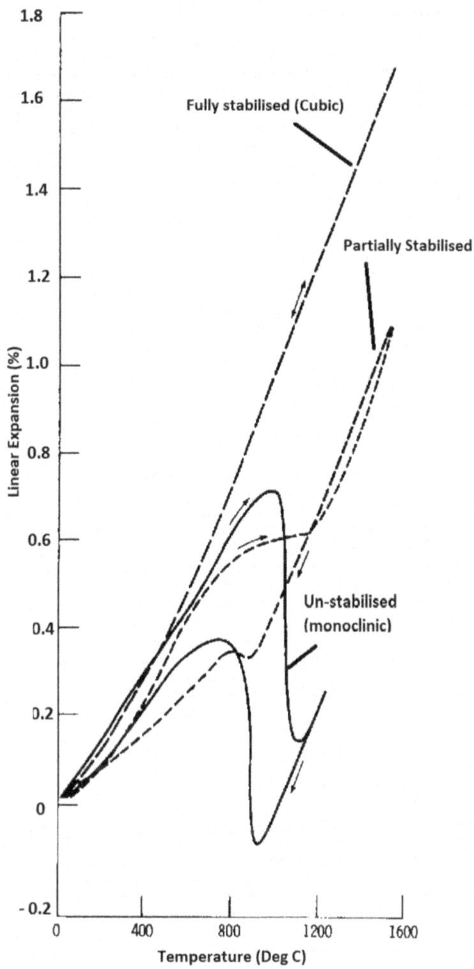

Fig. 1.19 Comparative thermal expansion of different quality ZrO_2 [19]

corrosion resistance zone. Further, combination of ZrO_2 with Al_2O_3 reduces the MOE, resulting in high thermal shock resistance. The well-known use of this material is with addition of carbon in manufacturing alumina-zirconia-carbon slide gate plate refractory in steel ladle. ZrO_2 refractory is used as insert in slide plates as flow control of Ca-treated steel and tundish nozzle for its excellent corrosion resistance. The tentative properties of zircon, zirconia and zirconia-mullite refractory are summarized in Table 1.19.

Table 1.19 The tentative properties of zircon, zirconia and zirconia-mullite refractory

Properties	Unit	Zircon (nozzle for tundish)	ZrO_2	Zirconia-mullite (Iso-statically pressed)
Chemical composition				
ZrO_2	%	62–65	94–96	18–20
Al_2O_3	%	0.5–1.0	0.1–0.2	65–70
SiO_2	%	25–35	0.4–0.6	10–15
MgO	%	nil	2.0–2.5	
Fe_2O_3	%	<0.5	<0.5	<0.2
Physical properties				
Bulk density	gm/cm^3	3.5–3.75	4.2–4.5	3.4–3.8
Porosity	%	16–18	12–14	1.0–2.0
Cold crushing strength	MPa	>80	>100	>300
RUL (ta)	oC	>1600	>1700	>1650
CTE at 1000 °C	%	0.4	0.8	0.65

1.5 Monolithic Refractories

Monolithic refractory is the name generally given to all unshaped refractory products, the word monolithic coming from the word *monolith* meaning "big stone". These are materials which are installed in some form of casting or ramming processes and ultimately harden to form a solid jointless mass. Monolithic refractories have no definite shapes or dimensions. Those refractories can be moulded or can be given any shape as per requirement. Because of this reason, these products are also known as the ready-to-use refractories and can be used as such by adding water or other binding agents, on site to repair or build a new lining in ladles, tundishes or critical domains in other furnaces. They consist of graded refractory aggregates mixed with different binders and special additives for modification of properties. The lining made of monolithic is usually jointless. Monolithic lining is gaining popularity during the last five decades worldwide due to the following advantages:

- It is a jointless lining which eliminates the possibility of hot metal and liquid slag passage through joints (in case of brick lining).
- Absence of firing for manufacturing monolithic saves energy.
- Process time for manufacturing monolithics is less as they do not need moulding and firing.
- Very fast installation reduces the downtime of equipment.
- Less man-power required.
- Easy for refractory maintenance jobs.
- Effective lining possible in complicated lining areas.

- Any types of shape can be made by the process of pre-casting.
- Lower refractory consumption.

1.5.1 Types of Monolithic Refractories

Classification of monolithic refractories depends on several aspects. Depending on the application purpose, monolithic refractories are divided in the following groups:

- Castables for monolithic construction and repair by gunning or grouting.
- Mortar for jointing shaped refractory.
- Ramming masses for filling gaps and providing expansion allowance.
- Plastics for monolithic construction.

Depending on the bond systems, monolithic refractories are divided into the following three groups:

- Hydraulic bond with hardening and hydraulic setting at room temperature.
- Ceramic bond with room temperature setting or heat setting.
- Chemical bond with setting by chemical reaction at room temperature or high temperature.

A broad classification of monolithics is shown in Fig. 1.20.

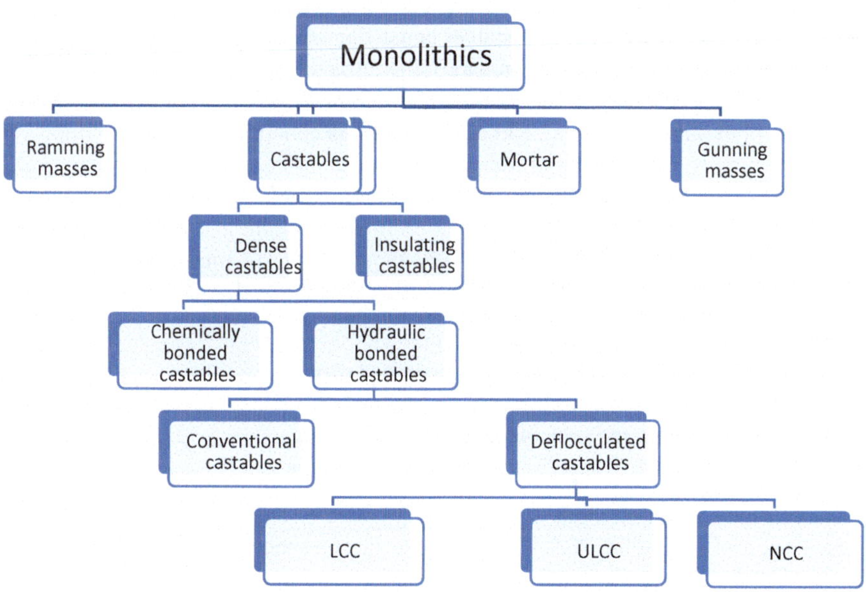

Fig. 1.20 A broad classification of monolithics

1.5.2 Castables

Castable refractories are the most common material type of monolithic refractories. Castable is literally combination of two words, "cast" and "able", in which refractory materials should have "fluidity" after mixing with water or any other liquid binder. In present time, castables are playing a leading role in the increased application area in the steel industry due to adaptability to automated means of installation, labour saving and reduced time for installation, which resulted in high productivity and reduced cost. Installation methods involve conventional casting, vibration casting, pumping, injection and troweling. The castable is generally recipe of graded aggregates and high-alumina cement as bond. With the development of low-cement castable and other high-performance castables, other different types of bonds have been used. The following properties are desirable in a castable:

- Sufficient fluidity with water content as low as possible for casting operation.
- Segregation of grains must be avoided by controlling water content.
- The casted body should be as strong as possible.
- Development of strength with temperature increase.
- Good durability in operating condition.

1.5.2.1 Conventional Castable

In the early development period, alumina cement was the primary bonding agent while phosphate and silicate bonds were also in use to a lesser extent in particular application areas. Conventional castables consisting of graded refractory aggregates and high-alumina cement have been part of the refractories since many decades. The properties of this product are largely dependent upon the selection of aggregates and hydraulic high-alumina cement, and improvement is generally restricted to improve aggregate quality and grading of the products.

Conventional castable contains 15–25% of high-alumina cement; this amount is necessary to achieve adequate strength at room temperature and medium temperature range. About 8–15% water is usually required during mixing depending on types of aggregate and quantity of cement added. The utilization of water in castable mix is:

- 0–8% water is taken up by the grain open porosity
- 6–10% water is needed for developing hydraulic bond by hydration of cement
- 2–6% water causes the castable mix to flow.

Compared to the fired bricks with same raw material used, the conventional castable shows lower strength in medium temperature range of 250–700 °C due to dehydration of hydraulic phases formation in the presence of cement. After firing above 1100 °C, ceramic bond such as mullite started to form and develop the strength of castables. Other disadvantages are higher porosity after heating, low-melting phases formed in the presence of high CaO reacted with Fe_2O_3 and SiO_2 content in high-alumina cement, and reduced hot strength above 1500 °C.

1.5.2.2 Low-Cement and Ultra-Low-Cement Castable

The introduction of low-cement and ultra-low-cement castables has been one of the most impressive developments in the field of monolithics in recent years. Castable with low cement contents was first mentioned in a French patent in 1969 [20]. Reduction of cement content to 5–8% without reduction of strength with temperature increase was achieved through the addition of a fine-grained refractory materials and de-flocculent, promoting a homogeneous distribution of the cement and micro-fine material added and reducing the water content for homogenous mixing of the castable. During mid-1970s, micro-silica with average grain size of 0.15 μm has come into use. This is amorphous SiO_2 added with cement in low-cement castable (LCC) composition. Refractory castables thus produced have high mechanical strength, high erosion resistance, good thermal chock resistance and low shrinkage.

Further improvement was made by introducing ultra-low-cement castable (ULCC) with the addition of high-alumina cement which is only 1–3 wt%, and it shows excellent high-temperature properties with temperature increase. The presence of low-melting phases such as anorthite ($2CAO.Al_2O_3.SiO_2$) and gehelanite ($CaO.Al_2O_3.2SiO_2$) is very less, and as it contains higher amount mullite at high temperature, the thermal shock resistance is significantly improved with higher strength at elevated temperature, compared to low-cement castable. The comparative composition of those different types of castables is shown in Table 1.20.

The high-performance castables, low-cement and ultra-low-cement castable described are based on a common bond system that incorporates a careful proportioned blend of fine components which, when mixed with selected aggregates, give a range of products to cover a wide range of maximum service temperature and different applications. The different products under LCC and ULCC are described below:

- 40–45% Al_2O_3 castable based on calcined clay grog
- 55–65% Al_2O_3 castable based on andalusite, kyanite, sillimanite, spinel
- 70–85% Al_2O_3 castable based on andalusite, mullite, spinel, bauxite and fused alumina and tabular alumina.

Table 1.20 Compositions of castables

Composition	CC	LCC	ULCC	NCC
Base materials	70–80%	80–90%	80–90%	80–90%
High-alumina cement	15–18%	5–6%	1.5–3%	0–1.5%
Reactive/calcined alumina	5–10%	5–10%	5–10%	5–15%
Micro silica	Nil	4–5%	5–6%	6–8%
Other additives	Nil	Nil	0.5–1.5%	4.5–6.5%

- 90–95% Al_2O_3 castable based on tabular alumina and fused alumina.

In steel plants, 55–65% Al_2O_3 and 70–85% alumina castables are widely used in backup lining of steel ladles and tundishes. The comparison is made with similar alumina content and service limit is shown in Table 1.21.

1.5.2.3 Castable Manufacturing Process

The main processes in the manufacturing of castables are mixing the graded aggregates with powders and various additives. The processes involve storage of raw materials, drying of raw materials, grinding, grading, and addition of different binders for final mixing and packaging. Castable raw materials may be shipped in bulk in the form of lumps of graded aggregates. Various fine materials used in castable readily absorb moisture, and therefore exposure in moisture or in direct sunlight must be avoided. No wet materials to be used, and if the lumps or graded aggregates contain moisture, it must be dried before crushing and grading. The coarse and medium fractions of aggregates are produced by crushing and grinding process. The lumps are to be crushed in jaw-crushers, gyratory or cone crusher and then grinded in grinding rod-mills or any other types of mills.

The final stage of grinding operation produces the fine fractions. The typical equipment used are different types of ball mills. Batches of aggregates are made by the combination of coarse, medium and fine grains.

Castables are a mixture of various raw materials with specified grain size distribution and formulated to a wide variety of compositions. Precise weighing, mixing and distribution are highly important in manufacturing castables, and especially critical to produce LCC and ULCC, which involves addition of micro-fine powders and additives in very small quantity. For effective mixing, mixers such as V-type, double cone mixer or high intensity mixers are used. An effective process for mixing fines in small quantity is to prepare blend consisting of fines and add the mixer for final mixing with the aggregates.

1.5.2.4 Testing of Castables in Laboratory

Sample preparation is one of the most important activities for testing of high-strength castable. Addition of water or other binders, type of mixer machine used, mixing time, use of vibration during casting, curing and drying influence the tested properties. Any deviation in processes deliver erroneous results. After drying for 24 h and heat treatment in recommended temperatures, samples are tested following the same procedure for testing shaped refractory. The procedure for sample preparation of high-performance castable (LCC/ULCC) is explained below:

- Use of high-intensity Hobert-type mixer and addition of recommended quantity water. Drinking cold water to be used to maintain the temperature of castable within 15–22 °C, during mixing and casting.

Table 1.21 Classification and competitive castable properties with respect to alumina content

Properties	Unit	55–65% Al_2O_3 castable		70–85% Al_2O_3 castable		90–95% Al_2O_3 castable	
Chemical analysis		Conventional (CC)	High-performance castable (LCC/ULCC)	Conventional (CC)	High-performance castable (LCC/ULCC)	Conventional (CC)	High-performance castable (LCC/ULCC)
Al_2O_3	%	55–65	55–65	70–85	70–85	90–95	90–95
SiO_2	%	35–42	40–50	15–22	8.0–10.0	1.5–2.0	2.2–3.0
CaO	%	6.5–7.5	1.5–1.8	5.5–6.5	1.5–1.8	4.5–5.5	0.3–1.5
Fe_2O_3	%	1.0–2.5	1.0–1.5	1.0–1.5	0.8–1.2	0.8–1.0	0.2–0.8
Physical properties							
Max. service temperature	°C	1450–1550	1450–1550	1600	1600–1700	>1700	>1700
Bulk density at 110 °C,	g/ cm^3	2.2–2.4	2.3–2.6	2.2–2.4	2.45–2.55	2.5–2.65	2.5–2.65
at 1000 °C		2.0–2.1	2.2–2.4	2.3–2.4	2.40–2.60	2.6–2.7	2.6–2.75
Cold crushing strength							
At 110 °C		50–60	65–75	5 5–65	65–120	5 5–65	65–120
At 1000 °C	MPa	35–40	60–70	3 5–40	60–100	45–55	60–80
At 1200 °C		34–36	75–85	3 8–45	75–100	45–50	70–80
At 1450 °C		40–45	80–95	50–60	80–120	50–55	80–120
Permanent linear change							
At 110 °C		0–1	0–1	0–1	0–1	0–1	0–1
At 1000 °C	%	−0.2–0.4	0.0–2.6	−0.2–0.4	0.0–2.6	−0.2–0.4	0. 0–2.6
At 1200 °C		−0.8–1.0	0–1.8	−0.8–1.0	0–1.8	0.2–0.5	0–1.8
at 1450 °C		−1.1–1.5	0.1–0.2	−1.1–1.5	0.1–0.2	0.2–0.5	0.1–0.2
H MOR at 1450 °C	Mpa	NA	4.0–5.0	NA	5.5–6.0	NA	4.5–5.5
	%	10.0–14.5	5.5–6.0	10.0–12.5	4.5–5.5	10.0–12.0	4.5–6.5

(continued)

Table 1.21 (continued)

Properties	Unit	55–65% Al_2O_3 castable		70–85% Al_2O_3 castable		90–95% Al_2O_3 castable	
Chemical analysis		Conventional (CC)	High-performance castable (LCC/ULCC)	Conventional (CC)	High-performance castable (LCC/ULCC)	Conventional (CC)	High-performance castable (LCC/ULCC)
Water required for casting							
Application method		Trouling/casting	Vibro casting/self-flow	Trouling/casting	Vibro casting/self-flow	Trouling/casting	Vibro casting/self-flow

- Dry mixing to be done for 2–3 min before adding water and final mixing for another 3–5 min to be followed on adding 4–6.5% water in LCC and ULCC and 10–16% in conventional castable. Flowability of castable under vibration is to be checked before moulding, the flowability to be maintained more than 80%.
- Generally, samples of cubes of 50 mm, 75 mm, or 100 mm side and rectangular shapes of 25 × 25 × 160 mm, 40 × 40 × 160 mm or 230 × 114 × 75 mm or any other shapes as recommended in standards is to be casted using the process of vibro-casting. Excessive vibration may cause segregation of grains from hydraulic matrix; hence, precautions should be taken during casting.
- De-moulding is to be done after 3–4 h of casting depending on setting behaviour of castable. The samples are to be air-dried (curing) for 24 h and ensure slow moisture removal in dense castable within temperature range of 15–22 °C and > 96% humidity. Presently, humidity-controlled drying oven is used for proper curing.
- After curing, the samples are to be placed in drying oven and to be dried at 110 °C for 24 h. In this stage, complete moisture removal takes place and the samples are then ready to fire at different temperatures as mentioned in product data sheet or as recommended by suppliers.

The following tests are to be done for complete evaluation of physical properties.

- Cold crushing strength of samples fired at different temperatures.
- Permanent linear change at different temperature. The 110 °C dried samples to be pre-fired at 950–980 °C before starting the test.
- Cold modulus of rupture after firing the samples at different temperatures.
- Hot modulus of rupture after preheating the dried sample at high temperature.
- Thermal conductivity: sample size of 230 × 114 × 75 mm to be used and testing to be done by hot ware method or by conventional calorimeter method. Samples to be pre-fired at 960–980 °C before starting the test.

The sample preparation is to be made according to ISO/DIS 1927-8. Few important procedures are explained below:

Test methods for carbon-containing refractories: Few of the high-alumina castables contain SiC and carbon, such as castable for blast furnace troughs and runners. The physical properties of products containing carbon are determined before and after removal of volatile components by carbonization and after removal of all carbon. Volatile components are required to be removed according to the carbonizing procedure defined in EN 993-3 which also entails the determination of the following properties:

- Carbonization mass loss
- Residual carbon content
- Carbon Yield (Total carbon removal from the carbonized test pieces followed EN 993-3)

1.5.3 Calcium Aluminate Cement (CAC)

Portland cement used in civil construction contains lime-silica compounds as a principal component. In calcium aluminate cement (CAC), lime-alumina compounds of high refractoriness form the principal components. CAC is manufactured either by fusion or by sintering processes. More than 70% Al_2O_3-containing cement, also called high-alumina cement, containing high-purity raw materials, are manufactured by sintering process, and CAC of 40–50% Al_2O_3 cement are usually manufactured by fusion process. After sintering in rotary kilns, the cement clinker is milled, either with or without additives, which modify the rheology and setting properties of cement. 70% Alumina cement is widely used in producing high-performance castables, and for some pure and special castables, 80% Al_2O_3 cement is usually used.

CaO and Al_2O_3 containing pure raw materials are used in high-alumina castables, and during sintering process CaO reacts with Al_2O_3 to form $12CaO.7Al_2O_3$(C12A7), $CaO.Al_2O_3$(CA) and $CaO.2Al_2O_3$(CA2) phases and free CaO. The characteristics of those phases are shown in Table 1.22. Among those phases, C12A7 and free lime are undesirable due to very fast setting; their content is kept low, <4–6% in 70–80% Al_2O_3 cement. The major phase is CA, which favours the high–temperature-strength development and control in setting of castable. The reactivity of calcium–aluminates decreases with increase of Al_2O_3 content with the formation of $CaO.6Al_2O_3$ (CA6) phase, which is non-hydraulic (does not react with water).

1.5.3.1 Hydration of Calcium Aluminate Cement

The calcium–aluminate phases hydrate in water addition and form hydrated bond. All the hydrated phases have specific crystal structure with individual crystal water content, depending on temperature and time of hydration reactions. The hydration process is shown in Fig. 1.21.

At low temperature hydrates (20–35 °C) with high water content are formed (CAH_{10} and C_2AH_8). However, these phases are metastable, and during drying, those phases will transform to the stable phases C_3AH_6 and gibbsite (AH_3). The release of water during this transformation leads to a shrinkage of the binding phases. The hydrate volume decrease is about 53%, which reduces the strength of

Table 1.22 Characteristics of calcium-aluminate phases in CAC

Minerals	CaO %	Al_2O_3%	Melting temperature (°C)	CCS (kg/cm^2	Density (gm/cm^3)	Setting characteristics
C12A7	48.6	51.4	1360–1390	150	2.69	Fast setting
CA	35.4	64.6	1600	600	2.96	Moderate setting
CA2	21.7	78.3	1750–1765	250	2.91	Slow setting

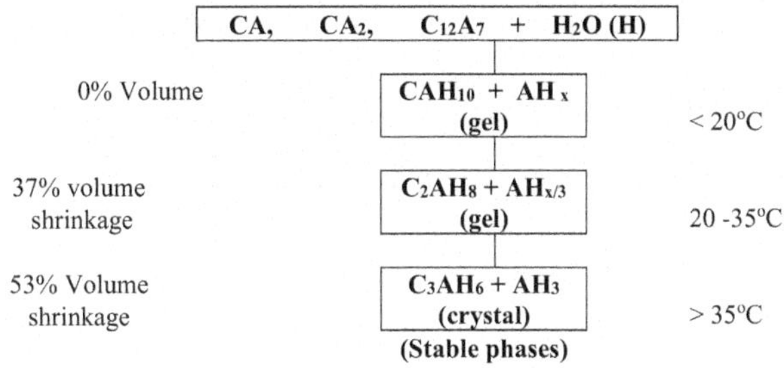

Fig. 1.21 Volume change with respect to phase formation at different stages. Sequential phase transformation depends on temperature and time of hydration

the bond matrix. The hydrating reactions are exothermic and these reactions are less in the case of LCC and negligible in ULCC. The setting time of alumina cement is influenced by temperature during hydration reactions and taking long time below 30 °C. Compounds formed during hydration of alumina cements dehydrate at 550 °C. After dehydration, the mineral phases are formed during preheating in high temperature. In the process of dehydration, C12A7 first formed and is stable up to 900–1000 °C. At 800–900 °C, CA starts forming and CA2 is also formed in more than 70% Al_2O_3-containing cement. CA and CA2 are stable up to 1400–1500 °C, and above 1400 °C, CA and CA2 react with free alumina to form calcium hexa-aluminate (CA_6), and leads to a theoretical volumetric change of +3.01% at above 1400 °C based on the following reaction:

$$CA2 + 4A = CA6 \tag{1.47}$$

CA disappears at the same time. During dehydration, gibbsite (AH3) dehydrates at 500–550 °C. In the presence of SiO_2 in castable composition, gehlenite (C_2AS) forms at 900 °C, which is eventually a low-melting phase.

1.5.4 *Spinel-Containing Castable*

High corrosion resistance and thermal shock resistance spinel-containing castables are nowadays widely used in steel industry worldwide. Spinel-containing castables exhibit less slag penetration and less corrosion when compared to spinel-free high-alumina castables. Spinel can be added as preformed or can be developed as in situ spinel high-alumina castables during heating at high temperature. Preformed spinel castables present high volume stability and in situ spinel castable has the advantage of gradual volume expansion with increasing temperature and high corrosion resistance towards FeO- and MnO-containing steel slags. However, there is a challenge relating to magnesia hydration and processing technology.

1.5.4.1 Preformed Spinel-Containing Castable

Preformed spinel is added in high-alumina castables for the following purposes:

- To increase the slag corrosion resistance.
- To improve mechanical strength at high temperature.

One of the important factors is the grain size of the preformed spinel. Spinel must be added to fine fraction of the castable formulation to attend the best slag and liquid steel penetration resistance. Preformed spinel contain matrix improves the corrosion resistance of a castable. Improvement of corrosion resistance depends on total amount of spinel added and its distribution in castable matrix. Nowadays, most of the spinel-added castables are low-cement castables.

The thermo-mechanical properties of castables, such as hot modulus of rupture (HMOR) and refractoriness under load (RUL) of high-alumina castables, are considerably improved by adding high-purity preformed spinel. It is reported that adding spinel in castable, HMOR at 1500 °C is improved from 8 to 17 MPa and RUL is improved from 1300 to 1500 °C [21]. Spinel addition in castable improves thermal shock resistance due to differential thermal expansion coefficient between alumina and preformed spinel and the differential expansion leads to micro-crack formation in castable matrix and improves thermal shock resistance. Castable with preformed spinel shows good thermal shock resistance in application of purging plugs, well blocks and seating blocks, used in steel ladles.

1.5.4.2 In Situ Spinel-Containing Castables

There are several examples where in situ spinel formation is formulated as part of the castable, possibly to reduce cost; however, there are significant disadvantages to this approach. The formation of spinel from alumina and magnesia leads to a remarkable volume expansion. Theoretically, this volume expansion will be 13%, based on a relative density calculation. In practice, however, this expansion is around 5%, still excessive for the microstructure to accommodate without cracking. Addition of fine silica, such as silica fume, is often used to promote liquid-phase sintering and to allow some local deformation to overcome the volume expansion. Excessive amount of MgO addition for higher quantity of spinel formation leads to cracking of castable due to the volumetric expansion in spinel formation, and hence the MgO addition is restricted to around 2.5–4.5%, which results in the formation of around 10–15% spinel formation. For the requirement of higher quantity spinel, additional preformed spinel is to be added.

In a ladle sidewall, spinel-forming castables are used for the advantages of volume expansion due to in situ spinel formation. It helps to prevent the crack formation due to shrinkage of castable in repeated heating and cooling. Also, the liquid phase formation due to silica addition in low-cement castable increases the mechanical flexibility of the castable, which can be beneficial to avoid cracking of

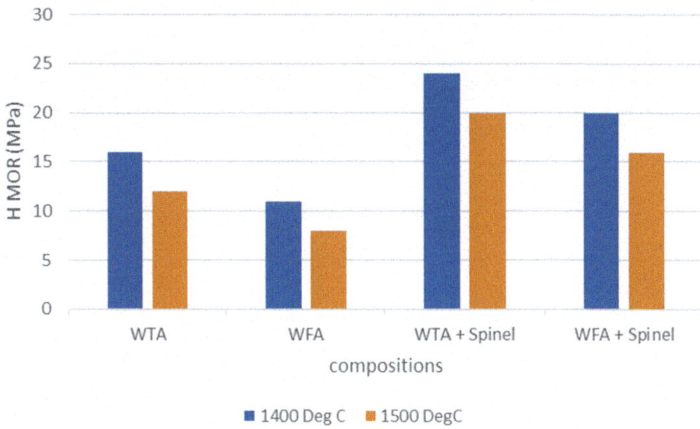

Fig. 1.22 Effect of spinel addition on HMOR

the lining during operation. In general, ladle sidewalls are less subject to erosion than a ladle bottom; thus, lowering the hot strength of the in situ castable is less of a factor.

One of the problem faced in spinel forming castable is the hydration of MgO in castable mix, when water is added. Hence, for pre-cast shapes, spinel-forming formulations are not preferred due to hydration of the agnesia additive. Also, many pre-cast shapes are used in areas of high erosion such as impact areas, purging plugs and well blocks. The reduced erosion resistance of in situ spinel formulations will reduce the life and increase operating cost. The disadvantages of using in situ spinel-containing castable are summarized below:

- Low erosion or abrasion resistance.
- Low flowability due to hydration of MgO.
- Restricted use due to volumetric expansion.
- Maintain raw materials purity to avoid formation of low-melting compounds.
- Fast setting of castable due to hydration of MgO.
- Preformed spinel-containing castables do neither have this expanding behaviour nor any unwanted hydration and thus be formulated with higher percentages of spinel addition or without adding fumed silica.

Tentative comparative hot strength of preformed spinel-containing castables with white tabular alumina (WTA) and white fused alumina (WFA) is shown in Fig. 1.22. Spinel added was 15–30%.

Properties of castables containing preformed and in situ spinel are shown in Table 1.23.

Table 1.23 Properties of castables containing preformed and in situ spinel

Properties	Unit	Low-cement castable (WTA)	Low-cement castable (WTA + in situ spinel)	Low-cement castable (WTA + preformed spinel)
Chemical compositions				
Al_2O_3	%	98	90	90
MgO	%		4.0–5.0	6.5–8.0
CaO	%	1.5	1.5	1.5
SiO_2	%	< 0.05	2.5–3.5	2.3–3.5
Physical properties				
HMOR at 1500 °C	MPa	15–17	20–22	13–15
RUL (T0.5)	°C	1690	>1700	1550
Mixing water	%	5.0–5.5	4.5–5.0	5.5–6.0

1.5.5 Ramming Masses and Plastic Monolithics

These are materials supplied in friable dry mix state, moist or ready to use. They are rammed into stable mould or behind stable brick work or templates. In special cases they are also applied by vibration. Sometimes the liquid binder is supplied separately and is added to obtain its ready-to-use consistency. The ramming mixes usually set above room temperature once subjected to heat.

Plastic mass are delivered in plastic rapped casing, in the form of pre-compacted blocks, and are installed by ramming into final position. It is possible to install them without moulds or templets. They set and develop strength, above room temperature during heating.

1.5.5.1 Ramming Mixes

Various commercial ramming mixes, such as alumina based ramming mixes including alumina–chrome, alumina–zirconia, as well as basic ramming mixes containing magnesia based mixes, MgO-Chrome and carbon- containing mixes are predominantly use in steel sector. Usually water or special liquid binders, such as gel-bonded, phosphate-bonded, acid-bonded and resin-bonded mixes, are used. Ramming mixes can have a ceramic and chemical bond system and the ramming mixes are rammed properly for compaction, and can be an excellent refractory lining equivalent to brickwork. The ramming mixes are used satisfactorily in the presence of slag and liquid metal and it also shows excellent thermal shock resistance. The properties of different quality ramming mixes are shown in Table 1.24.

Installation of ramming mix requires stable forms and it is important to note that ramming work is always conducted vertically on the ramming surface. Ramming mix is to be poured into the mould or template in instalment of every 50–150 mm

Table 1.24 Chemical composition and physical properties of different ramming masses

Properties	Unit	Fireclay	Andalusite	Mullite	Bauxite	Corundum	MgO—Dolo	MgO	MgO—Chrome
Chemical analysis									
Al_2O_3	%	45–50	55–62	70–74	80–85	>90	0.5–0.8	0.3–0.5	0.5–1.5
MgO	%						55–75	85–90	55–65
CaO	%						20–45	5.0–7.0	6.0–7.5
SiO_2	%	45–55	30–35	25–28	10.0–15.0	4.0–6.0	0.6–0.8	0.5–1.0	8.0–10
Fe_2O_3	%	1.0–1.5	0.8–1.1	0.2–0.5	1.5–2.5	0.2–0.35	3.8–4.5	6.0–9.0	5.5–6.5
Cr_2O_3	%								25–30
Physical properties									
Service temperature	°C	1450	1500	1700	1650	> 1750	>1700	> 1700	>1700
Bulk density	g/cm^3	2.3–2.35	2.4–2.45	2.30–2.35	2.60–2.65	3.05–3.07	2.75–2.9	2.8–2.85	2.75–2.9
Cold crushing strength	Mpa	40–50	50–55	65–70	50–60	100–120	50–80	65–110	80–110
Cold MOR	MPa	9.0–12.0	8.0–12.0	12.5–13	7.0–9.0	10.0–12.0	6.5–8.0	8.0–10.0	9.5–10.5

height, rammed properly by pneumatic or hydraulic rammer to obtain uniform ramming density. Continuous filling of ramming mass after completing one layer of ramming is to be maintained to avoid laminations. The ramming mixes are widely used in steel industries. The trough and runners of small and medium size blast furnaces are continued to line with special quality carbon-bonded ramming mixes. The other uses are in backup lining of steel ladle and vessels and bottom of EAF hearth and side wall. It is also used around the under-hearth cooling system and around the side wall plate coolers of large size blast furnaces.

1.5.5.2 Plastic Mass

The plastic masses are made of clay bonded or chemical bonded. Water is used in clay-bonded masses. Chemical-bonded products have wide application in steel ladles and side walls of re-heating furnaces. Phosphoric acid and mono-aluminium phosphate are used as binders in producing chemical-bonded plastic mass, whereas water is used for ordinary clay-bonded plastics. Properties of plastic refractories are summarized in Table 1.25.

Table 1.25 Properties of plastic refractories

		Clay bonded			Phosphate bonded		
Properties	Unit	Fireclay/ andalusite	Bauxite	Corundum	Fireclay/ andalusite	Bauxite	Corundum
Chemical analysis							
Al_2O_3	%	55–59	82–84	> 90	55–59	82–84	> 90
SiO_2	%	42–45	10.0–12.0	5.5–7.5	42–45	10.0–12.0	5.5–7.5
Fe_2O_3	%	1.1–1.5	1.1–2.0	0.4–0.6	1.1–1.5	1.1–2.0	0.4–0.6
Physical properties							
Service temperature	°C	1600–1650	1700	1750	1600–1650	1700	1700
Prefiring temperature	°C	1000	1000	1000	1000	1000	1000
Bulk density	g/cc	2.18–2.22	2.60–2.65	2.80–2.85	2.30–2.35	2.72–2.80	2.9–2.95
Cold crushing strength	Mpa	25–30	30–35	40–45	65–75	70–85	85–100
Cold MOR	MPa	2.5–3.5	4.5–5.5	5.0–5.5	8.0–10.0	12.5–14	18–25
permanent linear change	%	(−) 1.4–1.6	(−) 1.0–1.4	(−) 1.1–1.3	(−) 0.4–0.6	(−) 0.4–0.6	(−) 0.4–0.6

Characteristics of Plastic Masses

Compared to clay-bonded plastics, phosphate-bonded plastics have the following characteristics:

1. They are generally heat setting refractories.
2. They develop high cold setting strength after setting.
3. They are highly resistant to abrasion.
4. Their hot strength is higher than clay-bonded plastics.
5. They are more volume stable at high temperature compared to clay-bonded plastics.
6. They show good adhesion with old refractory in repair work, both in hot and cold condition; hence plastic refractory has a wide application in refractory maintenance and repair jobs.
7. Strength generally develops above 350 °C; below this temperature, strength drops due to moisture absorption, hence heating up to 350 °C is mandatory immediately after installation.

1.5.6 Application Methodology

Application methodology of monolithics plays a vital role for a successful installation of monolithics and their performances. Improper installation causes the failure of monolithic lining and shorten the campaign life. Utmost care needs to be taken during installation processes, which involve casting and forming, setting, curing and preheating. Application methods depend on types of monolithics and castables in particular. It also depends on type of bonds and their hydration characteristics and development of strength.

1.5.6.1 Installation of Castables

Conventional castables that also include insulating and hydraulic setting castables and gunning masses have a typical strength pattern depending on the degree of pre-firing. Generally, strength drops during pre-firing, within the temperature range of 800–1100 °C due to the dehydration of calcium aluminate hydrates, and changes the pore size distribution. Increasing strength above 1100 °C is the result of formation of new minerals from cement and ceramic bond formation. For LCC and ULCC the strength drop at intermediate temperature is negligible due to the presence of lesser amount of high-alumina cement and they show gradual increase of strength with rise of temperature.

Amount of water addition in LCC and ULCC plays a vital role in strength development at 110 °C and higher temperature; it also increases porosity and high-temperature shrinkage. Insulating refractory castable and conventional castables can get very hot after mixing water and during hydration. The temperature peak will depend on the quantity of castable used and ratio of surfaces of the cast surrounded by mould to the open surface for releasing water. During this time (around 15–20 h), the refractory castable is to be protected from fast releasing moisture from casting surface to prevent cracking. These refractory castables prefer to allow around 48 h soaking time before starting further processes of drying and firing. As LCC and ULCC contain lower cement content, the heat generation become very less and the soaking time may be maintained for 24 h.

Installation Methods of Castable

Installation method	Application area	Properties of materials required	Installation equipment	Important to check
Casting	Blast furnace, blast furnace iron troughs and runners, Torpedo mouth casting, Steel ladle, tundish, vacuum degassers, reheating	Flowability, setting time, strength after setting	Forma, Mixer, Core	Mixing time, water addition, Setting time
Pumping	Blast furnace iron troughs and runners, Torpedo mouth casting, Steel ladle, tundish, vacuum Bull nose of reheating furnaces	Self-flow pumpability, strength after setting	forma, Mixer, Pump	Continuous castable feed, water addition, setting time
Injection	Blast furnace stack, BOF convertor, vacuum degassers,	Self-flow pumpability,	Mixer, Pump	Injection pressure, castable flow, Back pressure
Gunning	Blast furnace stack, Blast furnace iron troughs and runners, Torpedo, Steel ladle, tundish, reheating furnaces	Powder flow, setting time, Hardening time	Gunning machine	Water and powder flow, stickability, Rebound loss

Features of the different casting methods of installation are:

- Extended life of the remnant lining.
- Formation of a homogeneous refractory structure.
- High-efficiency installation and reduced cost.
- Quick installation by automation.
- Pollution-free environment by less dust and noise.
- Reduce man power.
- Working environments.

1.5.6.2 Installation of Plastics and Ramming Mixes

The factors involved in the installation of plastics are as follows:

(a) Moulds: Moulds made of wood or steel or templates against the back-up lining are assembled to have enough strength to prevent buckling due to ramming pressure. The inner dimension of the mould must be proper to have accurate lining thickness.
(b) Anchors placing: For lining large area in furnace wall or roof, ceramic anchors fixed in the metallic clips of hangers are to be installed as per design. The length of the ceramic anchors is to be such that the top surface would be in the same level of the plastic rammed surface, as shown in Fig. 1.23a, and the clips are to be kept out side of the plastic lining.
(c) Ramming: Ramming begins with the placing of slabs or cakes of plastics and the cakes are rammed uniformly and ramming is to be done vertically. After ramming the first layer of cakes completed, another cakes to be added thereon and the work continues with a uniform pressure.
(d) Trimming: After completion of ramming, the forms are removed and the lining is finished by trimming the surface to the specified thickness and the top face of the ceramic anchors would be visible.
(e) Score lines: Score lines are cut over the trimmed surface for shrinkage and high-temperature expansion allowance. They are usually 1–3 mm width and 30–50 mm deep, and located around 1.0–1.5 m apart. The lines to be cut both vertically and horizontally and special cut around the burner ports as shown in Fig. 1.23a.
(f) Venting holes: Venting holes are provided to allow escape of moisture and volatiles from plastic lining during drying and heating up. These are 3–4 mm

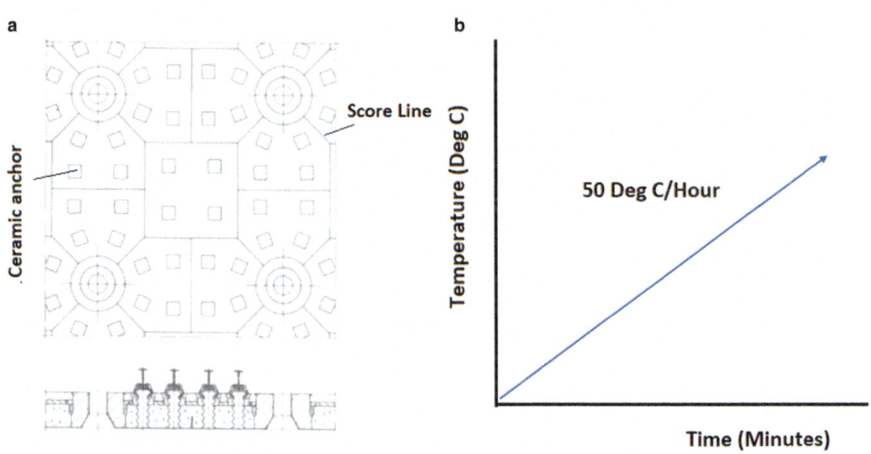

Fig. 1.23 (**a**) Schematic diagram on arrangement of ceramic anchors and score lines, (**b**) typical heating schedule of plastic mass

holes driven at 100–150 mm intervals all over the lining surface. The holes penetrate about 2/3 of the plastic refractory thickness.

(g) Heat up: Refractory lining installed with plastics or ramming masses is typically heating up with a faster heating rate of 50 °C/h and this is one of the advantages over castable lining to reduce the preheating time. A typical heating schedule is shown in Fig. 1.23b.

Characteristics and Installation Methods of Plastics and Ramming Masses

Types of monolithics	Plastics	Ramming mixes	Patching materials
Physical conditions	Mouldable/ramming	Granular with or without moisture	Ready mix
Type of bond	Chemical/clay	Hydraulic/ chemical	Hydraulic/chemical
Installation methods	Ramming by rammers	Ramming or gunning	Manual patching
Characteristics			
Lining thickness	Thick lining	Thick lining	Thin for repair
Setting characteristics	Air and heat setting	Heat setting	Heat setting
Shrinkage	High for clay bonded and low for chemical bonded	High	Low in drying and expansion on heating
Strength	High	High	Low
Spalling resistance	High for clay bonded and excellent for chemical bonded	High	High
Abrasion resistance	High	Excellent	Low
Corrosion resistance	High	Excellent	High

1.6 Corrosion of Refractory

1.6.1 Basic Corrosion Concept

Refractory corrosion is not only confined by molten contact in service condition. Continuous thermo-mechanical degradation expedites severe refractory damage mechanism through molten metals, oxides (slags), salts and gases that limit their performance. In this consequence, few unavoidable circumstances including

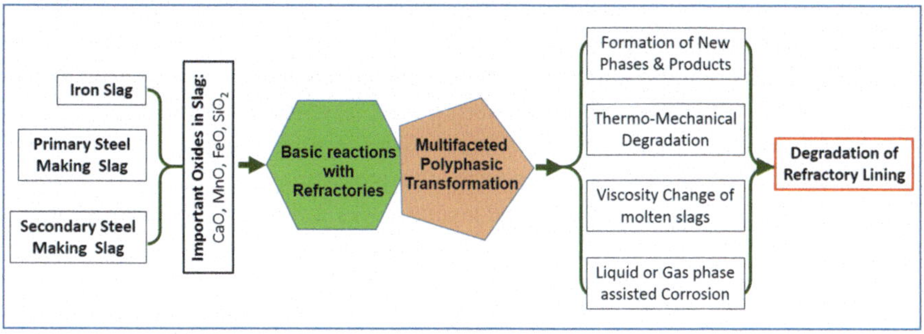

Fig. 1.24 A representative "corrosion model" influence by the different class of slags and their interaction with refractories

refractory phase changes, reduction of their lining thickness, and changes in the chemistry of slag and molten metals are common phenomena during application. Thus, basic target is to develop a quality refractory for particular grade of steel in order to minimize the unwanted incidents and premature failure. Despite these affects, viscosity of molten media influences the refractory corrosion kinetics. Probable basic reactions followed by complex reactions are prime reasons for the formation of new phases as well as corrosion that eventually reduce the lining thickness and enhance the liquid viscosity. A schematic representation of different reactions, important zones, effect of oxide constituents and subsequently their effect on refractory and molten liquid is represented in Fig. 1.24.

In obvious the new phase transformation is not a single step process; rather it is the result of different basic reaction mechanisms including dissociation, reduction, oxidation-reduction, volatilization, and dissolution/precipitation indecently occurring in the initial stage of slag–refractory interaction that expedite the virgin phase transformation of applied refractories. Despite phase transformation around working lining, thermo-mechanical degradation, viscosity change of slag followed by liquid- and gas-assisted corrosion facilitate the refractory lining wear. All these aspects are elaborated to justify the corrosion behaviour of different class of refractories during processing of iron to steel.

1.6.2 Slag Viscosity and Penetration

Most of the molten metal experience noncompetitive wetting behaviour with ceramics while almost all slags (ceramic oxides), glasses and fluxes react with ceramic oxides in slag lining refractories. In this constrain, carbon-based refractories restrict high degree of wetting characteristics during interaction of slag. In the perspective of refractory-metal-slag interaction, wetting predicts the probable liquid permeation through available open pores without any chemical reaction, whereas the corrosion expedites both chemical reaction and penetration. Thus, reactive molten slag is in forefront in order to describe the refractory corrosion behaviour. Initiation of the slag–refractory interaction depends on the viscosity of slag. Despite corrosion and dissolution, different phase formation and crystallite growth, rearrangement and precipitation are common phenomena during corrosion of refractories.

Fluid slag interact chemically, penetrate and dissolve the refractory that gradually increase the viscosity of slag and restrict the mass transport through refractory surface and progressively form saturated layer. However, low-viscous slag facilitates rapid diffusion and minimize the probability of formation of saturated layer on refractory surface. There are three popular techniques to measure the slag viscosity, concentric cylinder method, falling body method and oscillational method. Accurate slag viscosity measurement is an expensive and time-consuming incident, and thus wide range of slag viscosity (η) on the basis of composition and temperature (T) has been measured through a theoretical model, known as Frenkel equation. The Eq. (1.48) implies:

$$\eta = AT \exp (B/T) \tag{1.48}$$

where A and B depend on the slag composition, extensive properties as a function of constituents including viscosity is published in recent literature [22]. The predicted theoretical value varies within the range of 30% of experimental viscosity data.

In advance, the slag viscosity is not only the controlling parameter to penetrate refractory, rather necessary to consider Poiseuille's law to describe the penetration through available pores. The equation describes that the penetration rate directly depends on cube of capillary radius (r), predominately open pores and microcracks, and capillary pull on pressure, and inversely proportional with slag penetration depth (l) and dynamic slag viscosity (η).

$$\frac{dl}{dt} = \frac{r^3 \Delta P}{8l\eta} \tag{1.49}$$

The "r" value closely interlinked with the starting raw materials and processing conditions of refractories in manufacturer end, in other words microstructure and total pore volume. The pressure difference (ΔP) can be further expressed by:

$$\Delta P = 2\gamma \cos (\theta/r) \tag{1.50}$$

where γ is the slag surface tension and "θ" is the wetting or contact angle. In combination with Eqs. (1.49) and (1.50), followed by integrating results

$$l^2 = r \cos \frac{\theta}{2} \left(\frac{\gamma}{\eta}\right) t \tag{1.51}$$

The resultant equation predicts slag penetration depth can be suppressed by increasing slag viscosity and contact angle or by decreasing surface tension. This phenomenology clearly indicates the limitation of molten slag penetration from hot working zone to cold backup lining. Furthermore, the viscosity measures the mass transport diffusion after penetration into the refractory and that diffusivity (D) can be expressed by Stokes–Einstein relation:

$$D = \frac{kT}{6\pi\eta a} \tag{1.52}$$

where k is the Boltzmann constant (1.38×10^{-23} JK^{-1}), "a" is the radius of diffusing species. Thus, lowering viscosity enhances the diffusivity and results in formation of both solid and liquid during penetration. In this circumstance, the possibility of solid, liquid and gas (or combination) formation is more, and eventually different appearance of reaction products including strong adhesion with solid and viscous liquid, presence of fluid or gas, or combination of both are noticed. In brief, the slag penetration facilitates both active and passive corrosion. Active corrosion of refractory is predominating when the reaction product is soluble or like to dissociate in the liquid slag. However, gradual adherence of reaction products on refractory reduces the corrosion of original refractory surface and reduces the overall rate of corrosion, called passive corrosion.

1.6.3 Slag–Refractory Interaction

Slag-influenced refractory corrosion is a more complex phenomenon compared to sudden metal penetration in refractory lining followed by the vessel puncture. Brief on some common occurrence is summarized as a background of in situ change of refractories that eventually expedite the corrosion. Prior to discussing the corrosion mechanism, it is necessary to understand the classic difference between acid–base concept at room temperature and elevated temperature. Acid substance contributes proton (H^+) or accept electron (e^-) pair to form a covalent bond, whereas a basic substance contributes hydroxyl ions (OH^+) or accepts protons (H^+), which results in acid and base substance making the system acidic and basic, respectively. Before identification of acidic or basic oxides, a basic understanding comprehends that the high electronegativity of central atom (e.g. Si in SiO_2) provides more acidic oxide. In consideration of periodic table, electronegativity character increases from left to right and decreases down the column. A typical electronegativity data for central atom Ca = 1.00, Mg = 1.31, Mn = 1.55, Fe = 1.83, Al = 1.61, Cr = 1.66, Si = 1.90, P = 2.19.

In fact, Lewis theory suggests more oxidation state is capable to more electrons, means more acidic, however, same oxidation state (+3) of Al, Cr and Fe behave as amphoteric (both acid and base nature) oxide. In order to simplify and analyse the effect of oxides in corrosion, a baseline defines the neutrality ratio of $CaO/SiO_2 = 1$. Thus, acidic refractories experience less corrosion to acid slags ($CaO/SiO_2 < 1$) and basic refractories more compatible with basic slags ($CaO/SiO_2 > 1$). Beyond this compatible limit, the equilibrium between two mineralogical phases transforms to new class of phases and expedites the corrosion. In application, it is very difficult to attain refractories in chemical equilibrium in a microscopic scale as it is developed from different mixture of oxides; however, localized volume elements at the immediate refractory–slag interface may attain chemical equilibrium during formation of new phase.

Refractory corrosion is not only the phenomena of solid phase dissolving, rather momentarily it aggravates the composition change of contact liquid up to saturation level. This incidence eventually nucleate new phases from the saturated solution and subsequent their growth with respect to temperature profile. Solid-state diffusion of slag species into refractory grain phase helps exsolution (precipitation) at lower temperature. A particular phase diagram predicts the requisite temperature and probable phase formations when system has more than 3 or 4 oxides after slag–refractory interaction. Highest solubility oxide dissolves first and forms new phase using Gibbs energy minimization. In spite of material solubility, smaller and irregular shaped particles have high specific surface areas and exhibit higher dissolution rate.

Particle–particle necking is another weak region and rapid dissolve speed up new phase. These three aspects, including oxide solubility, particle size and shape, and weakest zone like particle–particle necking are essentially to encounter for the analysing of dissolution and phase formation. In this consequence, the shape of the new crystal depends on the interfacial energy of solid and liquid phase. Isotropic interfacial energy prefers to form spherical crystals while anisotropic envisages to form equilibrium shape manipulated by the Wulff construction [23]. This proposition elucidates the formation of equilibrium morphologies of different phases forming in refractory–slag interaction. However, local composition change including phase separation or impurity segregation alters the growing surface and certain degree of change of the equilibrium crystal morphologies during complex slag-refractory interaction.

Interface chemical reaction and reacting species diffusion governs the dissolution of refractory-slag interface. In order to limit the dissolution of refractory, the slag environment should be similar in nature. To accomplish such fact, acidic refractories and basic refractories are preferable for acidic and basic slag, respectively. The dissolution is controlled by either congruent (homogenous) dissolution when the diffusivity of reaction product is faster than the chemical reaction at the interface or incongruent (heterogeneous) dissolution where the rate of removal of reaction products by diffusion is slower than the rate of chemical reaction. In former case, the dissolution process is directly controlled by first-order reaction and expressed by:

$$J = K\left(\frac{A_c}{A_o}\right)C_m \quad (1.53)$$

where J is the dissolution rate (g. cm^{-1}. s^{-1}), K is rate constant, A_c is actual area of refractory (cm^2), A_o apparent area of refractory (cm^2), and C_m concentration of reactant species in the melt (g.cm^{-3}). Surface irregularities, grooves and pores increase the A_c/A_o ratio and result in high dissolution rate.

The latter case implies the formation of solute rich boundary layer, and refractory interface is saturated with reaction products, termed as indirect dissolution. If the boundary layer leads to formation of a solid interface, the rate of corrosion can be expressed by Nernst equation:

$$J = D\,\frac{C_s - C_m}{\delta} \qquad (1.54)$$

where D is the diffusion coefficient ($cm^2.\ s^{-1}$), C_s is saturation concentration of refractory in the melt, C_m the concentration of reactant species in the melt, and "δ" is effective boundary layer thickness (cm). In this circumstance, C_s and saturation of liquid in solid both are indeed effective to understand the new phase formation. For example, if the solid is unsaturated with respect to any one component of the liquid, then solid solution may be the result.

Despite static interaction, the stirring has effect on refractory-slag interaction analogous to the floor condition. Interestingly, stirring has no apparent effect on the direct dissolution rate. Herein, the atoms diffuse from the interface at a rate proportional to $t^{1/2}$ as reactants are depleted and dissolved species grown up without liquid convection or stirring. The indirect dissolution rate enhances when the boundary layer thickness reduces or breaking up under stirring convection. In actual the layer thickness persistence depends on degree of convective flow caused by density and thermal gradients, the liquid viscosity, mean diffusion coefficient and the vessel size. In several vessels, turbulence flow is unavoidable circumstance where it severely affects the refractory as it tends to pull out the loosely bound fine grains in the bricks by abrasion and erosion wear. In practical situation, the dissolution phenomenon is often incidence of combined first-order reaction and diffusion control mechanism.

The refractory degradation due to slag attack is observed in two ways. The first one is uniform corrosion that involves dissolution and the entire area in contact with slag has a loss in thickness of bricks. The second one is more severe impact containing slag penetration into the pores of refractory, localized corrosion to enlarge the pores that increase the penetration of slag and finally disintegration of the bricks. Pore size is larger, more surface tension and low viscosity are all known to intensify the rate of penetration and subsequent structural refractory degradation. The experimental correlation for the rate of refractory corrosion in slag is much useful to explore the role of different parameters on the rate of corrosion.

$$R = CL_0 \times \left(\frac{T^{\frac{2}{3}}}{\mu^{\frac{8}{9}}}\right) \times (fHA)^{\frac{1}{9}} \qquad (1.55)$$

where R is the rate of refractory attack by slag (cm/s), C is the constant for the furnace geometry, L_0 is the refractory solubility in slag (g/g), T is the absolute temperature of the hot face of the refractory (K), μ is the viscosity of slag (poise), f is the fraction of slag adhered to the refractory wall, and H is the heat discharge rate in the furnace chamber ($kJ/m^3/h$). A = ash content of the fuel in grams per gram of ash.

The correlation gives a strong dependence rate of refractory corrosion on the solubility in slag, slag viscosity, and temperature. Very common and interesting constituent MgO solubility in slag is encountered to establish this fact. A basic fundamental correlation can be implemented for the situation where the MgO mass transfer in slag controls the dissolution rate of refractory (see Eq. 1.54). It is clear that both kinetic factors (diffusivity and thickness of the boundary layer) and the thermodynamic system (solubility) properties influence the rate. MgO addition to the slag plays two major roles. First one shifts the composition towards saturated slag, lowering (C_s–C_m), and the second one changes the rate of mass transport by changing the viscosity of slag. A slag saturated over with MgO consists of solid phases such as MgO-Al_2O_3 spinel or magnesiowüstite MW ((Fe.Mg) O) solid solution that eventually increase the slag viscosity and lower its rate of degradation. Figure 1.25 shows, MgO concentration increasing leads to MW formation that is expected to reduce the rate of dissolution. MgO solubility in FeO-SiO_2-CaO-MgO slags equilibrated with pure iron at 1600 °C. The limit of solubility is induced as a function of FeO and CaO/SiO_2 content in the slag, and thus one can optimize the MgO content in a slag to control the refractory degradation [24, 25].

1.6.4 Primary and Secondary Slags

1.6.4.1 Iron-Making Slag (BF Slag)

The slag from iron making is predominately found in blast furnace and torpedo ladle. This slag facilitates the prime alumina-silica refractory corrosion mechanism through two interactive substances, namely FeO (+2, basic oxide) and alkaline vapor. The BF slag composition is identified by that of the impurities in the coke, fluxing stone, and ores are charged into the blast furnace is shown in Fig. 1.26a. Blast furnace slag is composed of several oxides, and their brief chemical analysis CaO (32–45%), SiO_2 (32–42%), Al_2O_3 (7–16%), MgO (5–15%), S (0.7–2.2%), FeO_3 (0.1–1.5%) and MnO (0.2–1.0%) predicts an overall idea about the metallurgical process and the possibility use of this by-product. However, the composition range may vary from source to source, and iron content reduces in modern blast furnace [26].

Depending upon the process of cooling, there are three kinds of slags are produced, they are (1) air-cooled slag, (2) granulated slag, and (3) expanded slag. In brief,

1. Air-cooled slag is generated by empowering the molten slag to cool under ambient condition in a pit. Porous and low-density aggregates are the result during slow cooling condition, and this slag can be used for several domains.
2. Granulate slag is generated by extinguishing the molten slag by means of water or water sprays with high pressure. Crystallization prevents by extinguishing, thus resulting in glassy aggregates, granular. This slag is crushed, fined and screened

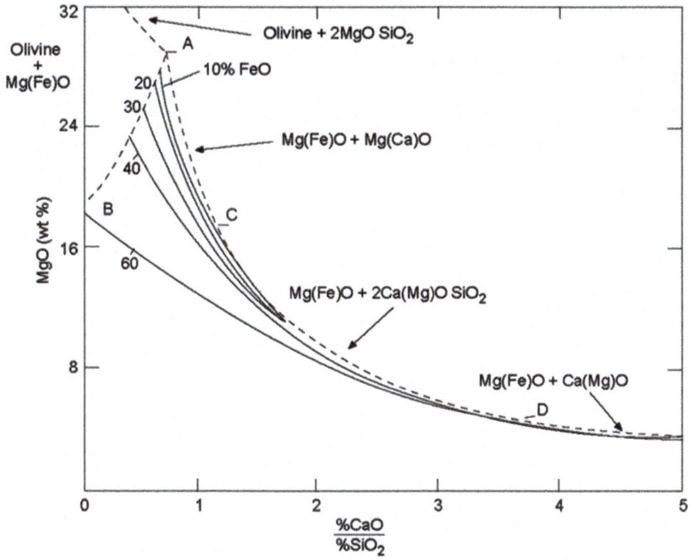

Fig. 1.25 The solubility of MgO in FCSM slags at 1600 °C [24, 25]

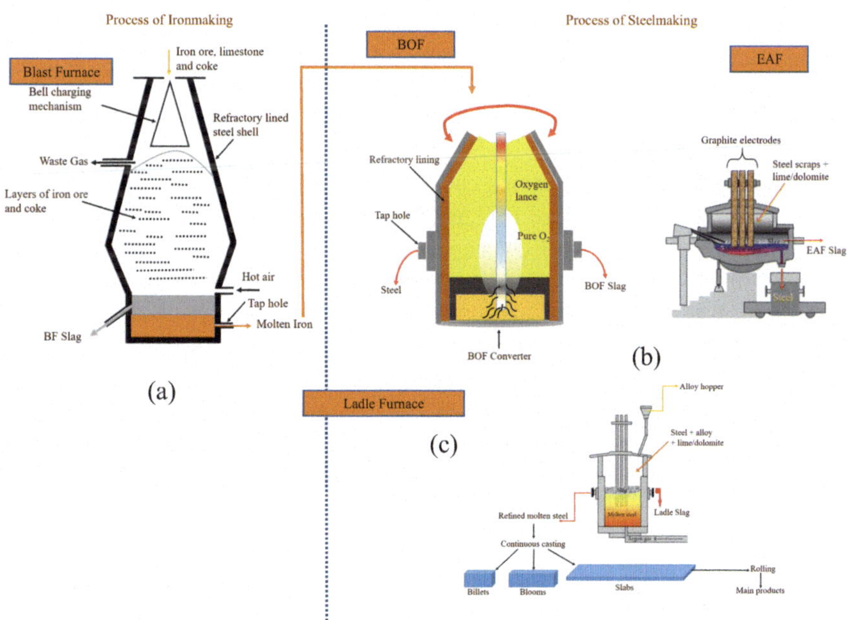

Fig. 1.26 (**a**) Iron-making slag, (**b**) primary steel-making slag, (**c**) secondary steel-making slag

for use in different utilizations, especially in the production of cement due to its characteristics of pozzolanic.

3. Expanded slag is produced through governed by making colder of molten slag in water or water with the addition of steam and air under pressure, steam generation and extra gases improve the porosity and vesicular nature of slag, results in aggregate of light weight applicable for use in concrete [27].

1.6.4.2 Primary Steel-Making Slag

BOF Slag

BOF (Basic Oxygen Furnace) is for making steel and associated slag is produced from charging steel scraps, molten iron produced from blast furnace, refining and alloying agents and refractory dissolution. Generally, the BOF charge consists of approximately 10–20% of scrap steel and 80–90% molten iron. The continuance scrap steel charging in the BOF plays a major task in cooling down the temperature, and thus the furnace temperature is necessarily maintaining near to 1600–1650 °C in order to complete the chemical reactions during conversion of iron to steel, and by-product as slag. The resulting slag from the process of steel making on the upper surface (due to density difference) of the molten steel is shown in Fig. 1.26b. In practice, the BOF is turned in one direction in order to pour the steel into the ladles. After removal of complete steel from the BOF, it is turned again in a counter direction to spill the liquid slag into ladles. The slag coming out from the steel-making process is referred as the BOF slag. The chemical reactions that occurred during impurities removal are the controlling factor to predict the chemical constituents of the BOF slag.

Generally, the mineralogical and chemical compositions of both BOF and EAF slags are almost same. CaO and FeO are the main chemical elements of both slags. The chemical elements of the BOF slag are SiO_2, FeO and CaO; while the transformation of molten iron into steel, some amount of the Fe (iron) cannot be regained into the produced steel. The oxidized iron was identified in the chemical composition of BOF slag. In consideration of the furnace performance, the iron oxide content can be as high as 38% in the BOF slag. Substantial amount of SiO_2 (7–18%), MgO (0.4–14%), Al_2O_3 (0.5–4%) and CaO as high as 35% is noticed in BOF slag. The high lime is formed due to addition of large quantity of lime/dolomitic lime during the transformation of molten iron into steel. BOF slag has porous and heterogeneous morphologies of both sub-angular and spherical shaped particles.

EAF Slag

EAF (Electric Arc Furnace) use electric arc with high power to generate the required heat to melt the steel scrap of recycled and transform into the desired composition of steel. The steel-making process on EAF is not dependent on the BF production since the actual input is scrap steel and certain quantity of pig iron. EAFs are having the graphite electrodes and giant kettles of resembling with a spill or a weird notch on one side. The electric arc furnace dome can pin and swivel to facilitate the raw materials under loading. The charging of different kinds of scrap steel to the furnace by using receptacle during steel-making process followed by the graphite electrodes let down into the furnace. Thereafter, an arc leads to passing electricity through the electrodes and metal itself. The heat is generated due to the resistance of metal and electric arc to the flow of electricity. Once heat is generated, the scrap melts and electrodes are driven extending far down through the layers of scrap. The process of melting initiated, a liquid steel pool is developed at the foot of the furnace. Few iron, concurrently, with impurities additionally in the metal, including carbon, manganese, silicon, aluminium, and phosphorus, get oxidized during the process. The oxidized elements mingle with CaO (lime) to obtain slag. The carbon powder is also needed to injected through the slag phase floating on the surface of the molten steel to refine the steel, which leads to the formation of CO that causes the slag to foam, resulting in increasing heat transfer energy efficiency. The required chemical composition of steel is achieved once the EAF is turned and the steel and slag are poured out of the furnace into separate ladles as shown in Fig. 1.26b.

Chemical composition of EAF slag is almost similar to BOF slag. However, because of the feedstock steel scrap composition variation, the resultant EAF slag composition may vary from BOF slag to some extent. Despite common oxides CaO, FeO, Al_2O_3, MgO and SiO_2, certain amount of oxide impurities like MnO and sulphur oxide are noticed. This slag preferentially forms irregular platy-particles and consists of extreme rough and porous texture [28].

1.6.4.3 Secondary Steel-Making Slag (Ladle Furnace Slag)

The steel produced from the BOF and EAF is again refined to get the required chemical composition; these are secondary refining processes of steel making. The major purposes of secondary refining process are degassing of oxygen, final desulphurization, hydrogen, nitrogen, removal of impurities and final decarburization. Ladle furnace similarly like smaller versions of electric arc furnace, which also have electrodes of graphite, are fastened to an arcing transformer used to heat the steel in the furnace. The desulphurizing agents like Ca, Mg, CaSi, CaC_2, etc., injecting through a lance, and the sulphur absorption in the steel can be decreased to 0.0002%. During deoxidation, the addition of aluminium and silicon forms Al_2O_3 (alumina) and SiO_2 (silica); later the oxides are absorbed by the slag obtained by the refining processes shown in Fig. 1.26c.

During the steel-refining processes, the ladle slag is produced, in which many alloys are included to produce different grades of steel from the ladle furnace; because of this chemical composition is different from those of BOF and EAF slags. Herein, many alloys are added into the ladle furnace in order to get the steel with different grades. Therefore, the ladle slag chemical composition is majorly variable; generally, the content of FeO in the ladle slag is very low almost ($<10\%$)

Table 1.26 Iron and steel slags utilization around the world

Slag	Technique of production	Applications
BF slag	Air-cooled	The aggregate of concrete, base, subbase, filter aggregate, construction fill, rockwool, scour protection, etc.
Granulated blast furnace slag (GBFS)	Molten slag quenched with high-pressure water sprays	Construction sand, subbase, construction fill, grit blasting, glass, cement manufacture, tiles, stabilizing binder, etc.
BOF slag	Air-cooled and watered	Asphalt aggregate, base, subbase, sealing aggregate, construction fill, grit blasting, adsorbent for phosphate, subsoil drains, etc.
EAF slag	Air-cooled and watered	Asphalt aggregate, base, subbase, grit blasting, subsoil drains, etc.
Ladle furnace slag (LFS)	Air-cooled and watered	Adsorbent for phosphate, base, subbase, asphalt aggregate, sealing aggregate, subsoil drains, construction fill, etc.

compared to BOF and EAF slags. In other words, the contents of CaO and Al_2O_3 are generally higher for ladle furnace [29].

1.6.5 Effective Use of Iron/Steel slags

By-product iron/steel slag is presently used and sold for different applications, mainly for road base, asphalt, and concrete aggregate [30]. Research on the iron and steel slags utilization has been carried out in various fields as follows: (1) cement industry, (2) aggregate in asphalt or concrete, (3) glass making, (4) construction of roads, (5) agriculture soil conditioner, (6) aquaculture, (7) land amendments and (8) ceramic tiles as tabulated in Table 1.26.

References

1. Source: World Steel Association
2. Retrieved from: https://www.marketwatch.com/press-release/refractory-market-2019global-industry-analysis-size-share-trends-market-demand-growth-opportunities-and-forecast-2025-2019-03-18.
3. Retrieved from: https://www.mordorintelligence.com/industry-reports/refractories-market
4. Retrieved from: https://imformed.com/refractory-mineral-trends-irefcon16/
5. Retrieved from book: John B. Wachtman, *Mechanical Properties of Ceramics*, Wiley.
6. Retrieved from book: EP Popov, *Engineering Mechanics of Solids*, (1999), Upper Saddle River, NJ: Prentice Hall, pp. 111–115.
7. Retrieved from book: C. Kittel, *Introduction to solid state physics*, Wiley, p. 122.
8. Retrieved from book: W. D. Kingery, *Introduction to Ceramics*, Wiley, p. 593.
9. RTE, Houldsworth (1923).

10. T. J. Lu and N. A. Fleck The thermal shock resistance of solids. *Acta Mater.* (1998) Vol. 46, No. 13, pp. 4755–4768.
11. T.K. Gupta, J. Am. Cera. Soc. (1972), 55, pp. 249.
12. Cristobal Feliciano et al., "Rapid and Durable Repairs of Blast Furnaces Shafts with Jetcasting®: Assessment and Industrial Utilization" -ISSN 2176-3135.
13. Retrieved from book: C. A. Schacht, *Refractories Handbook* (2004), CRC Press.
14. Retrieved from: http://www.expertsmind.com/questions/thermal-properties-30137289.aspx
15. Emad Mohamed, M. Ewiais, "Carbon based Refractories", *Journal of the ceramic society of Japan* (2004), 112 [10], pp 517–532.
16. D. Mohapatra, D. Sarkar, "Effect of in situ spinel seeding on synthesis of MgO-rich $MgAl_2O_4$ composite", *Journal of materials science* (2007), 42 (17), pp. 7286–7293.
17. Retrieved from book: Budnikov
18. "Evolution of Thermal Microcracking in Refractory ZrO_2-SiO_2 after Application of External Loads at High Temperatures". *Materials* (2019), *12*(7), 1017; DOI: https://doi.org/10.3390/ma12071017
19. Retrieved from book: Routschka G, Barthel H, Refractory Base Materials, *Pocket Manual Refractory Materials* (1997), Vulkan-Verlag, Essen/Germany.
20. Prost, L. and Pauilliac, A. "Hydraulically setting refractory compositions", French patent application no. 6934405 (1969).
21. E Y Sako, M A L Braulio, P O Brant, V C Pandolfelli, "The impact of pre-formed and in-situ spinel formation on the physical properties of cement bonded high alumina refractory castables", *Ceramic International* (2010), 36, pp. 2079–2085.
22. "Melt Corrosion_Oxide_Non Oxide_Refract", Lexington, MA, SciGlass'; SciVision. Ref –06.
23. Retrieved from book: W. A. Tiller, *The science of crystallization: microscopic interfacial phenomena*, (1991) Cambridge, Cambridge University Press.
24. Mansoor Barati, "Application of Slag Engineering Fundamentals to Continuous Steelmaking", DOI: https://doi.org/10.1016/B978-0-08-096984-8.00015-X
25. Retrieved from book: G. Tro¨mel, K. Koch, W. Fix, N. Großkurth, Arch. Eisenhu¨ttenwes. 40 (1969), p. 969, (Fig. 2).
26. Leonard W. Bell et al., "Ground Granulated Blast-Furnace Slag as a Cementitious Constituent in Concrete".
27. Retrieved from book: *SLAG-IRON AND STEEL Indian Minerals*, Yearbook 2015.
28. Irem Zeynep Yildirim et al., "Chemical, Mineralogical, and Morphological Properties of Steel Slag", July 2011, Volume 2011.
29. Farshid Maghool et al., "Laboratory Evaluation of Ladle Furnace Slag in Unbound Pavement-Base/Subbase Applications", *American Society of Civil Engineers*, (2016) DOI: https://doi.org/10.1061/(ASCE)MT.1943-5533.0001724
30. Pen-Chi Chiang et al., "Iron and Steel Slags", DOI: https://doi.org/10.1007/978-981-10-3268-4_11

Chapter 2
Iron- and Steel-Making Process

2.1 Introduction

The earth crust contains about 8.0% iron. But it is present in large proportion as its compound, mostly in the form of oxides. Iron is extracted in the form of metallic iron (Fe) by reducing the oxide ores with a reducing agent. The principal agents commonly employed to serve this reduction process are the naturally occurring forms of carbon such as coke, charcoal and coal. In order to extract the iron, Fe (55–60%)-rich ore is usually used. The most abundant oxide mineral is found as hematite (Fe_2O_3), followed by magnetite (Fe_3O_4). Those minerals are contaminated with sand, clay and other rocks, which are the sources of impurities, such as silicon, phosphorus, aluminium, manganese and sulphur, and those impurities to be removed below the acceptable quality limit during extraction. Extracted iron can be classified into three categories with respect to carbon (C) content, and these are wrought iron, cast iron and steel. In present application point of view of iron, steel is the most important product, followed by cast iron. Wrought iron is used in very small quantity for ornamental purposes. Blast furnace iron making is the most popular process for extraction of iron from its ores. Other well-known processes are EAF and Direct Reduction Iron (DRI) processes. However, blast furnace will continue to be the most efficient route to produce pig iron for its high productivity and cost optimization, for many years to come [1, 2].

Pure iron has a very high melting point (1536 °C), but in presence of carbon, the melting point lowers significantly, the lowest melting temperature (eutectic point) is 1150 °C with an alloy of 4.2% C. Steel is an alloy of carbon and iron with C content of less than 4.2%. Iron has the ability to form alloys with many other elements, such as chrome, nickel, and manganese, and thus addition of those elements modifies the properties of steel, to fulfil the requirement of different applications, and coined as special steel. Based on alloy formation, thousand grades of steel are produced through primary and secondary steel-making processes. The level of impurities

S. Biswas, D. Sarkar, *Introduction to Refractories for Iron- and Steelmaking*,
https://doi.org/10.1007/978-3-030-43807-4_2

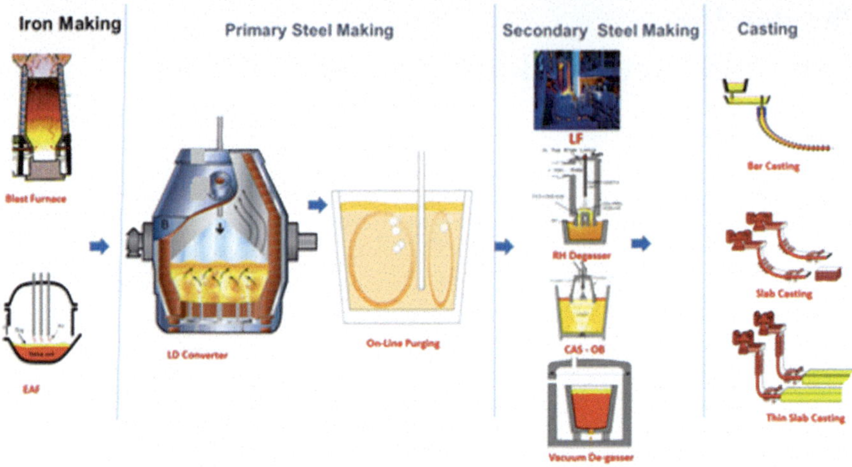

Fig. 2.1 Modern process of iron making and steel making

including carbon is further reduced by purging oxygen or under vacuum, to produce different grades of steel.

Globally, steel products have been mainly manufactured via primary steel-making process that constitutes two major routes. These are integrated Blast Furnace–LD Process/Basic Oxygen Furnace Route (BF–LD/BOF) and Directly Reduced Iron–Electric Arc Furnace (DRI–EAF) steel-making practice. Alloy steels are generally made in EA and electric induction furnace. The most commonly used is BOF process as it delivers very high productivity and flexibility to produce high-quality conventional steel. BOF steel making rapidly begun to replace the other existing open-hearth process worldwide since 1940. In the BOF process, pure oxygen at supersonic speed is blown vertically through a lance on the surface of molten hot metal to reduce carbon and other impurities, and the total process is completed within 45–55 min.

High alloy special steels with stringent properties are produced through secondary refining process. Secondary steel making is primarily concerned with the refining of crude steel in terms of both composition and cleanliness before casting. This secondary processing of liquid steel involves a wide variety of techniques, majorly including deoxidation or killing with Al/Si additives, vacuum degassing, desulphurization, homogenization, alloying and control of inclusion chemistry and content. These ladle metallurgy operations are commonly performed using the electric arc equipped gas stirred ladle furnaces. Tight control of ladle metallurgy practices in terms of chemistry and consistency to yield high-grade steels is rather adequate. The processes of iron and steel making up to casting protocol have been schematically illustrated in Fig. 2.1.

2.2 Overview on Blast Furnace Iron Making

A modern blast furnace plant consists of the following important sections:

1. Blast furnace proper.
2. Hot blast stoves.
3. Hot blast carrying system and tuyere assembly.
4. Raw material storage and handling.
5. Liquid slag and hot metal disposal and transport.
6. Gas cleaning system.

The schematic arrangement of the various section of the blast furnace is shown in Fig. 2.2. Among several components, the most important in the perspective of iron manufacturing and refractory consumption zone is "blast furnace proper", and thus more focus has been given on it.

2.2.1 *Basic Construction of Blast Furnace*

A blast furnace is a circular in cross section with height of approximate 30 m. Outer shell is made of steel and inside is lined with refractory. The old mantle supported structure is replaced by modern design of "free standing" with advantages of uniform load distribution, larger volume, easy to upgrade, as shown in the following

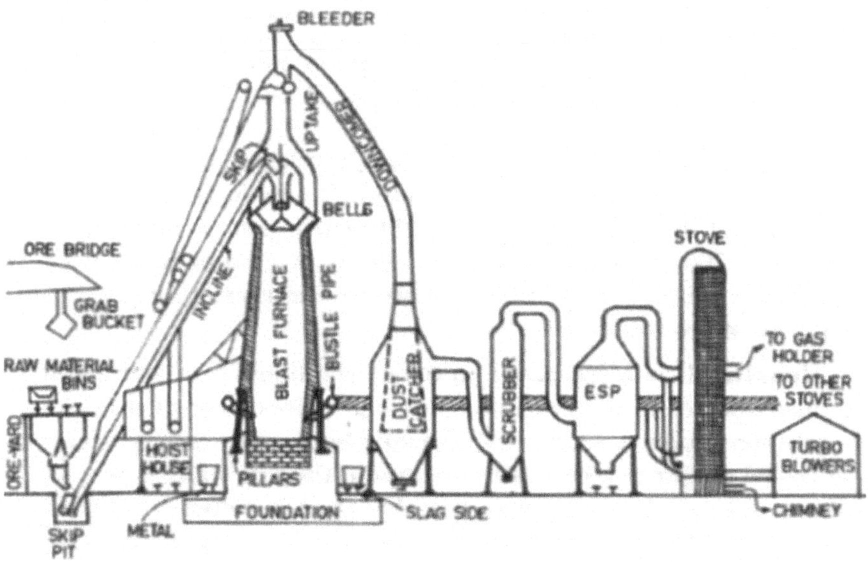

Fig. 2.2 Schematic diagram of blast furnace iron making including raw material feeding to metal output

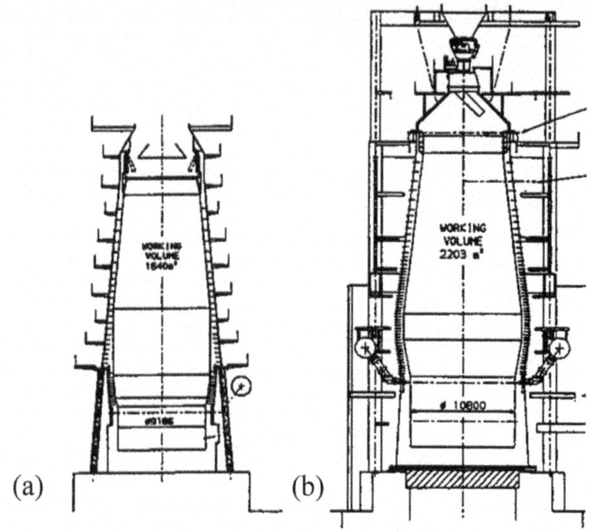

Fig. 2.3 Schematic diagram of blast furnace structure. (**a**) left—pre-1992 rebuild, and (**b**) right—post-1992 rebuild or modern blast furnace [2]

schematic diagram in Fig. 2.3. Different parts of the modern blast furnace are further elaborated in Fig. 2.4.

In consideration of iron-making processes, chemical reactions and temperature profile, the furnace is divided in the following zones, along the vertical length, as shown in the Fig. 2.4.

- Stack: It is usually considered from stock line to mantle level. In this zone, burden is completely solid at the periphery wall. The charge is heated around 200 °C at stock line to nearly 1100–1250 °C at the bottom of the stack.
- Belly: It is the cylindrical portion at lower stack. The chare materials begin to soften and start reacting with ascending burden.
- Bosh: In this zone, the melting of burden takes place except coke in this zone. Direct reduction starts in this zone. The gangue and flux combined to form slag, called bosh slag. Desalinization and desuphurization are the major reaction takes place in this zone.
- Hearth: Most of the feed materials are melted, and it is a receptacle to collect the liquid slag and metal. Coke burns in the tuyere level and a small fraction descends into hearth and solid coke mixed with slag accumulated at the centre of the hearth, called "Deadman", which either sits on the hearth or floats in hot metal, just above hearth bottom.

In order to increase the output of a blast furnace, hearth diameter thus working volume needs to increase. The cumulative capital and operational cost of a large blast furnace is lower than smaller blast furnace, and thus demand of large blast furnace is progressively increasing to fulfil the production target

(continued)

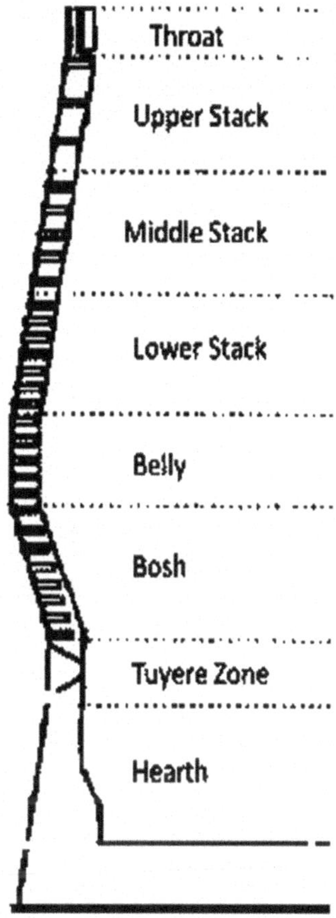

Fig. 2.4 Different parts of the furnace, from raw material charge to output region sequence: throat, upper stack, middle stack, lower stack, belly, bosh, tuyere zone and hearth

Table 2.1 Different sizes of blast furnace (BF), their hearth details, working volume and production per day

Type of BF	Hearth diameter (m)	Working Vol (m^3)	Production/day (tons)
Mini blast furnace	3–4	250–350	< 500
Small size	5–8	450–900	500–2500
Medium size	9–11	1000–2500	3000–7500
Large size	>12	>3500	>9000

[2]. The size of the blast furnaces can be classified with respect to Hearth Diameter and Working volume. Based on the capacity, the blast furnaces are of the following types, shown in Table 2.1.

2.2.2 *Blast Furnace Reactions to Produce Metallic Iron*

The major charge materials in blast furnace are sinter or pellets iron ore and limestone, and the ascending gas CO is a product of combustion of carbon (coke) in tuyere region. Iron ore (Fe_2O_3, Fe_3O_4) is reduced to FeO in lower stack and bosh area [3]. All the raw materials are charged from furnace throat and hot air is purged through the tuyere, where the hot air helps for the combustion of C follows the reaction:

$$2C_{(s)} + O_{2\,(g)} = 2CO_{(g)} \tag{2.1}$$

Therefore, thus released CO gas acts as prime resource of thermal energy for the smelting operation and production of metallic Fe. This ascending gaseous matter (CO) preheat the feed material and reduce the major amount of iron oxide in descending burden at stack portion of the blast furnace. Hydrogen gas (H_2) is another major source of heat for the reduction of iron oxide and is usually evolved by an endothermic reaction between moisture present in the hot air blast and carbon. This chemical reaction (Eq. 2.2) attributes to the evaluation of more reducing gaseous mixture (i.e. CO and H_2).

$$C_{(s)} + H_2O_{(g)} = CO_{(g)} + H_{2(g)} \tag{2.2}$$

Depending on the temperature, the reduction process of iron oxides in the descending burden by the ascending CO and H_2 gaseous mixture can be categorized into the following two stages.

Stage I: Reduction of hematite (Fe_2O_3) and magnetite (Fe_3O_4) at a critical temperature below 970 °C takes place in the stack region of blast furnace. The relevant stoichiometric and sequential chemical reactions are as follows:

$$Fe_2O_{3(s)} + 3CO_{(g)} = 2Fe_{(s)} + 3CO_{2\,(g)} \tag{2.3}$$

$$Fe_3O_{4(s)} + 4CO_{(g)} = 3Fe_{(s)} + 4CO_{2\,(g)} \tag{2.4}$$

$$Fe_2O_{3(s)} + 3H_{2(g)} = 2Fe_{(s)} + 3H_2O_{(g)} \tag{2.5}$$

$$Fe_3O_{4(s)} + 4H_{2(g)} = 3Fe_{(s)} + 4H_2O_{(g)} \tag{2.6}$$

Stage II: Further reduction of iron oxide (FeO) to metallic Fe in the temperature range of 970 °C–1000 °C is happening at the lower portion of the blast furnace. This specified temperature is supposed to be higher than the critical temperature of reactions. The corresponding chemical reactions have been given.

$$FeO_{(s)} + CO_{(g)} = Fe_{(s)} + CO_{2\,(g)} \tag{2.7}$$

$$FeO_{(s)} + H_{2(g)} = Fe_{(s)} + H_2O_{(g)} \tag{2.8}$$

However, the evolved CO_2 gas and water vapour (H_2O) are relatively unstable in the presence of coke at this temperature, stimulated to react with the carbon at a higher rate to regenerate CO and H_2 gaseous mixture. The respective chemical reactions are as follows:

$$CO_{2(g)} + C_{(g)} = 2\,CO_{(g)} \tag{2.9}$$

$$H_2O_{(g)} + C_{(g)} = H_{2(g)} + CO_{(g)} \tag{2.10}$$

Therefore, the overall reduction reaction facilitated by either CO or H_2 gas in this zone can be represented as above.

2.2.3 *Gaseous or Indirect Reduction of Iron Oxides*

Carbon monoxide gas (CO)-assisted reduction of hematite (Fe_2O_3) to metallic Fe usually transforms above 570 °C through formation of the intermediate compositions magnetite (Fe_3O_4) and wustite (FeO).

In this circumstance, a reversible and equilibrium attainment can be noticed at constant pressure with respect to variable temperatures. A typical equilibrium relationship within low-temperature reduction of magnetite (Fe_3O_4) to iron (Fe) is illustrated in Fig. 2.5. The reduction process completion at stack zone maintains the equilibrium between CO and CO_2 at a particular temperature as per the Boudouard reaction, which is named after Octave Leopold Boudouard.

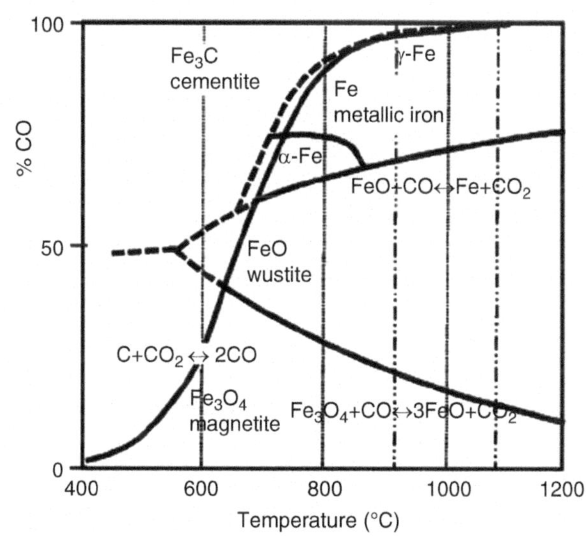

Fig. 2.5 Equilibrium diagram of Fe–O–C system

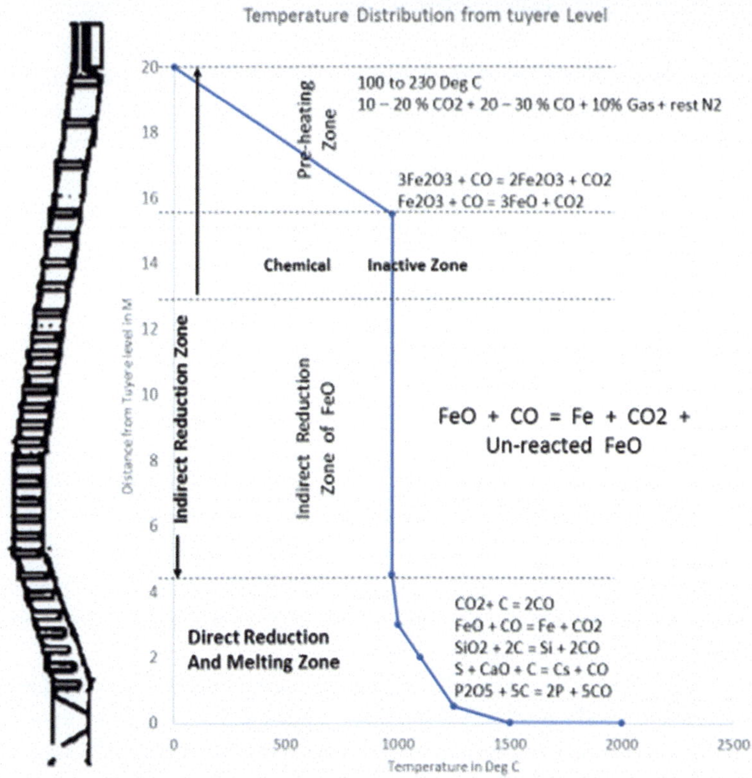

Fig. 2.6 Reduction of iron ore inside blast furnace

$$2CO = CO_2 + C \tag{2.11}$$

The redox reaction follows chemical equilibrium within CO, CO_2 and C at a definite temperature. The available gases interact with coke in preference below 800 °C, and develop substantial amount of CO, may be concentration raises to 90%, and hence the process leads to the reduction of FeO to metallic Fe, such phenomenon is known as indirect reduction. The reducing agent CO is generated from the combustion of fuel carbon, and the indirect reduction process is completed within the isotherm of 970 °C, as shown in Fig. 2.6.

2.2.4 *Direct Reduction of Iron Oxide by Solid Carbon*

The direct reduction process of iron oxide by solid carbon is highly endothermic in nature and occurs when the molten slag flows over incandescent coke in the bosh region at a temperature above 1000 °C [4]. Therefore, the corresponding reaction can be written as follows:

$$FeO_{(s)} + C_{(s)} = Fe_{(s)} + CO_{(g)} \tag{2.12}$$

Herein, a competitive discussion on the reduction phenomenon of FeO by carbon or gaseous substances provides a better insight in order to understand the thermal behaviour since the carbon-assisted reduction is endothermic while CO-assisted is exothermic in nature, thus actual situation in the furnace is relatively complicated. While discussing such aspect, need to remember that in the beginning the CO develops from coke, and equilibrium limitation restricts the complete combustion to CO. Gaseous reduction yields excessive thermal surplus, whereas the thermal-deficit carbon reduction follows only one-third of the total carbon to produce each mole of iron. Therefore, both of the reduction processes are in competition to balance the thermal history and thus fuel saving during process. In brief, the content of oxygen in the range of 60–65% present in the burden is being removed by gaseous reaction and the remaining 35–40% by carbon reduction.

2.2.5 *Other Reactions in Blast Furnace*

Despite formation of Fe, there are several other processes are involved to complete the iron-making process. Most predominate transformations including MnO to Mn, SiO_2 to Si, removal of sulphur, P_2O_5 to P and slag formation are noticed.

2.2.5.1 Reduction of MnO

It is a multistage process occurring in the blast furnace stack region. In the beginning, the higher oxides are reduced by CO; however, the MnO is only reduced in the presence of C at the temperature above 1500 °C. The later-stage reaction (Eq. 2.15) absorbs substantial amount of heat, thus final-stage reduction demands high temperature, and conversion efficiency varies 65–75% of the charged manganese compounds.

$$MnO_2 + CO = MnO + CO_{2(g)} \tag{2.13}$$

$$Mn_3O_4 + CO = 3MnO + CO_{2(g)} \tag{2.14}$$

$$MnO + C = Mn + CO_{(g)} \tag{2.15}$$

The resultant reduces the dissolution of manganese (Mn) into hot metal while the unconverted 25–35% remains in the slag. Such incidence reflects the manganese partitioning phenomenon and indicator of the thermal state of the hearth.

2.2.5.2 Reduction of SiO_2

Silica (SiO_2) to Si reduction follows two-step process before transferring into hot metal. In tuyeres region, silica in ash form silicon monoxide (SiO) gas when come into contact with coke, and volatize, which eventually react with molten iron and form FeO. The FeO is subsequently reduced to Fe by the existence of enormous CO. Despite such phenomena, direct reduction of SiO_2 to Si is also taking place at very high temperature.

$$SiO\ (g) + Fe = Si + FeO \tag{2.17}$$

$$SiO_2 + 2C = Si + 2\ CO_{(g)} \tag{2.18}$$

The reaction rate is expedited at high temperature, and thus silicon content of the hot metal is proportional to the hot metal temperature for a particular burden and slag composition. The resultant Si percentage in hot metal can synchronize through higher silicon resource in charge as well as coke feeding rate.

2.2.5.3 Removal of Sulphur

Coke is the prime resource to enter sulphur in blast furnace, and released either as H_2S or a gaseous compound mixture of carbon monoxide and sulphur (COS) in burning of coke. Upward gas movement through the stack enable the interaction with lime in the flux and certain combination with iron. A plausible reaction mechanism can be pointed in Eqs. (2.19) and (2.20).

$$FeO + COS = FeS + CO_2 \tag{2.19}$$

$$FeS + CaO + C = Fe + CaS + CO_2 \tag{2.20}$$

The iron sulphide (FeS) is preferentially removed in the presence of basic flux lime (CaO) at very high temperature in the hearth region. However, the sulphur removal depends on three parameters referred as hearth temperature, slag basicity $(CaO + MgO)/(SiO_2 + Al_2O_3)$ and slag volume.

2.2.5.4 Reduction of P_2O_5

High-temperature reduction of P_2O_5 to phosphorous (P) is followed by the Eq. (2.21). Unlike Mn and Si, herein, the complete reduction of P_2O_5 and formation of phosphorous prefers to dissolve in the hot metal.

$$P_2O_5 + 5C = 2P + 5CO \tag{2.21}$$

2.2.5.5 Slag Formation

Continuous and vigorous reactions of charged materials in blast furnace produce metals and remaining unconverted materials considered as blast furnace slag. The prime constitute of slag is CaO, MgO, SiO_2 and Al_2O_3, and their cumulative content usually exceeds 95% of total slag weight. Despite these oxides, preferentially other oxides MnO, K_2O, Na_2O and FeO are present in the slag. Owing to the compositional behaviour variation, the slag melting temperature varies from 1250 to 1450 °C. However, the slag basicity index [(CaO/SiO_2) or (CaO + MgO)/(SiO_2 + Al_2O_3)] depends on the SiO_2 content, and it is maintained in the range of 0.9–1.1, which is acidic in nature. Eventually, this index influences the elemental distribution between slag and hot metal, thus final hot metal composition. To maintain the hot metal quality, it is required for complete slag and metal separation, which takes place in iron trough.

2.2.6 *Cooling System*

The main function of the cooling system in a blast furnace is to cool the furnace shell and prevent it from over-heating and burning that causes for the stoppage of furnace operation. To accomplish this, the cooling system must be able to extract the excess heat generated into the furnace and loaded on the shell that finally help to maintain the desired isotherm inside the furnace. Campaign of blast furnace depends on the life of refractory lining that severely affected attribute to the chemical attack and mechanical wear (erosion, chemical attack and oxidation) during iron processing [3]. These preferentially occur at elevated temperature and within certain temperature range as shown below:

- CO-attack on stack refractory: 450–850 °C.
- Alkali attack: 800–950 °C.
- Zinc vapor attack: 750–900 °C.
- Oxidation by O_2 on C refractory: >400 °C.
- Oxidation by CO_2 & H_2O: >700 °C.

Hence, from the refractory performance point of view, the lining refractory to be kept lower than the temperature of reaction by designing external cooling system and the best available cooling media is water to reduce the shell temperature. Process of enhancing blast furnace campaign life by external cooling is known as "Thermal solution", and the process is further discussed in details in Chap. 3.

Cooling plate and staves are developed and introduced to achieve high cooling efficiency where heat flux is very high. Both systems developed in parallel and the cooling medium in a cooling system placed inside the furnace, designed to sustain long campaign life. Apart from cooling in stack, belly and bosh area are essentially encountered for effective cooling. Cooling of hearth wall is also introduced by stave

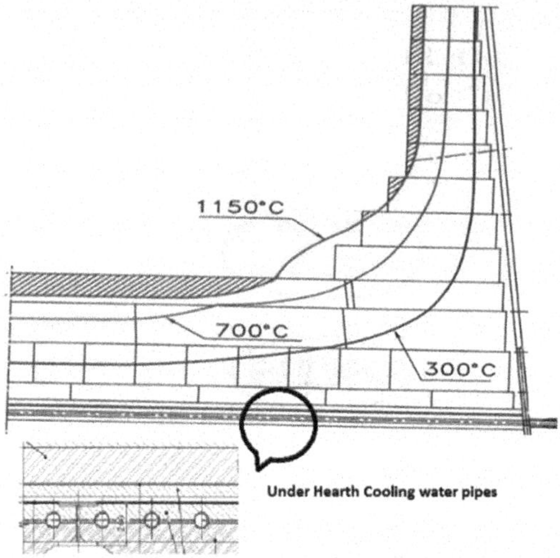

Fig. 2.7 A typical cooling arrangement below the refractory hearth of blast furnace

cooling. Besides the hearth wall, under-hearth cooling assembly installation is a common practice for large size blast furnace. Generally, water through pipes are used in under-hearth cooling, as shown in Fig. 2.7. For small size furnace, under-hearth cooling is taken care by hearthside wall stave cooling.

The main bosh cooling and stack cooling systems used in a modern blast furnace are as follows:

- Cast iron and copper cooling plates,
- Cast iron and copper cooling staves,
- Box coolers,
- Cigar coolers,
- Spray cooling,
- Combination of stave and plate cooler for intense cooling.

In small- to medium-size mantle-supported blast furnaces, cast iron staves, box coolers and copper plates are widely used in bosh and stack areas. In few furnaces, spray cooling is used in lower stack and bosh area. Cigar coolers are installed in cooling to improve cooling in few selective areas, like tap-hole area.

2.2.6.1 Cooling Plates

Since last more than a century, plate coolers are in use in middle stack, lower stack and belly area. The cooling efficiency depends on quality of cooler plate body, water flow rate, inlet water temperature, spacing of coolers (pitch) and quality of refractory used.

The thermal shock resistance and thermal conductivity are the most critical properties of refractory to optimize the effective cooling, which control the isotherms inside the furnace and erosion of refractory lining. A schematic diagram of plate coolers is shown in Fig. 2.8a. The plate coolers are embedded in the brick lining, as shown in Fig. 2.8b.

2.2.6.2 Cigar Coolers

In few selective areas, such as tap hole spool area and slag notch area, intensive cooling is required and cigar coolers are the best choice for these areas. Cigar coolers are machined from a solid copper bar to form a cylindrical shape and channels are made inside the coolers for continuous and uniform water circulation. It is placed between the plate coolers assembly. The typical schematic diagram of cigar cooler and its assembly is shown in Fig. 2.9. Two types of plate coolers are in use, coolers made of cast iron and made of copper. Utilizing the superior conductivity of copper, the copper coolers are used in high heat zones, such as in lower stack, belly and bosh areas. Use of cigar coolers could also improve the cooling of adjacent refractory and thus reduce the corrosion effect on lining.

2.2.6.3 Stave Coolers

In small- and medium-size furnaces, stave coolers are used in high heat zone area, such as in bosh and hearth wall. In those areas, cast iron staves are used. The main advantages of using stave coolers over plate coolers are uniform cooling throughout the refractory lining surfaces. In modern large-size blast furnaces, the plate coolers in lower stack and belly area are replaced by high-conducting copper plate coolers. One of the disadvantages of using plate coolers is very low cooling surface area compared to stave cooling, which covers very large cooling surface. In plate cooling system, maximum cooling effect is experienced around the plates and insufficient cooling was observed in between the spaces (pitch) of coolers assembly, as shown in Fig. 2.10.

Hence, in modern large-size blast furnaces, the cooling assembly in belly and stack zone is also replaced by staves. With the increase of size of blast furnaces, the

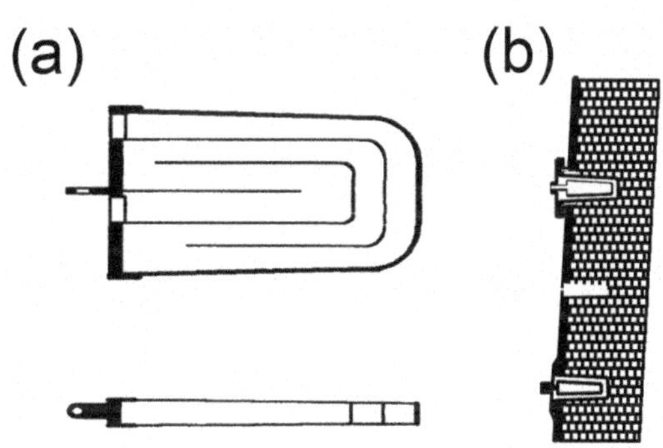

Fig. 2.8 (**a**) Plate cooler and (**b**) assembly of coolers in brick work

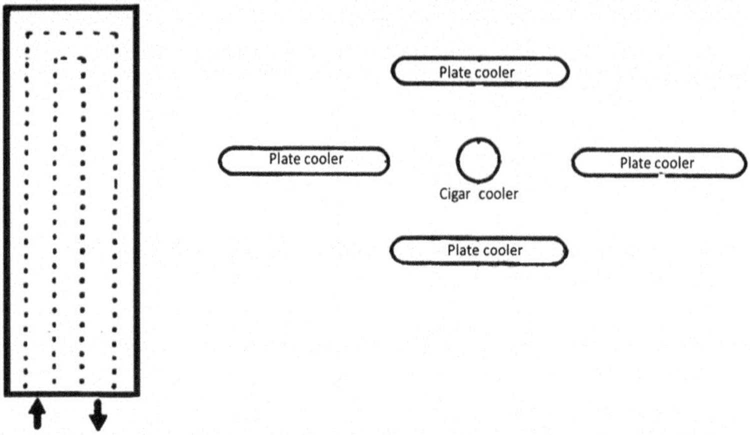

Fig. 2.9 Cigar cooler and its assembly with plate coolers

Fig. 2.10 A typical assembly of stave plate cooling consists of different efficiency zones

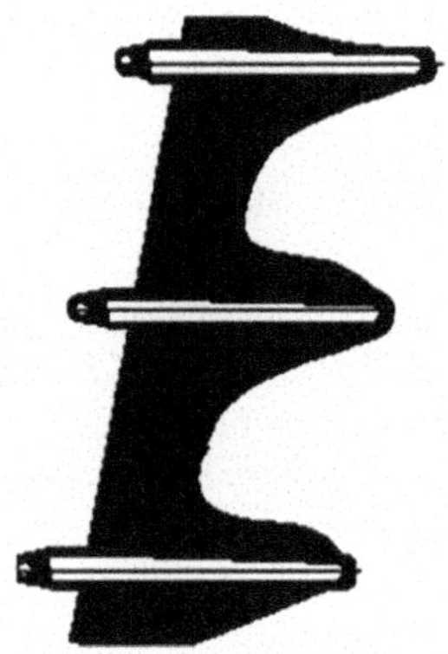

heat load inside the furnace has been increased. In highly eroded area at lower cooling zone, the shell temperature would increase to as high as more than 550 °C and the shell became red-hot, and cracking of shell occurs. One of the process is to reduce the distance (pitch) between the coolers, but it would not eliminate the occurrence of over-heating.

To maximize the indirect reduction and optimize the direct reduction iron oxides, proper temperature control is mandatory to design by continuous improvement in cooling efficiency, and this was achieved by upgradation of cooling system through development of stave cooling system. Stave cooling developed in parallel to plate cooling. These are flat surface cooling members with large cooling surface area, parallel to the shell embedded

(continued)

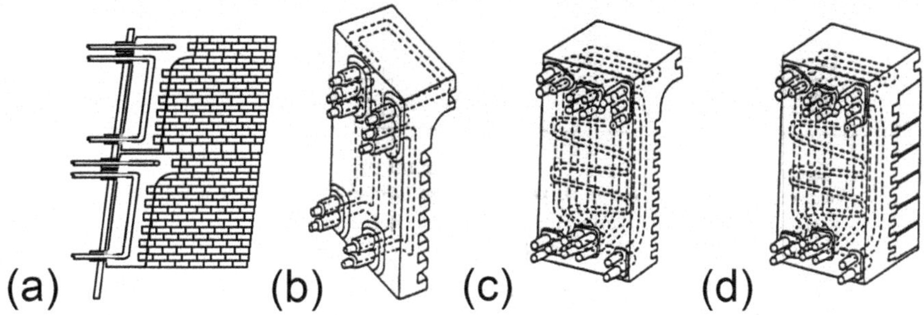

Fig. 2.11 (**a**) Assembly of stave coolers with refractory lining, (**b**) second generation, (**c**) third generation and (**d**) fourth generation stave plate cooler

between refractory work and shell. The main application areas are as follows:

- High heat load area—bosh and belly.
- Hearth wall.
- Tuyere area.
- Area requires uniform cooling—in modern furnaces, hearth wall, bosh, belly and stack.

Initial staves are made of cast iron. A schematic diagram of a stave cooler assembly and the continuous upgradation of its design is shown in Fig. 2.11. Until the design of third-generation staves, the thick refractory lining was used as shown in Figure 2.11c, but fourth-generation stave is appreciated as a breakthrough cooler design system in blast furnace as shown in Figure 2.11d. The thinner layer of refractory, 230–360 mm width, is only embedded into the stave, and no additional brick lining is required. The main advantages of using fourth-generation coolers are as follows:

- Reduction of heat loss,
- Intensive and uniform cooling,
- Reduced installation time,
- Larger volume above tuyere,
- Long campaign.

The latest lining in front of stave coolers at blast furnace stack area is shown in Fig. 2.12.

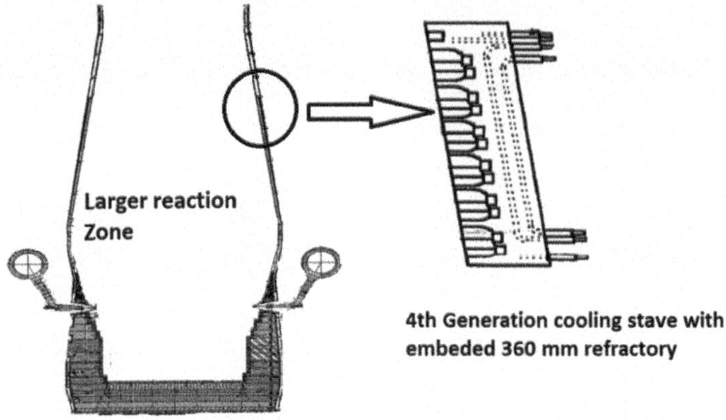

Fig. 2.12 Latest cooling stave design

2.2.6.4 Fourth-generation SiC Refractory for Cooling Stave

SiC refractory was first used in blast furnace about more than 40 years ago. In continuation of that, the SiC was developed as a most suitable refractory material to use in high heat zone [5]. The most important properties of refractory used in stack at high heat zone embodied into fourth-generation staves are as follows:

- High corrosion and abrasion resistance,
- High alkali attack resistance,
- High thermal conductivity,
- High toughness and resistance to crack growth,
- Uniform and finer pore-size distribution,
- Oxidation resistance at intermediate temperature.

Silicon carbide (SiC) refractories have almost all above properties. It has excellent thermal shock resistance due to its very high thermal conductivity, and it is resistant to alkali attack. Development went from oxide bonded to oxinitride-bonded bricks and ended to nitride (Si_3N_4)-bonded bricks. Apart from corrosion resistance, development of new bond system had improved oxidation resistance and durability in blast furnace operating conditions. Depending on the bond system, the four types of SiC bricks are available: (a) oxide bonded SiC, (b) direct bonded SiC, (c) Si_3N_4 bonded SiC and (d) SiAlON bonded SiC, details of such bricks are given in Table 2.2.

In the late 1970s and early 1980s, there was considerable interest in beta-SiC bonded, direct bonded SiC refractory to use in blast furnace stack and bosh. But gradually the use of those bricks was declined. In plate cooling system, the coolers are used as anchors or supporting members for thicker brick work around the coolers. In large-size furnaces, the main cause for failure of refractory around such plate coolers is thermal stress generated due to high-temperature and high-pressure operation. The other cause is the corrosion due to chemical attack of alkali and zinc

Table 2.2 Different grades of SiC refractory used in BF cooling system

Properties	Oxide bonded	Direct bonded	Nitride bonded	SiAlON bonded
SiC (%)	86	>90	84	71
Al_2O_3 (%)	NA	NA	NA	13
N_2 (%)	NA	NA	7	6.5
Fe_2O_3 (%)	1.1	0.25	0.5	0.2
Porosity (%)	20	16	17	16
Bulk density (Gm/cc)	2.52	2.58	2.6	2.69
Bending strength				
At 20 °C (Mpa)	20	35	45	40
At 1400 °C (Mpa)	15	40	44	42
Thermal conductivity at 1000 °C (W/mK)	11	17	16	15
Thermal shock resistance at 1000 °C, (change in C MOR after 5 cycles in water quenching) (%)	−16	0	−28	0
Refractoriness under load (RUL) ta (°C)	1480	1650	1650	1640
Oxidation test (mass gain after 20 h at 1300 °C) (%)	0.25	0.42	0.79	Nil
Alkali test (H MOR)				
Before alkali attack (Mpa)	15	40	44	28
After alkali attack (Mpa)	6	35	36	13

(Zn). Si_3N_4 bonded SiC bricks are the superior in corrosion resistance and high temperature strength, considered as the prime advantages of non-oxide refractory over alumina refractory.

In stave cooling system, cooling members cannot support lining bricks, and hence thicker lining in front of coolers is not possible. In the fourth-generation cooling system, the staves were cast together with bricks, and the width of the bricks was maintained in the range of 230–360 mm. Hence, from the beginning, Si_3N_4 bonded SiC bricks consist of high CO attack resistance, high corrosion resistance and high thermal conductivity which are considered for new class of design. In few plants, the combination of stave coolers and plate coolers are in practice for intense cooling as well as adequate support to the thick refractory bricks lining in front of coolers and the arrangement as shown in the Fig. 2.13.

In early 1987, the SiC bricks with SiAlON bonding had been introduced with superior oxidation resistance, with other favourable properties and at present, it is accepted in wide horizon with stave cooling system. SiAlON is a continuous solid solution of Si, Al, O and N as an end member of Si_3N_4. Its chemical formula is written as $Si_{(6\text{-}z)}$. Al_z O_z $N_{(8\text{-}z)}$ when $(0 < Z < 4.2)$, as shown in Fig. 2.14. In the beginning, the prime raw materials graded silicon carbide added with aluminium metal powder, and co-fired in the presence of nitrogen gas in order to develop in situ oxinitride bond. SiAlON solid solution may be formed in the matrix of SiC bricks by the reaction of Si_3N_4 and Al_2O_3 during sintering at elevated temperature under reducing atmosphere, as shown in the equation: $Si_3N_4 + Al_2O_3 = \beta\text{-SiAlON}$.

Fig. 2.13 Assembly of stack coolers, embedded in the lining

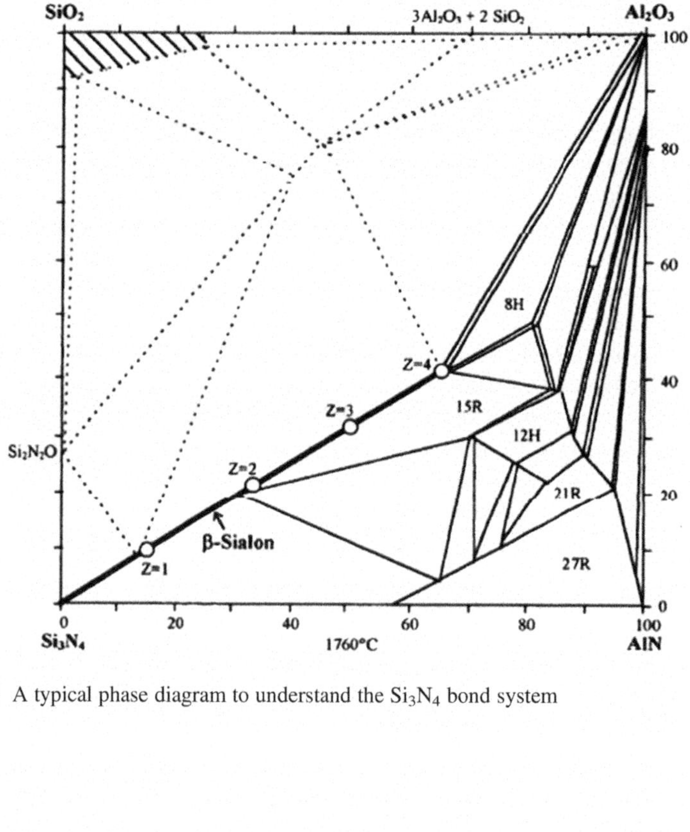

Fig. 2.14 A typical phase diagram to understand the Si_3N_4 bond system

The strength of the bricks is enhanced as the bonding force increases between SiC particles by formation of SiAlON, until the Z number reaches 2, and when the Z reaches 3, Al_2O_3 became excess for the formation of sialon and remains as Al_2O_3 in the matrix of sialon bonded SiC bricks, which forms whiskers, resulted in further development of bonding strength.

2.2.7 *Cast House Practice*

Blast furnace tap hole mix and trough materials are generally known as cast house refractories. In the past, inexpensive low grade alumino-silicate refractories were used in cast house of small and medium size blast furnaces. With the increase in working volume, productivity and high top-pressure, phenomenon changes in high-quality refractory and trough design around the globe. Today, those improved refractories have an excellent bearing capability during iron-making operation. The good cast house practice has substantial contribution to fulfil the high productivity for blast furnace having the internal volume of more than 5000 M^3. This has resulted increase in tapping rate, tapping temperature and duration of tapping. To cope up with this stringent operating condition, however, the proper drainage of hot metal to be ensured.

2.2.7.1 Tap Hole Practice

Molten iron accumulated in the hearth and is periodically removed from the furnace through a hole located at the hearth side wall. This opening is known as tap hole or iron notch. This tapping hole is normally closed with the help of a clay gun. The gun was filled up with carbonaceous refractory mass, called tap hole clay or mud-gun clay. The clay is pushed into the hole by gun. For tapping hot metal, the hole is further opened by drilling through the enblocked tap hole clay by a drilling machine. The construction of tap hole is shown in Fig. 2.15. The pre-requisite for trouble-free tapping are as follows:

- Constant tap hole length,
- Uniform discharge speed of hot metal through tap hole,
- Complete drainage of hot metal during casting,
- Smooth opening and closing of tap hole.

To achieve the above results consistently, it is required to have a total cast house system approach. It should start right from the tap hole design, tap hole clay material design and proper application technology.

Tap hole clay is used for plugging of tap hole during iron processing followed by drilling machine opens the tap hole clay in making a path for discharging iron and slag, and further plugging of tap hole after tapping. The mud gun injects or extrudes the clay into the blast furnace tap hole where it hardens and checks the hot metal and

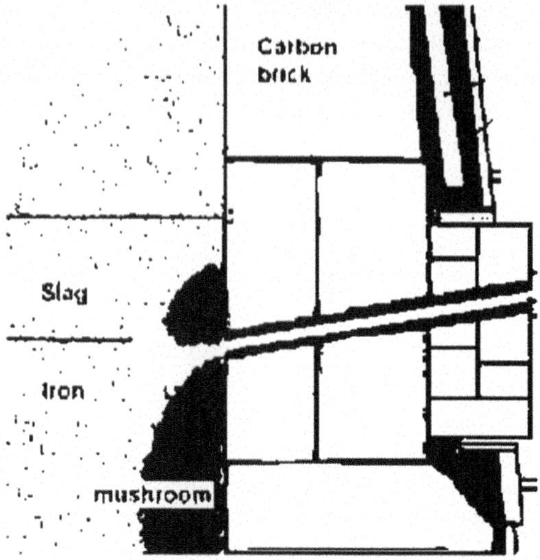

Fig. 2.15 The tapered cylindrical passage represents tap-hole region, pig iron and slag mixture flow from top to down and form mushroom near to tapping zone

slag from coming out of blast furnace. The performance of tap hole clay is of fundamental importance to the successful running of a blast furnace. The tap hole length, tapping duration and tapping rate, the hot metal and slag balance are directly related to uniform furnace performance. Tap hole length indicates hearth side wall condition near tap hole. Tap hole clay is expected to protect hearth side wall by forming a protection layer of clay known as "mushroom." The longer cast duration contributes to not only reduction of clay consumption but also reduction of workload at cast house such as gun up and drilling.

2.2.7.2 Tap Hole Clay Mix

The tap hole clay at ambient temperature or sometimes heated at around 80 °C is rapidly pushed into the blast furnace, where iron temperature is of 1500 °C. The clay is capable to absorb the sudden thermal fluctuation without deterioration of physico-thermal properties. Binders used in the tap hole clay play a very important role such as plasticity, polymerization speed and texture inside the tap hole. Previously, many plants were using conventional hydrous tap hole mixes made from clay, coke, pitch and other refractory aggregates. These proved to have inadequate erosion resistance in blast furnace environment. In order to improve the setting characteristics and erosion resistance, anhydrous (tar-based) tap hole mixes of similar composition were developed. Anhydrous tap hole clay was found to perform superior to hydrous tap hole clay. Application of anhydrous tap hole clay not only increases the furnace productivity but also results in significant decrease in maintenance cost. Starting from those innovation, many improvements have been carried out (e.g. application of silicon carbide, silicon nitride, fused alumina, ultra-fine carbon, etc., as a refractory aggregate) with anhydrous tap hole clay and still development work is going on.

Table 2.3 Tap hole clay mix

Process characteristics	Tar	Tar/Resin	Resin
Polymerization speed	2	4–5	5
Plasticity	5	2	1
Texture at rapid heating	2	4–5	5
Texture at ramming	4	2–3	2
Oxidation resistance	4	2	1
Price	5	1–4	1
Environment	1	2	4.

The provided numerical indicates the lower to upper limit, 1–5, respectively

In general, tar binder is employed for its high plasticity at hot condition. Plasticity is one of the most important characteristics of tap hole clay at the time of ramming as it provides excellent extendibility inside the tap hole at high temperature. As binder gives great influence on tap hole clay plasticity, same time it is important to understand the volatilization characteristic of each binder. Tar bonded clay has broad range of volatilization up to 450 °C while resin bonded clay shows rapid volatilization before 200 °C.

In present scenario, blast furnace tap hole mixes can be classified into three major groups according to the class of binder; (1) tar bonded mixes, (2) resin bonded mixes, (3) mixes bonded by blends of above two binders. The characteristics of binders are shown in Table 2.3. In convention, the resin binder is employed for the capability of faster polymerization or hardening speed at hot condition after extrusion. In order to create good texture with having minimum cracks inside the tap hole, faster hardening speed is believed to be important in order to prevent defects caused by volatilization of binders. Resin bonded tap hole clay has characteristic of restoration of crack inside tap hole. In addition to tar and resin binder, by changing the mixing ratio of tar and resin, it becomes possible to adjust the plasticity and polymerization speed in accordance with the furnace requirements.

For blast furnace, getting tap hole length is a paramount factor preferred to use tar bonded tap hole clay, whereas in case of relatively short tap hole rapid hardening resin bonded tap hole clay is preferred. Amount of binder also plays a vital role for deciding clay mix property. Volatilization of binder in clay mix not only provides required plasticity and hardening speed but at the same time causes defect. So it is important that the amount of binder in clay mix should be as low as possible. However, the reduction of binder results in deterioration of tap hole clay properties in terms of achieving required plasticity at certain extrusion pressure. In this context, the use of sintered and fused aggregate is found to be useful, thus ultra-fine powder (carbon, silica) results in significant reduction of binder amount while keeping extrusion pressure in accepted level. Tap hole clay with low binder content will have lower apparent porosity and higher corrosion resistance.

2.2.7.3 Tap Hole Mixes Design

As mentioned in tap hole mixes must have the following main characteristics to confer good performance:

- Good workability in the mud gun and in the tap hole to guarantee an adequate injection;
- Quick drying and hardening after injection, in order to seal the furnace and shorten the mud gun's holding time in front of the tap hole;
- Good permeability in order to ensure the escape of the forming gas (tar or resin volatiles);
- Good adhesion to the remaining old mud from the tap hole's internal regions to form "mushroom";
- Good resistance to erosion and corrosion against hot metal and slag;
- At the same time, relative easy drilling of the set or sintered tap hole mix expedite the tapping process;
- Good shelf life, without hardening caused by the binder over a prolonged period in storage;
- Chemical compatibility with the internal refractory structure of the blast furnace.

2.2.8 *Drainage of Hot Metal Through Trough and Runners*

First, the molten iron comes out from the blast furnace through tap hole and flows on the cast house runner, known as blast furnace trough. Herein, the molten metal along with slag (produced during iron processing in blast furnace) is to be separated by a barrier called skimmer. It separates the slag from metal by allowing the slag leave from the top into slag runner as slag has low density than molten iron. The iron passes through a passage in the skimmer at the bottom. Thus, extensive interaction and sever corrosive environment affect the trough refractory that demands excessive care for interruption and continuous operation. Operational practice discloses that the performance of trough refractory not only depends on the refractory composition, rather proper cast house design is a very critical parameter to enhance the refractory life. Previously, the contact in between trough and molten iron for a prolonged time was not allowed, and maintained sharp inclination to move fast attribute to limited quality of refractories. However, the scenario has been changed, and now trough enables to withstand the molten iron for a longer period without metal solidification because of improved quality of refractories and design of trough. The schematic diagram of iron trough is shown in Fig. 2.16.

The runner design is of primary importance in the cast house sector. The main criteria are as follows:

- Reduction of the kinetic flow energy of the tapped stream in the main runner;
- Optimized segregation of slag and hot metal;

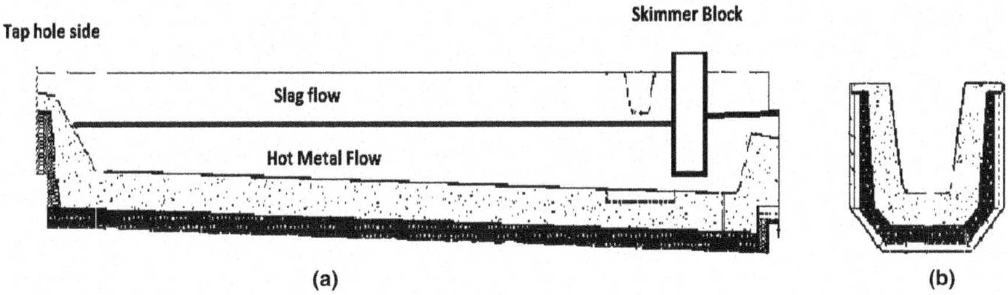

Fig. 2.16 Schematic diagram of iron trough. (**a**) Side view; (**b**) cross-sectional view

- Minimized thermal loss in the main runner;
- Reduced consumption of refractory material by an optimized runner geometry.

The different designs of trough and refractory practice are further explained in detail in Chap. 3, Sect. 3.5.

2.3 Modern Steel-Making Practices

2.3.1 Bessemer Process

Today's modern steel-making practice had been initiated about 200 years ago, when Henry Bessemer patented his process in 1856, and the process was introduced as Bessemer process of steel making [6]. The Bessemer process follows bulk steel production in large quantities and made it suitable for various applications. Thus, the introduction of Bessemer process is looked upon as an industrial revolution. This process involved the removal of impurities and lowering of carbon level by blowing of air through bottom of the tuyeres.

Oxygen in the air blown would react with Si, C and Mn present in hot metal to form oxides like SiO_2, MnO, etc. In this process, Fe in metal was also oxidized to FeO, and thus the presence of FeO, MnO and SiO_2 makes the slag acidic and hence it is also called *acid Bessemer process*. Acidic refractory was used as the lining material of the vessel. Carbon was oxidized to CO and CO_2 gases and escaped out through the mouth of the vessel, as shown in the schematic diagram (Fig. 2.17). As the oxidation reactions are exothermic, the refining process was carried out without any external heat input which eventually boosts up the refining process in 20 min. Once refining was completed, the slag and metal were tapped out separately and the liquid steel cast as ingots.

However, this process experience major limitations including removal of sulphur and phosphorous from hot metal, and both oxygen and nitrogen content is extremely high. In spite of the disadvantages, the acid Bessemer process was predominant up to around 1910. Later, the inventor S. D. Thomas enabled the phosphorus removal in 1879, and the process is coined as the Thomas process or *Basic Bessemer process*.

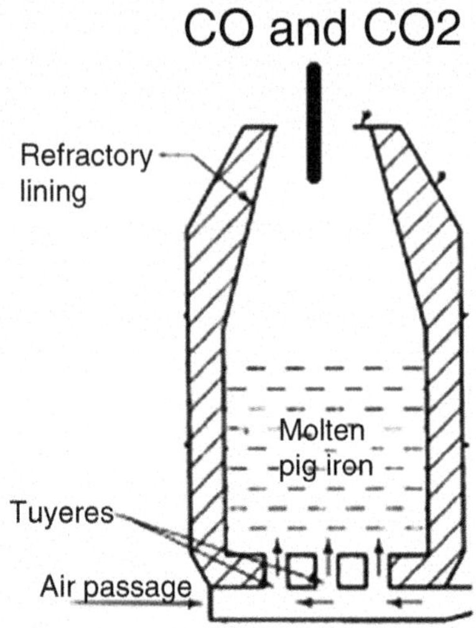

Fig. 2.17 Typical schematic representation of Bessemer convertor

The basic Bessemer process involved the addition of basic oxide CaO as flux during the blow to form a basic slag with high CaO content, which was capable to remove phosphorus as calcium phosphate. The vessel was lined with basic refractory to protect the slag with high basicity.

2.3.2 Open-Hearth Process

The open-hearth process of steel making began in 1868, and this process dominated the world steel production for more than 100 years. However, the process was very slow, and it required external heat input. Heat was provided in open-hearth process to the furnace by the combustion of gaseous fuel or gas. The air used for combustion was recycled by reheating through regenerators that helped to attend flame temperature more than 1600 °C. In this process, the charge was a mixture of scrap and molten hot metal. The scrap was initially heated near to its softening point and molten pig iron from the blast furnace was poured onto it.

The basic slag was prepared by adding lime with iron ore. The atmosphere of the furnace was maintained always oxidized by adding oxygen through lance. This process could make almost all grades of steel, but it is usually avoided because of very slow process kinetics as 6–8 h minimum is required for each heat or cycle. The incremental development in steel making took place with the introduction of Linz and Donawitz (LD) process in 1850 and today worldwide, maximum steel is produced by that protocol. The trend of world steel production from 1950 to 2015 is shown in the following Fig. 2.18.

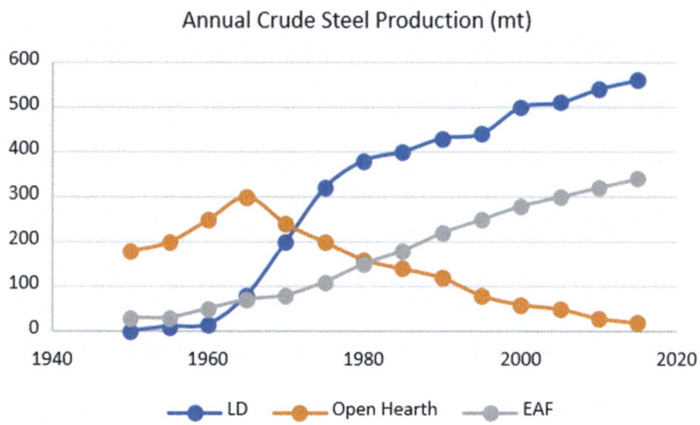

Fig. 2.18 Comparison of steel production through different routes

2.3.3 Primary Refining Process Through BOF

A modern steel-making process has three broad groups, primary steel making through basic oxygen furnace (BOF) followed by secondary steel-making process and electric arc furnace (EAF) route, and those group of processes together account for almost the total liquid steel production in the world. The LD convertor is a refined version of Bessemer convertor, where blowing of air was replaced by blowing oxygen. Around the year 1950, the Linde division of Union Carbide Corporation devised a more efficient air separation process, which made bulk oxygen available at affordable price. This made a remarkable change in bulk steel-making process by introducing modern oxygen steel making. The first commercial plant using the effective process of oxygen blowing begun at Linz and Donawitz in Austria in 1952–53, giving the process a popular name of LD steel making, which later became widely known as BOF (basic oxygen furnace) steel making. However, it had limitation of handling high-phosphorus hot metal, bath homogeneity in terms of temperature and composition. To get around this problem, at first pure oxygen bottom-blown steel-making process (using tuyeres) and later combined top- and bottom-blown converter processes had been developed.

The vessel (or reactor, or converter) comprises a steel shell with an internal lining of refractory bricks (magnesite or dolomite), supported by a stout steel ring equipped with trunnions, whose shaft is driven by a tilting system. The internal volume of the vessel is 7–12 times greater than that of the total volume of the steel to be treated, in order to confine the majority of metal projections entrained by the oxygen blast, together with swelling of the slag during periods of foaming.

This typical converter geometry shows the convertor mouth, oxygen lance, trunnion ring, trunnion, tilting mechanism, and tap hole (this is for steel different from BF tap hole), as shown in Fig. 2.19. Typical capacities are 200–350 tonnes of liquid steel, and the tap-to-tap cycle is about 30 min with a 15 min oxygen blowing

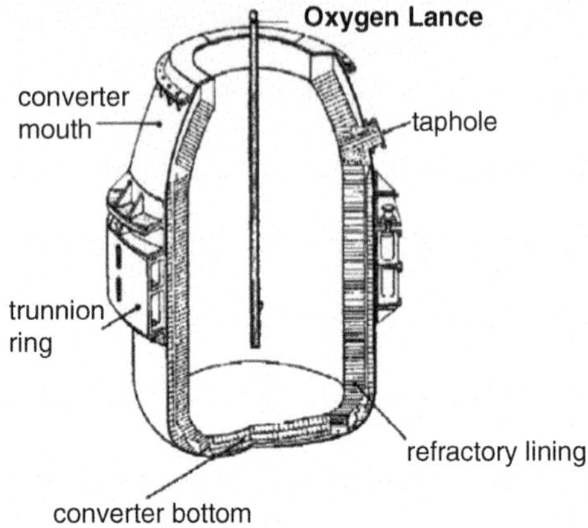

Fig. 2.19 Schematic diagram of LD convertor

period. Several classes of operational features including raw material feeding, their homogenization, maintaining equilibrium and stirring can facilitate to obtain the desired quality of the steel output. The brief has been given in order to understand how to control the input–output in converter.

2.3.3.1 Input–Output in LD Converter

Effective output of converter is a critical incidence to regulate the steel quality and productivity.

Thus, the following important features are essential to be encountered in the perspective of both input and output features of converter [7].

1. The prime input liquid hot metal (output of blast furnace) has to be desulphurized or dephosphorized before charging in converter.
2. Other iron-containing materials, especially scrap and ore, are also needed to add for the adjustment of the thermal balance and required steel temperature.
3. Lime (CaO) and dolomitic lime (CaO-MgO) in the size range of 20–40 mm lumps are usually charged to form the slag consisting of appropriate composition and viscosity.
4. In order to refine and reduce the carbon, pure oxygen is injected either through bottom tuyeres or through multi-hole lance.
5. During such process, there are three outputs: major liquid steel and other two valuable by-products, exhaust gas rich in CO (about 80–90%) and slag.
6. Gas is recovered through the closed-circuit gas recovery system, say combustion hood, and often used in the burners of reheating furnaces.
7. Slag is poured and collected followed by steel tapping.

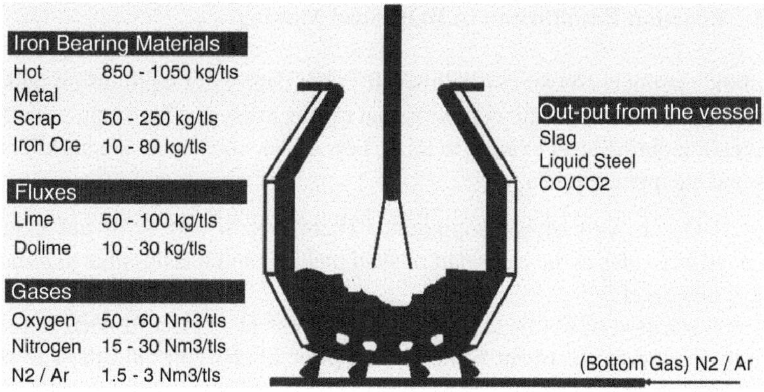

Fig. 2.20 Input and output in an LD converter

The brief of input–output parameters of LD converter is highlighted in Fig. 2.20.

2.3.3.2 Reactions in BOF

In the vessel-operating condition, sulphur removal is not feasible, hence sulphur to be reduced in hot metal prior to pouring the hot metal into the LD vessel. During refining by oxygen purging, rapid oxidation reactions of C, P, Si, Mn and Fe take place, and their oxides finally discard either in exhaust gas or slag without maintaining thermodynamic equilibrium. Equation (2.22–2.27) represent different stages of chemical reactions in LD vessel:

$$C + \frac{1}{2} O_2 = CO \tag{2.22}$$

$$2P + (5/2)O_2 = P_2O_5 \tag{2.23}$$

$$Si + O_2 = SiO_2 \tag{2.24}$$

$$Mn + \frac{1}{2}O_2 = MnO \tag{2.25}$$

$$Fe + \frac{1}{2}O_2 = FeO \tag{2.26}$$

$$2Fe + (3/2)O_2 = Fe_2O_3 \tag{2.27}$$

The produced CO is preferred to be oxidized into CO_2 above the melt (post-combustion), and evacuated through exhaust hood, and the ratio $CO_2/(CO + CO_2)$ is known as post-combustion ratio (PCR). The developed solid oxides combine with charged basic oxides (CaO, CaO-MgO) and form liquid slag that float on the molten metal bath surface.

2.3.3.3 Reaction Equilibrium in BOF Steel Making

Most of the chemical conversion follows different degrees of equilibrium under the same reaction temperature and condition that results in diverse by-products and level of purification during steel making in BOF. Few classic incidences are highlighted to understand such phenomena.

1. SiO_2, TiO_2 are very stable compound. Therefore, Si and Ti in hot metal are removed as oxides in the early part of steel making, and it is possible to remove to a very low level.
2. Desulphurization of hot metal occurs to some extent in blast furnace and subsequently during pretreatment of hot metal, prior to charging into the LD vessel. Sulphur can be transferred from liquid metal to slag, only under reducing conditions. Since, primary steel making is a highly oxidizing process, and the slags are highly oxidizing in nature, sulphur removal is very limited.
3. Removal of carbon, phosphorus and manganese along with oxidation of Fe to FeO takes place throughout the blowing and those impurities can be removed significantly.
4. Decarburization is the extensive and important reaction during oxygen steel making. Carbon is reduced through the reaction of [C] + [O] = CO (g). Removal of carbon depends on the dissolved oxygen in liquid metal and reaction with blown oxygen. Carbon is also be reduced by reacting with FeO and MnO in slag as follows:

$$(\text{FeO}) + [\text{C}] = [\text{Fe}] + \text{CO (g)} \tag{2.28}$$

$$(\text{MnO}) + [\text{C}] = [\text{Mn}] + \text{CO(g)} \tag{2.29}$$

   The maximum rate of decarburization is in the range 0.20–0.28 wt% of C per minute. About 3.5–4.5% carbon in hot metal is oxidized to CO and CO_2 during the oxygen blow, resulted in steel production with less than 0.2 wt% of carbon. Carbon is reduced in three distinct stages. In the first few minutes of the oxygen blow, the rate of carbon removal is very slow as nearly all the oxygen supplied are consumed by Si metal to oxidize into SiO_2.

   The second stage follows continuous and rapid oxidation of high carbon content in the metal, and eventually the rate of decarburization slows down as the carbon becomes less available to react with the available oxygen and the excess oxygen reacts with Fe to form FeO in slag during third stage of oxidation. The change in melt condition and the slag composition during oxygen blowing time is shown in Fig. 2.21. Below about 0.4% C, the rate of decarburization decreases with a decreasing carbon content because of high degree of oxygen consumption during oxidation of P, Mn, Fe as well as an increasing amount of oxygen dissolution in the steel bath.
5. Removal of phosphorus is a very complex process. Although the boiling point of P is very low, a considerable amount remain devolve in liquid iron because of its

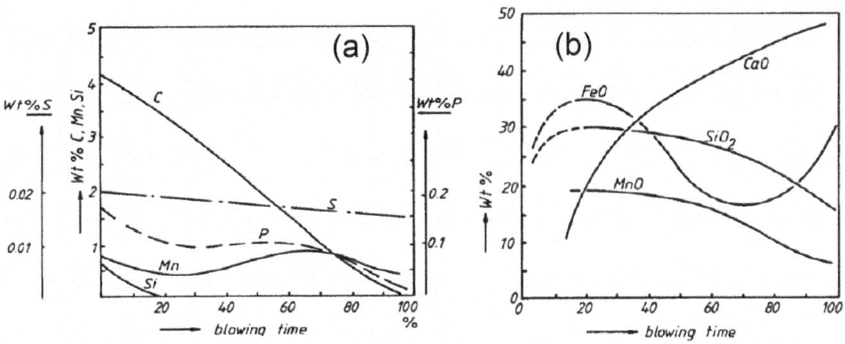

Fig. 2.21 Change in (**a**) composition change in melt condition, and (**b**) slag composition during blowing in LD convertor

strong interaction with iron. In Ellingham diagram, the iron and phosphorus lines are so closed to each other that the entire phosphorus in the burden is reduced with iron in blast furnace. This has made the oxidation of P very complex and challenging. However, in steel-making process, it is possible to reduce the activity of P_2O_5 by adding a very strong basic flux like lime (CaO). For effective removal of phosphorus, the steel making slags to maintain very high basicity.

2.3.3.4 Bottom-Stirring Practice in LD Converter

Chemical composition inhomogeneity and non-uniform temperature all along the vessel are common phenomena during top-blown oxygen purging, resulted in improper mixing. Thus, introduced bottom-stirring practices using inert gases like Ar and N_2 through permeable components or tuyeres to improve the resultant mixing efficiency. In a common industrial practice, N_2 gas is introduced in bottom-stirring practices using inert gases like Ar and N_2 through permeable components or tuyeres to improve the resultant mixing efficiency. In a common industrial practice, N_2 gas is purged in the first 60–80% of the blow, and argon gas is purged in rest of 40–20% of the blow.

The rapid evaluation of CO in the first stage of refining prevents nitrogen pickup in the steel. Some of the effect of bottom blowing and the resulted improved mixing include the following:

1. Decreased FeO content in the slag. Plant study showed that by bottom purging, FeO content in the slag is reduced by 5.0%. This results in better metallic yield and reduces corrosion and oxidation of refractory lining.
2. Reduced dissolved oxygen in the metal. This lowering of dissolved oxygen in steel, reduces consumption of aluminum metal in the ladle during deoxidation. Studies have shown aluminum savings of 0.15 kg/ton of crude steel, due to bottom blowing.

3. Higher Mn content in steel. Bottom steering reduces ferro-alloy addition, as Mn in steel increased by 0.3%.
4. Better phosphorus removal for the same amount of slag Fe.
5. Higher vessel life.

2.3.4 Secondary Refining Process

Primary steel making is aimed at rapid refining of hot metal (pig iron), produced in blast furnace. It is capable to refine in macro level to arrive at broad steel specification, but is not designed to meet the specifications of special grades of steels. In order to satisfy such a demand in quality of steel, the steel obtained from a primary steel-making unit needs to undergo further refining process, known as *Secondary steel making* [8]. The harmful impurities in steel includes sulphur, phosphorus, carbon, oxygen, nitrogen and hydrogen. The later three gaseous elements occupy interstitial sites in iron lattice and hence are known as *interstitials*. The effect of those impurities on product quality are loss of ductility, lower strength and poorer corrosion resistance. Oxygen and sulphur are also contributing in non-metallic inclusions in steel. The presence of those inclusions is harmful and to be maintained as low as possible. Carbon can also present as interstitial in iron lattice as well as in the form of cementite (F3C). Unlike other interstitials, some carbon is always required in steel, and hence carbon is specified in product specifications. Some special grades of steel, with stringent specification, are interstitial-free (IF) grade, very low-level carbon grade (Ultra-low carbon steel) and those grades are made in secondary refining processes. Few special grades of steel with their specification are shown in Table 2.4.

Thus, precise chemistry on deoxidation and decarburization of liquid steel is essential to encounter during secondary steel-making process. Solubility of oxygen in molten iron in equilibrium with FeO is very high, oxygen level of BOF steel typically ranges from 600 to 1000 ppm. On this contrary, the solubility of oxygen in solid steel is negligibly low. Therefore, during solidification of liquid steel, the excess oxygen is evolved out and causes defects by reacting with C, Mn, Si, etc., to form CO, MnO and SiO_2. Presence of CO creates blow holes and other oxides form inclusion during solidification. Therefore, level of dissolved oxygen in liquid steel to be reduced to an acceptable limit by addition of strong oxide formers, such as Al, Si, Ca and Mn metals. Those metals react with dissolved oxygen to form the oxides and transport to slag, as slag formers and the process is known as deoxidation or killing of steel. Hydrogen, nitrogen along with oxygen are also dissolved as atomic H, N, O in molten steel. Usually, the nitrogen and hydrogen both have also harmful influence to degrade the steel properties. Nitrogen is absorbed by molten steel during steel making from atmospheric air, and at the same time the hydrogen is picked up from the moisture in solid charges. Hydrides are thermodynamically unstable and hence entrapped hydrogen [H] is converted to H_2 gas and diffused to atmosphere forming pin holes in steel structure. Dissolved hydrogen results in loss of

Table 2.4 Different classes of steel and their maximum allowable impurities level and inclusion size

Steel product	Maximum allowed impurity fraction	Max. allowed inclusion size
IF steels	[C] $\leq$ 30 ppm, [N] $\leq$ 40 ppm, T. O. $\leq$ 40 ppm [C] $\leq$ 10 ppm, [N] $\leq$ 50 ppm	
Automotive and deep-drawing sheets	[C] $\leq$ 30 ppm, [N] $\leq$ 30 ppm	100 μm
Drawn and ironed cans	[C] $\leq$ 30 ppm, [N] $\leq$ 40 ppm, T. O. $\leq$ 20 ppm	20 μm
Alloy steel for pressure vessels	[P] $\leq$ 70 ppm	
Alloy steel bars	[H] $\leq$ 2 ppm, [N] $\leq$ 20 ppm, T. O. $\leq$ 10 ppm	
HIC resistant steel sour gas tubes	[P] $\leq$ 50 ppm, [S] $\leq$10 ppm	
Line pipes	[S] $\leq$ 30 ppm, [N] $\leq$ 50 ppm, T. O. $\leq$ 30 ppm	100 μm
Sheets for continuous annealing	[N] $\leq$ 20 ppm	
Plates for welding	[H] $\leq$ 1.5 ppm	
Bearings	T.O. $\leq$ 10 ppm	15 μm
Tire cord	[H] $\leq$ 2 ppm, [N] $\leq$ 40 ppm, T. O. $\leq$ 15 ppm	10 μm
Non-grain-orientated magnetic sheets	[N] $\leq$ 30 ppm	
Heavy plate steels	[H] $\leq$ 2 ppm, [N] = 30–40 ppm, T. O. $\leq$ 20 ppm	Single inclusion 13 μm; cluster 200 μm
Wires	[N] $\leq$ 60 ppm, T.0. $\leq$ 30 ppm	20 μm

steel ductility. Thus, degassing under vacuum is employed to remove low level of gaseous substances [9].

2.4 Type of Processes and Special Consideration

The history of secondary steel making went through three distinct stages of development. In the first stage, simple ladle metallurgy was developed in order to improve the following:

- Removal of inclusion by gentle bath stirring;
- Desulphurize steel by synthetic slag and injection metallurgy;
- Modify inclusion by primarily lime addition;
- Deoxidation control.

Normally the temperature drop of molten steel during tapping from primary steel-making vessels is around 20–50 °C, an additional temperature drop of around 40–60 °C occurs in the course of secondary steel making, and further temperature drop of around 10–20 °C is observed in the final stage of casting. Hence, it was most desirable to adjust temperature of liquid steel during secondary refining and in the next developmental stage. To solve this issue, ladle furnaces are developed to improve productivity and quality of steel by faster removal of impurities through high-temperature reaction kinetics. However, ladle furnace operation makes the refractory performance more challenging due to high-temperature corrosion and severe thermal shock. The main advantages of using ladle furnace are as follows:

- Reheating of steel and control of reaction temperature;
- Large alloying to produce high alloy steels;
- Halogenation of heat and composition by introducing bottom purging in ladles;
- Production of clean steel;
- Efficient deoxidation.

In the third stage of development, degassing under vacuum enabled to produce ultra-low carbon and very low interstitials steel, which has resulted in its wide range of applications. As a result, increasing proportion of steel is produced through vacuum treatment. Now-a-days, the secondary steel making is basically comprising of LF treatment and vacuum treatment. The main features of this process are as follows:

- More close to homogeneous chemistry.
- Ultra-low level of impurities;
- Faster production rates;
- Higher recovery of alloying elements;
- Deeper decarburization;
- Desulphurization for extra low-end sulphur;
- Very effective degassing and deoxidation;
- Micro-allowing;
- More flexibility to produce different stringent grades of steel;
- Improvement in cleanliness and flexible tundish metallurgy through continuous casting.

Reduction of Impurities in Steel

Level of gaseous impurity components is significantly reduced from 1960 to 2015, as shown in Fig. 2.22. The requirement of stringent properties, maintaining very low level of inclusions, proved that the secondary refining process is an essential part of modern steel making and it is universally accepted because of its excellent efficiency.

- Stirring treatment;

(continued)

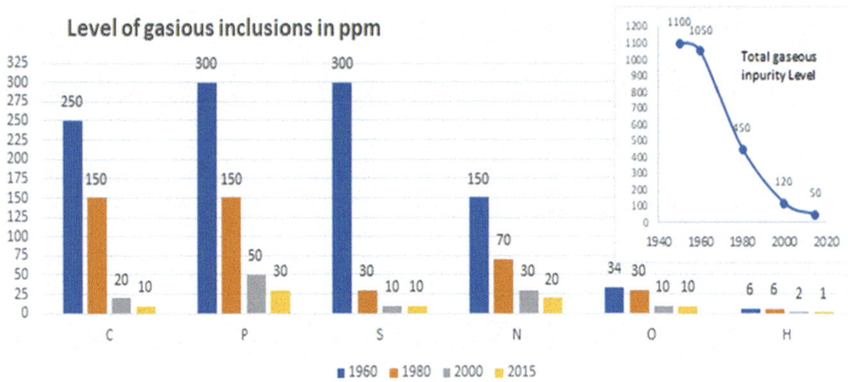

Fig. 2.22 Level of gaseous impurities from 1950 to 2015

- Synthetic slag refining with stirring;
- Vacuum treatment;
- Injection metallurgy;
- Closed casting by use of shrouds, mono-block-stoppers and sub-entry nozzles;
- Tundish metallurgy.

However, in most of the steel plants, the most well-known and highly efficient secondary refining processes adopted are as follows:

- Secondary refining in ladle furnaces (LF treatment);
- RH-degasser;
- RH-OB degasser;
- CAS-OB.

The essential operational activities of the secondary refining are as follows:

2.4.1 Ladle Furnaces

The steel refining operation is broadly considered as deoxidation, desulphurization and dephosphorization through controlled addition of alloying elements and simultaneous inclusion modification. Thus, different operational features eventually simulate the refractory selection for such vessel. A brief chemistry and relevant process protocol on steel making facilitate to develop the new generation of refractory.

2.4.1.1 Refining of Liquid Steel

Steel refining in ladle is initiated by a deoxidization process that expedite by Fe/Mn and/or Fe/Si, followed by the final-stage deoxidation by aluminium (Al). There are well-known three categories of steel deoxidation phenomena:

1. Steel deoxidized with ferromanganese to yield 100–200 ppm dissolved oxygen;
2. Semi-killed steels deoxidized with,
   (a) Si/Mn to yield 50–70 ppm dissolved oxygen.
   (b) Si/Mn/Al to yield 25–40 ppm dissolved oxygen.
   (c) Si/Mn/Ca to yield 15–20 ppm dissolved oxygen.
3. Killed steels deoxidized with Al to yield 2–4 ppm dissolved oxygen.

This protocol has several advantages including minimum N pickup, minimum P reversion from the carried-over furnace slag and minimum Al loss due to reaction with carried-over furnace slag. Despite this synergic effect, modern ladle metallurgy adapts to add synthetic slags to boost up the production of ultraclean steels in specific extra low sulphur content. Use of manganese and silicon together for deoxidation is found more effective than with either of those elements alone, as activities of the oxides is less than unity in the deoxidation process, $[Si] + 2(MnO) = 2[Mn] + (SiO_2)$. The symbols within the square brackets refer to species dissolved in the steel. Activities of MnO and SiO_2 are lowered further in the presence of the aluminosilicate phase. For example, dissolved oxygen content reduces to approximately 20 ppm by means of deoxidation with silicomanganese in the presence of aluminium. Aluminium is a very effective de-oxidizer since Al_2O_3 is more stable oxide than SiO_2. But their oxidized product remains solid even at steel-making temperature, and hence aluminium cannot be used alone rather used along with Mn and Si. The basic philosophy of phosphorous and sulphur removal has already been discussed in Sect. 2.3.3.3.

2.4.1.2 Steel Alloying

Ferro-alloys of different grades and carbon contents are used for alloying to obtain the necessary steel chemistry at best possible economy. Elements such as Al, Si and Mn are added primarily as deoxidizers. Elements such as Zr, B and Ti are added for deoxidation in making special grade steel. Cr, W, Mo, Ni, V and Nb are added to make high alloy steel, e.g. stainless steel. However, nickel and chromium do not act as a deoxidizer, they are added as an alloying addition in stainless steels. They can be added at any time during secondary refining process.

Several methods of alloy addition are practiced. Most common protocols include throwing of filled bags, adding with a shovel or via mechanized chutes, wire feeding, powder injection and bullet shooting. A special process for making alloy additions is the CAS process (Composition Adjustment by Sealed argon bubbling). In CAS process, a refractory lined snorkel is immersed in the steel bath and argon (Ar) gas is purged through a porous plug, fitted at the centre of the bottom of the ladle. Alloy additions are made onto the liquid steel within the area covered by the snorkel. Within the covered area into the snorkel, low oxygen pressure is maintained and thus alloy addition is done in the absence of oxygen, preventing oxidation of alloy elements. The process is known as CAS-OB, and it is adopted by many integrated steel plants. Another well-known technique of alloy addition is wire feeding by means of the cored wire, and the process is developed primarily for the addition of calcium to the steel. This process is also followed for adding elements that are less dense than steel or have a limited solubility, high vapour pressure and high affinity for oxygen. For example, ferroboron or tellurium additions can be made by wire feeding. Aluminium is also added by wire-feeding process. Advantages of Al wire additions comprise higher aluminium recovery, better control of aluminium content and improvement in cleanliness that is accompanied by the release of heat. The enthalpy released is sufficient to melt the compound, thus allowing rapid dissolution of the ferro-alloy into the liquid steel.

Stirring of steel bath is achieved by bubbling inert gas like nitrogen or argon in the steel bath held in steel ladle. The rising gas bubbles tends to lift up the non-metallic inclusions due to its surface tension effect. The lifted-up particles are supposed to be absorbed in slag which is present at the top of the molten steel bath. As the bubbles rises in the bath, their sizes increase due to reduction of ferro-static head and this helps to lift up the inclusion up to the slag layer. There is a limiting flow rate at which this would happen in optimum. Any higher rate of blowing reduces the inclusion removal by rigorous mixing of steel with slag at the slag–metal interface. The rising gas coming in contact with slag layer opens up the slag layer, often called "red-eye" formation (details in Sect. 2.4.3.1). Very wide red-eye formation reduces the efficiency of inclusion removal; however, very small eye formation is considered as an ideal situation for effective gas purging through bottom. Typically, argon gas is blown at 3.5–5.0 bar pressure, depending on bath size in tonnage and with a rate of 200–500 Ndm^3/min.

Gas is often purged through a porous refractory with continuous channels. The plug is made of magnesia, alumina, spinel containing alumina refractories. The several designs are available as shown in Fig. 2.23, conventional random type with porous refractory, directional type and combination of random and directional type. The pores need to be fine enough so that metal does not enter because of the surface tension effect. Similarly, they tend to get closed due to entrapped slag at the end of the cast. A good plug would readily give life of hundreds of heats. The efficiency of stirring is measured in terms of energy associated with quantity of bubble formation and their sizes.

The homogenization of bath temperature achieved by applying the buoyant energy of injected gas, which is calculated as follows:

$$E = 14.23 \left(\frac{VT}{M}\right) \log \left[\frac{1 + H}{1.48P}\right] \tag{2.30}$$

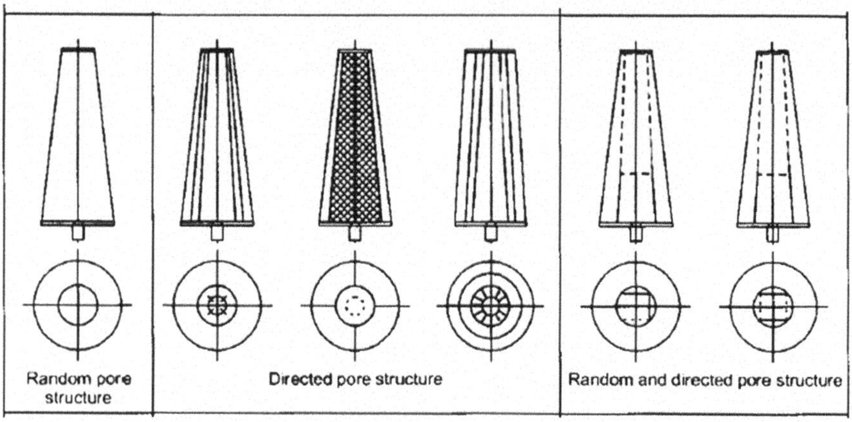

Fig. 2.23 Commercially available different designs of porous plugs

where, E = stirring energy, V = gas flow rate (Nm^3/min), T = bath temperature (K), M = bath weight (ton), H = depth of gas injection (m), P = gas pressure at bath surface. The mixing time, as defined as more than 95% hominization, is given as follows: Mixing time in second = $116 \times E^{-1/3} \times D^{5/3} \times H^{-1}$ (where D = ladle diameter in m).

2.4.1.3 Stirring Liquid Bath in Ladles

Stirring is an important criterion to make homogenize the liquid in content vessel, may be either making sugar solution in room temperature or steel in very high temperature. In obvious, a specific care has to be taken care during stirring by neutral gas like argon gas as the slag–metal interfacial area is affected by the degree of agitation in the bath commonly known as the stirring power. Stirring has now been employed universally, and the modern stirring processes now aim to fulfil the following criteria (Fig. 2.24):

- Homogenization of temperature and chemistry of the melt all through the refining process;
- Efficient effect of alloy addition in steel bath to adjust chemistry;
- Enable to produce clean steel.

2.4.1.4 Chilling Effect on Alloy Addition

One of the adverse effects of adding alloys and fluxes is the temperature decrement, as the kinetics of alloy dissolution is endothermic. The effect of various alloying additions, including coke, on the change in temperature of the steel for an average

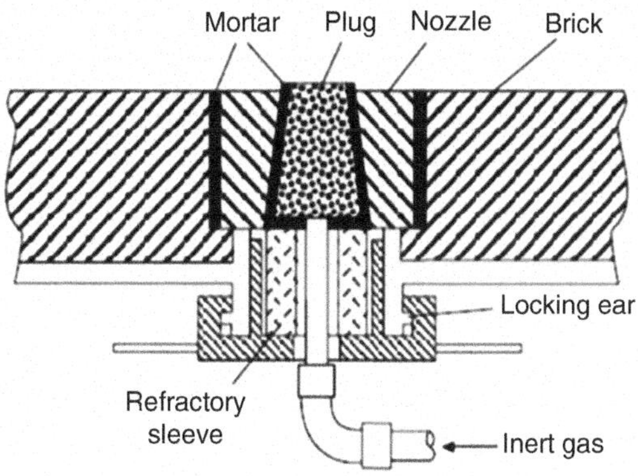

Fig. 2.24 Assembly of refractory porous plug at ladle bottom

Table 2.5 Effect of alloy addition in ladle temperature

Addition of 1.0% alloying element with 100% recovery	Change in steel temperature ΔT (°C)
Coke	−65
FeCr (50%), high °C	−41
FeCr (70%), low °C	−28
Fe.Mn, high °C	−30
Fe.Si (50%)	0
Fe.Si (75%)	14

bath temperature of 1650 °C is summarized in Table 2.5. It can be seen from the table that ferrosilicon is the only ferro-alloy that, upon addition, does not result in a decrease in steel bath temperature; in fact, the use of Fe-Si (75%) results in an increase in temperature. Dissolution of silicon into liquid iron is an exothermic process. Addition of Fe-Si (75%) usually expensive affairs than silicon source through Fe-Si (50%). However, the use of the former material is economically justified under certain conditions, particularly when relatively large quantities must be added and when the shop has no or limited reheating facilities such as ladle furnaces.

Addition of flux and slag conditioner decreases the temperature of the steel in the ladle. The effect of these additions on the change in steel temperature is determined from the heat capacity data as summarized in Table 2.6. In aluminium killed steel, the heat loss due to the flux addition is balanced by the heat generated during deoxidation reaction by aluminium.

2.4.1.5 Preheating of Steel Ladle

The steel ladle is lined with high alumina bricks (70–80% Al_2O_3) in metal zone and basic bricks, Magnesia-C or Magnesia chrome in slag zone. MgO-C bricks usually contains approximately 10% carbon and small amounts of metallic additions such as aluminium to minimize the oxidation of carbon. The thermal conductivity of lining

Table 2.6 Change in temperature due to flux addition

Flux addition 1.0 kg/ton of steel	Change in steel temperature ΔT (°C)
SiO_2	−2.5
CaO	−2.1
MgO	−2.7
CaO·MgO (dolomite)	−2.3
$CaO{\cdot}Al_2O_3$ (Ca-aluminate)	−2.4
CaF_2	−3.2

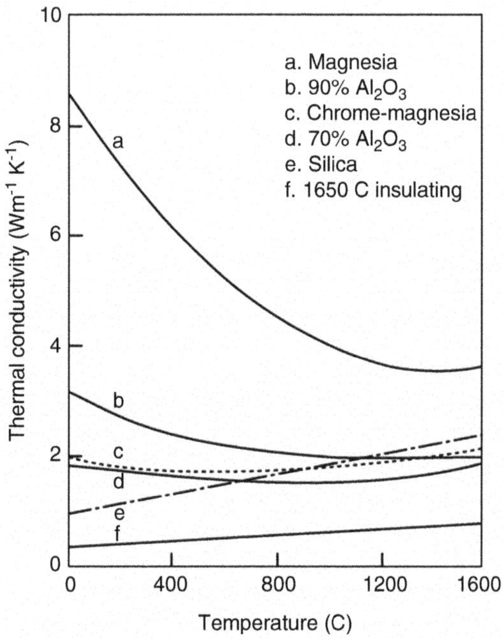

Fig. 2.25 Thermal conductivity of different refractory oxides

refractory is playing a very important role on the selection of ladle refractory as shown in Fig. 2.25.

It is seen that the conductivity of magnesia is significantly higher than other refractory materials. Magnesia also has a higher linear expansion, which made it poorer thermal shock resistance. Addition of carbon as graphite in magnesia-C bricks improves the thermal shock resistance, but it further increases thermal conductivity. Heat storage capacity of those bricks is also high. The high-temperature properties of dolomite are similar to those of magnesia.

Because of high thermal capacity and thermal conductivity, the temperature of MgO-C bricks in hot face increases rapidly compared to cold face lining as shown in Fig. 2.26. Hence, the rate of preheating is a critical operation in order to control and minimize the stress generated because of high thermal gradient between hot face and cold face.

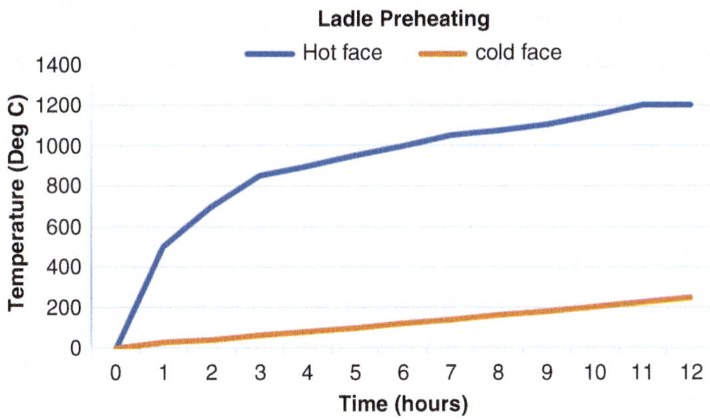

Fig. 2.26 Hot face and cold face temperature of ladle refractory

Faster heating rate causes cracking and oxidation of hot face bricks. At the end of preheating of ladle for 6–8 h, the cold face temperature would reach to 100–150 °C. A faster heating rate may cause the following problems:

- Generation of thermal stress within the bricks causes cracking.
- Cracking of steel shell due to hoop stress.
- Thermal shock resistance of high alumina brick in metal zone may not be high enough to withstand a rapid heating rate.

Usually the ladle is preheated up to 1400 °C and soaked in steady state for 6–8 h. After preheating is completed, the ladle is moved to the converter to receive primary steel for the refining as mentioned in early.

2.4.2 *RH-Degasser*

Vacuum degassing processes are traditionally classified into the following categories:

- Ladle degassing process (VD, VOD, VAD).
- Stream degassing process.
- Circulation degassing processes (DH and RH).

Currently, stream degassing is no longer in practice for most of the major steel plants, rather ladle degassing process is in forefront. Amongst the circulation during degassing processes, RH (Ruhrstahl Heraeus) degassing process and its variants is most popular. In order to avoid additional temperature – drop of around 20–50 °C during secondary steel making, RH is being employed for heating and temperature adjustment. Using both ladle degassing [Vacuum Arc Degassing (VAD) and Vacuum Oxygen De-carburization (VOD)] is a common practice for the refining of stainless steel through oxygen lancing under vacuum; however, in recent these are also adapted to produce ultra-low carbon steel. Similarly, RH-OB was developed with oxygen blowing in RH chamber to produce ultra-low carbon steel.

2.4.2.1 Basics of RH Process

The RH process was coined from the pioneer organization Ruhrstahl and Heraeus, where primary target was initiated to reduce the hydrogen in liquid steel. However, the process efficacy was very low due to low degree of vacuum in the vessel [10, 11]. However, in 1960, by using steam ejector vacuum pumps, sufficient low pressure was achieved to reduce the hydrogen content less than 1 ppm. In advance, the RH process was further used at the end of 1970s for the decarburization. It is an extremely successful protocol to minimize the final carbon content less than 20 ppm, important specification for automotive steel. Despite carbon minimization, alloy addition during degassing facilitates to achieve high yield for alloy and precious chemical composition of steel, as the process is carried out in the absence of air and very limited metal-slag reactions. Pickup of nitrogen in steel is thus avoided, preciously. A typical RH degasser system is shown in Fig. 2.27. It consists of upper vessel, lower vessel and two snorkels, one is called inlet snorkel, through which the steel is uplifted into the lower vessel, and the other is called outlet snorkel, and thus steel is circulated from ladle to RH vessel and flows back to ladle during degassing.

Usually, the snorkels are welded to the lower vessel bottom. There are facilities for introducing lift gas by means of 12 tuyeres at two levels into up-leg snorkel; this helps to lift the liquid steel and thermo-chemical halogenation. A three-stage steam

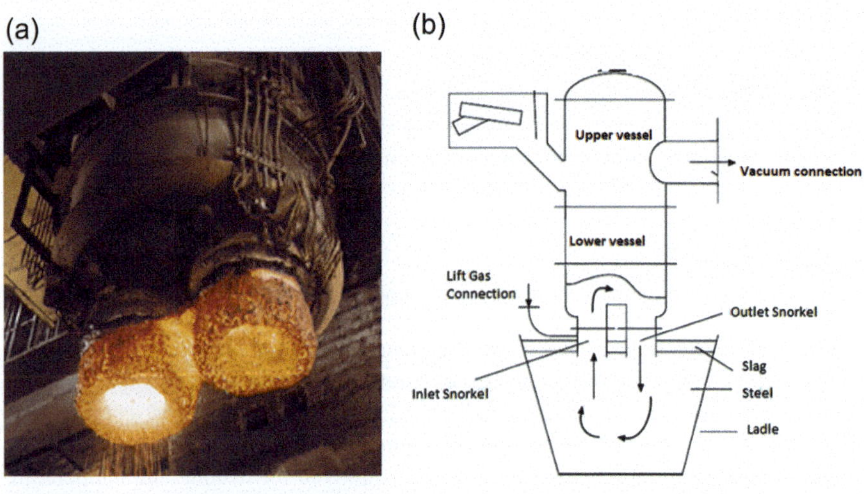

Fig. 2.27 (**a**) RH degasser, and (**b**) schematic representation of inner–outer snorkel, during operation

Table 2.7 Competitive degassing process and steel chemistry data for RF process and ladle refining process

Parameters	RH processes	Ladle refining			
	RH-OB, RH-MBF	RH	VOD	Tank	Ladle furnace
C content (ppm) in final product	<10	<20	<60	40–50	30–40
Rate of decarborization	Very high	High	High	20–30% less than RH	
Decarborization time in min.	<13	15	Depends on initial C	20–25	20
H content (ppm) in final products	<1	<1	Not as low as RH process		
Desulphurization	Powder injection by lance	No		Very efficient	
Chemical heating	Yes	No	Yes	No	No

ejector pump with triple ejectors connected in parallel at the third stage and three water ring pumps in parallel as the fourth operating stage in a barometric configuration generate the vacuum. The liquid steel is held in ladle. Typical parameters of a widely used RH degasser and the product chemistry are described as follows:

- inner snorkel diameter: 750 mm,
- deep vacuum: 1200 kg/h at 0.67 mbar,
- Suction capacity varies from 550 kg/h at 0.67 mbar,
- Lift-gas flow rate: 100–150 Nm3/H,
- TOP lance maximum oxygen blowing capacity: 3800 m^3/h (stp),
- Hydrogen content after degassing: max. 1.0 ppm,
- Nitrogen content after degassing: max. 50 ppm,
- Carbon content after decarburization: max. 20 ppm.

RH process has been further redefined as RH-OB and RH-MFB, to use oxygen blowing and temperature increase by multifunction burner. This upgradation has been done in consideration of four major areas:

- Introduction of multifunction burner (MFB),
- Enhancement of steel recirculation rate,
- Increase in the vacuum pumping capacity,
- Introduction of the micro alloy addition system.

The comparison of different degassing processes is shown in Table 2.7.

2.4.2.2 Multifunctional Burner (MFB) in RH-OB

The Multiple Functional Burner is introduced in RH-OB during 1997, where vessel temperature enhancement is carried out by heating with the burner and oxygen blowing by top lance [12]. The advantages of the modified process are;

1. Oxygen blowing under vacuum enables the RH-OB unit to produce low carbon steel from higher initial carbon levels and to decarburize faster than the existing rate at a normal start carbon level.
2. Oxygen gas burner, which helps in maintaining the hot face temperature of the vessel between two treatments and heating the vessel refractory lining at significantly higher rate than the present heating rod system. Burner heating is now being applied to all types of steel during and between treatments to maintain the refractory temperature around 1400 °C.
3. Oxygen blowing for chemical heating to reheat the steel by exothermic reaction between oxygen and aluminum.
4. Oxygen blowing for de-sculling of the vessel.

The introduction of MFB has improved the performance of lining refractory in lower vessel. It is enabled to maintain the brick hot face temperature of more than 1400 °C, reduce the thermal shock in lining bricks. Figure 2.28 shows the temperature drop experienced by lining bricks, between with and without MFB, during stand-by position within 10 min of the RH vessel. The difference is expected to increase with time, and hence it has a direct effect on refractory performances and thus the life of lower vessel was increased by 1.2–1.5 times in RH-MFB.

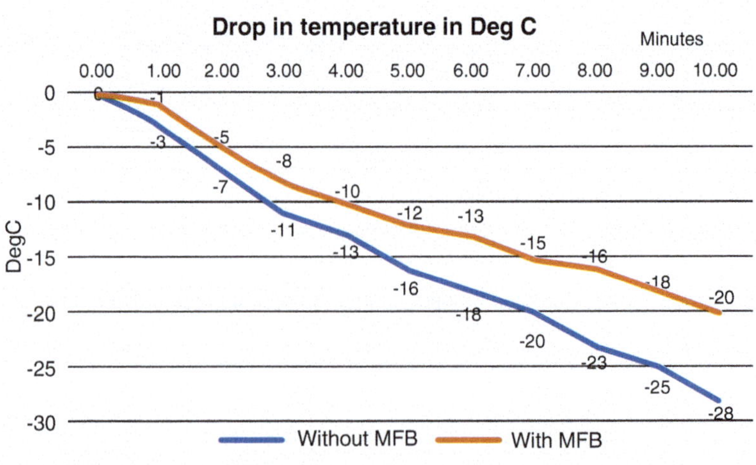

Fig. 2.28 Drop in temperature during processing in RH and RH-MFB degasser

2.4.2.3 Ultra-Low Carbon Steel in RH Vessel

The foremost requirement of RH process is the production of steel grades with very low hydrogen and carbon-level steel. The process involves vacuum decarburization, degassing of other gases including hydrogen and subsequent alloying. The heats for ULC grades are tapped at LD converter with dissolved oxygen level of 500–700 ppm. Forced decarburization is done by oxygen blowing in case of low input oxygen dissolved in steel. De-carburization reaction at RH is controlled by means of computerized pump-down program to avoid severe splashing inside the RH vessel. Indication of end of decarburization is available from waste gas analysis.

The snorkels are dipped into the liquid steel to start the process of degassing. Injection of inert gas started, commonly argon is used, to lift the liquid steel through the inlet snorkel. After attaining the optimum depth of immersion of both snorkels into steel in ladle, the vessel is evacuated by starting the vacuum pump and decarburization starts. The process of decarburization is very complex, and the efficiency depends on degree of vacuum created and circulation rate. After decarburization is completed, the balance oxygen is deoxidized either by aluminium or by silicon as per the requirement. Other ferro-alloys are being added after deoxidation. Since the degassed steels have lower dissolved oxygen, sufficient time is allowed after aluminium addition to promote inclusion floatation. The resultant products have hydrogen of <1.0 ppm and carbon of <10 ppm. Sampling and chemical analysing techniques become particularly important for ultra-low carbon steels.

2.4.3 CAS-OB

The composition adjustment by sealed argon bubbling-oxygen blowing (CAS-OB) facility is a part of secondary steel-making route aimed to have a good chemical temperature control of the liquid steel prior to the continuous casting at low capital cost [13]. The CAS-OB process was developed in the 1980s by Nippon Steel Corporation (NSC), Japan, where reheating of liquid steel can be done by chemical reaction heating.

2.4.3.1 Basics of CAS-OB

The important features of the CAS-OB process are shown in Fig. 2.29. Herein, the losses of deoxidizers and alloys through oxidation are low, and the dissolution become faster because of high turbulence and stirring at the plume eye. With additional facility of oxygen blowing, temperature can be raised because of exothermic reaction of aluminium oxidation.

As a result, arc heating is not required, and thus the process has an advantage of low capital cost and hence several steel plants around the world have adapted this

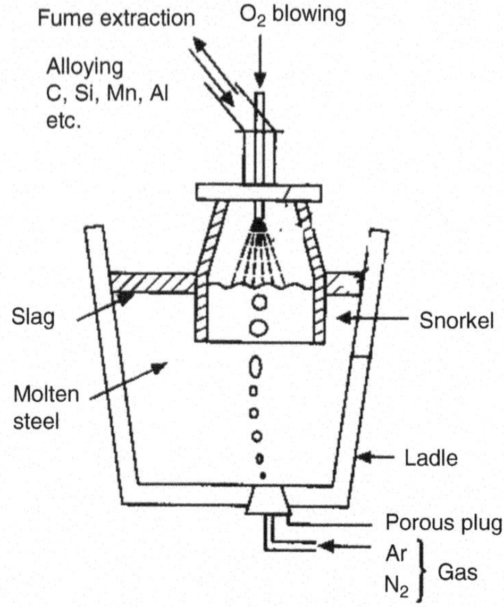

Fig. 2.29 A typical schematic representation of CAS-OB system

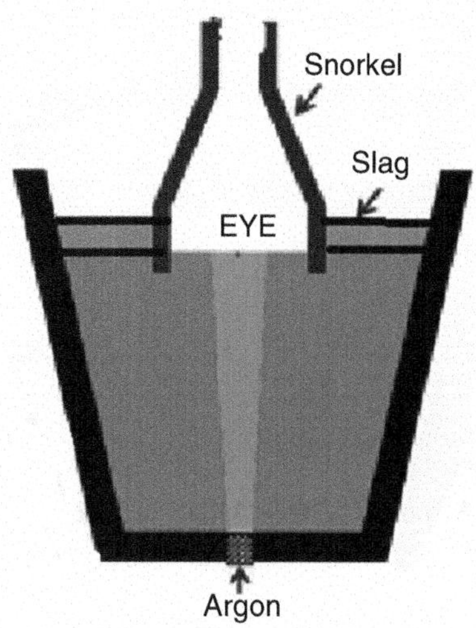

Fig. 2.30 Eye formation in CAS-OB

process and installed it. The secondary refining process is carried out in a steel ladle, and the top of the ladle is partially covered by a snorkel. The molten steel is stirred by bottom gas purging through refractory purging element and thus homogenous temperature and chemistry are achieved. In the same time, the dissolved impurities and gases in liquid steel float upward and are removed. After the slag over the liquid steel blown onto the skirt of the ladle top, the snorkel is placed over the "eye" opening onto the liquid steel, as shown in the Fig. 2.30. Then alloy and deoxidizers are added within the snorkel under an inert and slag-free region in order to adjust the chemical compositions.

If the melt needs to raise the temperature, aluminium is being added and oxygen gas is blown simultaneously through the top oxygen lance. The average refining cycle is around 35 min and refining temperature varies in the range of 1580–1650 °C.

2.4.3.2 CAS-OB Reheating

An increase in the steel temperature under treatment is achieved by utilizing either of the CAS-OB Al reheating or CAS-OB Si reheating processes. For CAS-OB Al reheating, the heating of the molten steel is achieved by utilizing the exothermic reaction between aluminium and oxygen. To reheat, aluminium (from the CAS-OB Al holding hopper) is delivered at a controlled rate into the steel via the immersed CAS snorkel. For the CAS-OB process, the snorkel immersion depth is increased compared to the basic CAS operation. At the same time, oxygen is blown onto the surface of the steel inside the bell through an oxygen lance. The oxygen blow rate is controlled and matched with the rate of aluminium addition. Reheat rates up to 10 °C/min are achievable by this method. The amount of reheating is restricted to between 5 and 50 °C on each reheat sequence. The volume of O_2 delivered and the weight of Al equate to the temperature increment. The CAS-OB reheat process creates alumina, which floats to the surface aided by the argon stirring, and is absorbed by the slag. Once the oxygen blowing and aluminium feed are completed, the O_2 lance is raised back to its top position, and the snorkel is raised back to the CAS position. Occasionally the steel under treatment can contain an excessive amount of aluminium. The CAS units can be used to remove aluminium by blowing oxygen alone. The oxygen reacts with the aluminium in the steel creating alumina as per the reheating operation. The reheat is achieved through the addition of O_2 and Al to the molten steel. The following exothermic reaction takes place:

$$2\mathrm{Al} + 3/2\mathrm{O}_2 = \mathrm{Al}_2\mathrm{O}_3 + 31.01\ \mathrm{MJ/kg\ of\ Al} \tag{2.31}$$

In this process, the consumption of Al and O_2 is controlled through following parameters:

- Oxygen flow rate;
- Required volume of oxygen (m^3);
- Aluminium feed rate (worthy to remember that this will be fixed during commissioning to achieve heating rate of 10 °C/min);
- Required amount of aluminium (kg).

Industrial practice follows the fixed reheating rate at 10 °C/min that eventually controlled by the consistent oxygen flow rate of 50m^3/min (3000m^3/h). Similarly, aluminium delivery rate is approximately fixed at 50 kg/min during commissioning and adjustment of CAS-OB set-up. During such operation, an optimum O_2 delivery time is essential to maintain, for example, O_2 delivery is essential to complete in 20 s starting from valve opening, otherwise the reheating sequence may be aborted. To achieve the desired temperature increment, the correct volume of oxygen and the equivalent weight of aluminium must be delivered in a calculative manner.

However, that to prevent the oxygen reacting with other alloys in the steel it is normal to begin the aluminium in advance of the O_2 delivery. If the Al delivery rate deviates from the set point by greater than 25%, the delivery will be aborted. At the start of the CAS operation, the operator must position the CAS-OB snorkel such that the bottom edge (which is refractory covered) is immersed approximately 200 mm (when reheating the immersion depth is adjusted to a different value usually around 400 mm) into the steel in the ladle. The steel level can vary significantly in the ladle; due to variations in metal weight, scull build, etc. The distance from the ladle's top edge to the metal level in the ladle is known as the freeboard, which is used to establish the treatment position for the CAS arm and thus the snorkel. The freeboard is measured by the operator lowering the arm, providing the relevant ladle car is stationary in the treatment position, until the bottom lip of the snorkel is at the top of the liquid steel surface. This needs to be seen by the operator via a CCTV camera positioned such that the snorkel position can be observed and monitored.

2.4.3.3 Refractory Erosion and Snorkel Life

The refractory performance of CAS-OB unit depends on the life of snorkel, which is lined with monolithic castable refractory and their erosion rate is very high. During operation, slag viscosity is increasing due to formation of excess Al_2O_3 in slag and causes jamming of snorkel, as shown in Fig. 2.31.

Thus, erosion rate of snorkel refractory (containing oxy-carbide alumina refractory) during steel treatment is severe due to interaction of CAS-OB slag [14]. The increased immersion depth with reduction in argon flow maintained to reduce the steel circulation rate inside the bell. As the aluminium is added and oxygen is blown inside the bell, the reduced circulation rate of steel helps to achieve high aluminium

Fig. 2.31 Snorkel erosion and jam formation

content in steel inside the bell for the reaction and thereby reduces the oxidation of other elements such as Si, Mn, C and Fe.

References

1. Principle of Blast Furnace Iron making – A K Biswas, SBA Publication, 1999
2. Design of 5500m3 Blast Furnace at Shougang Jingtang – Zang Fuming, Qian Shichong, Zhang Jian, Journal of Iron & Steel Research International, 16, 2, 2009,1029–1033
3. Current Blast Furnace Design Philosophies – Derek Macauley, Steel times International, Vol 19, No 2. February 1995, pp 20–26.
4. Chemical reactions in the Blast Furnace as it happens - W-K. Lu, Department of Materials Science and Engineering , McMaster University, AISTech 2004 Proceedings – Vol 1(395)
5. Application of refractories in modern Blast Furnaces – Efficient combination with cooling element, Nicolas Mousel, Paul Worth S.A Luxembourg – A report.
6. Richard J. Fruehan : The Making Shaping and Treating of Steel(Steel Making volume), The AISE Steel Foundation, 2004
7. Theoretical and Practical Aspects of Lime dissolution in laboratory experiments and in BOF , B. Deo, P K Gupta, M.Malathi, et .el, 5th European Oxygen steel making congress 26–28 June 2006, Steel Institute VDEH, pp. 202–209
8. A. Ghosh , Secondary Steel Making – Principle and application, CRC Press, 2001
9. Vacuum Degassing – RJ Fruehan, Iron & Steel Society, 1990.
10. RH metallurgy, Dieter Tembergen, Rainer Teworte, Robert Robey, Millennium Steel - 2008, 104–108.
11. Mathematical Modeling for RH Refining process of Molten Steel, Ji – WEI and Neng – Wen YU, Steel Making Conference Proceedings, 2002, pp. 687–704.
12. High Speed Decarburisation by modernized RH-OB And New decarburization Technology Under Reduced Pressure By CAS-OB, – Okitomo Kunitake et.el, Steel Making conference Proceedings – 2001, pp 625–633.
13. Physical Modelling and industrial Trials of CAS-OB refining process for low carbon aluminum killed steels, W J Ma, Y -P Bao, L -H Zhao, M. Wang, Iron Making and Steel Making, 2014, 41, 8, 607–610.
14. CAS – OB refining: slag modification with B_2O_3 – CaO and CaF – CaO, H – M Wang, G -R Li, et.el. – Iron Making abd Steel Making, 2007, Vol 34, No 4,pp 350–353.

Chapter 3
Blast Furnace Refractory

3.1 Introduction

The total campaign of a blast furnace depends on the performance of lining refractory. The large size blast furnaces require consistent and prolong campaigns with high utilization factors in order to produce economical pig iron. Most of the blast furnaces have specific output of 3000–6000 t/m^3 of inner volume and produce about 15×10^6 tons of hot metal in a total campaign. The durability of refractory lining depends on the stresses experienced by the refractory lining, which are of two types: physical stresses and thermal stresses. The main sources of physical stresses are gas flow through the stack, erosion by descending burden and flow of liquid iron in the hearth, which have a direct impact on refractory erosion. The gas flow through the stack is determined by the geometry of different zones of the furnaces and by the burden distribution. The stack walls above the granular zone, called cohesive zone, are exposed to gas flow at elevated temperature. Herein, the thermal stresses involve the heat load inside the furnace which is the cause for high-temperature reactions of refractories with gases in stack and belly area, corrosion reaction with slag at below cohesive zone at belly and bosh, and penetration of hot liquid metal in hearth. Apart from the thermal stresses, the alkali infiltration by the refractories and the possible reactions with their bonding phases, especially in the bosh and lower stack, have the greatest effect on the lining as the alkalis are predominately introduced mainly with the coke.

The use of dense refractories with higher thermal conductivity, thermal shock resistance and corrosion resistance counteracts these stresses. Introducing an effective cooling system, the chemical reactions within refractories/slags and refractories/gases can be reduced. Improvements in gas diffusion and liquid penetrations into refractories are also brought about by reduction of porosity and pore size. These requirements are also satisfied by using carbon refractory and SiC containing refractories in different zones of blast furnaces.

S. Biswas, D. Sarkar, *Introduction to Refractories for Iron- and Steelmaking*,
https://doi.org/10.1007/978-3-030-43807-4_3

3.2 Demand on Refractory Lining

Increase in furnace size and productivity has led to an increase in stresses exerted on the refractory lining. The Fig. 3.1 summarizes the essential demands made on the quality of refractories based on the stresses and wear processes indicated in the different zones of the furnaces, where different influences prevailing the different regions of blast furnace [1].

Wear Mechanism of Blast Furnace Refractories

1. Infiltration of gases in stack and bosh, leading to spalling.
2. Dissolution of refractory in hot metal and slag at operating temperature.
3. Thermal shock.
4. Mechanical abrasion.

However, there are two ways or approaches for preventing the chemical wear. These are referred to as

- Thermal approach.
- Refractory approach.

Both the abovementioned procedures to prolong the blast furnace campaign life are explained in Sects. 3.4.4 and 3.4.5. The critical properties to support the material as well as thermal approach of refractories are shown in Table 3.1.

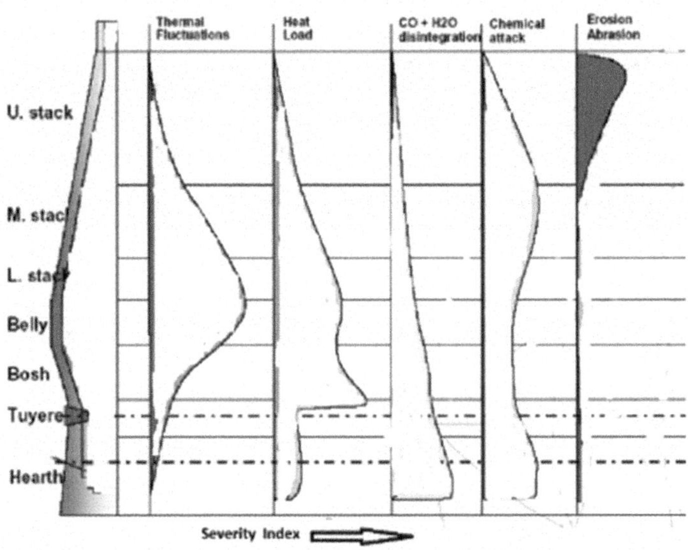

Fig. 3.1 Different stress and wear fascinated zones of blast furnace

Table 3.1 Critical properties of different refractories used in blast furnace [2].

Types of refractory	Alkali attack	Temperature of chemical reaction T°C	Thermal conductivity (?) in W/mK	Thermal shock resistance (°C/min)
High alumina (62% Al_2O_3)	Severe	550	1.5–2.2	5
Mullite	Medium	680	2.15–3.2	8
Oxide bonded SiC	Low	860	6–8	50
Si_3N_4 bonded SiC	Low	890	12–14	120
SAILON bonded SiC	Very low	980	16–18	80
Carbon refractory	Very low	800	18–22	400
Graphite refractory	Low	850	120–160	550

Till 60–70s, the lining refractory materials used in blast furnace construction including hearth were made of different qualities of high alumina bricks. All the zones including hearth were lined with those bricks. The main disadvantages to achieve higher campaign are due to poor alkali resistance, poor thermal shock resistance, and very low thermal conductivity, which affects the cooling of the lining as indicated in Table 3.1. Effect of slag attack and the alkalis on high alumina refractory causes the formation of low melting compounds, which melts below 1400 °C, resulted in severe corrosion of high alumina refractory during furnace operation. Primary effect of slag and alkali attack predominate to deteriorate the high alumina refractory, when major phases are encountered as mullite ($3Al_2O_3$, $2SiO_2$) and silica (SiO_2), and relevant reactions related to such phenomena are summarized below [3]:

Effect of slag attack:

- CaO (slag) + brick = Anezthite ($CaO{\cdot}Al_2O_3{\cdot}2SiO_2$).
- more CaO + mullite = Glelanite ($2CaO{\cdot}Al_2O_3{\cdot}SiO_2$),
- MgO (in slag) + brick = Cordierite ($2MgO{\cdot}2Al_2O_3{\cdot}5SiO_2$).

In this circumstance, liquidus phase appears at 1100 °C and facilitate 60% solubility of alumina refractory composition in contact with liquid blast furnace slag, resulted in severe refractory corrosion.

Effect of alkali attack:

- 2 K (g) + CO (g) = K_2O(g) + C (s)
- 2 K(g) + CO (g) = K_2CO_3(g) + 2C (s)
- K_2O + brick = Leucite ($K_2O{\cdot}Al_2O_3{\cdot}4SiO_2$)
- more K_2O + mullite = Kaliophilite ($K_2O{\cdot}Al_2O_3{\cdot}2SiO_2$)

Herein, the liquid appears at 980 °C. The campaign of blast furnace depends on the performance of hearth refractory, mid-term repair of hearth is not feasible and hence any deterioration of hearth refractory causes the end of campaign. So, complete shutdown of the furnace is mandatory till rebuilding of complete refractory lining. However, in other area, mid-term repair is common practice to prolong the campaign by adopting the following methods:

1. Robotic stack gunning.
2. Grouting in stack, belly and bosh area.
3. Mid-term repair by complete change of refractory from tuyere level to upper stack.

Till 50s and 60s, most of the small- and medium-size blast furnaces hearth was lined with high alumina refractory, and due to limitation on high corrosion and low thermal conductivity, the campaign life was 8–12 years. Introduction of carbon hearth in 1970 and later the use of SiC refractory in bosh, belly and lower stack area were the breakthrough for modern large blast furnace lining [4]. The advantages of using carbon in hearth are as follows:

- Resistance to alkali attack and resistance to metal penetration.
- Resistance to hot metal infiltration.
- High thermal conductivity and high erosion resistance.
- High volume stability and excellent thermal shock resistance under reducing atmosphere.
- Very high temperature of vapourization.
- High hot crushing strength.
- Non-wetting towards molten iron and slag.

The upgradation of refractory quality in blast furnace lining is shown in Fig. 3.2. Since entering the twenty-first century, there is no brickwork for new generation shaft and belly of blast furnace. The bricks used for embedded cooling staves were microporous corundum and silicon carbide brick. For hearth and bottom of blast furnace, semi-graphitized carbon blocks and microporous carbon blocks are adopted.

All of the carbon blocks used possess high heat conductivity, usually 20 w/m k. With a length of carbon block for hearth wall, the cooling staves are capable of withstanding a heat flux of 18,000 k cal/m^2·h (or 21,000 w/m^2·h) and maintaining the temperature of hot surface less than 1000 °C [5]. Apart from refractory quality upgradation, the quality of construction, new design of bricks and shapes for optimum stress release during operation are decisive factors to prolong the campaign. The most important factor for maintaining an efficient performance of blast furnace is to control the distribution of gas flow in blast furnace. The central gas flow should be kept stable, and the peripheral gas flow should be kept even and moderate in order to control the starting of cohesive zone within the region of lower shaft. In the early stage of carbon hearth lining, the carbon blocks were made of anthracite with high porosity, it was further upgraded to superior quality carbon blocks in micropore and super micropore quality with very low pore size and very high conductivity. In consideration of different upgradation, the present campaign life of large size blast furnace is more than 25 years.

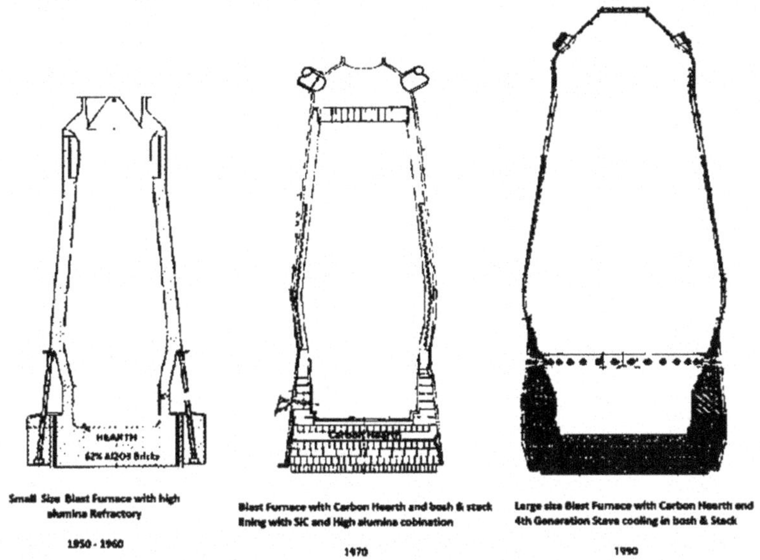

Fig. 3.2 Upgradation of blast furnace design and refractory

The design of the blast furnace profile has a direct impact, and thus the critical dimension of blast furnace is an essential consideration for higher campaign life and improved productivity. It also influences the performances of lining refractory. Determination of refractory thickness at different zones depends on furnace size and profile [1]. The blast furnace profile has evolved through the experience and therefore the calculations of dimensions are empirical in nature, depending on changing nature of the charge materials. The critical dimensions of a furnace profile are given as follows:

1. Useful furnace height and volume.
2. Diameter of hearth, belly and throat.
3. Hot heel height (height of tap hole from hearth bottom).
4. Shaft angle and bosh angle.
5. Height of bosh, belly and stack.

In a blast furnace, the total reduction of iron oxide occurs in the lower stack and below. The profile must be such that the stock remains the longest possible time in that zone, where the gas velocity should also be relatively sluggish. This can be achieved by widening of this zone by introduction of larger diameter and height of the belly and a shorter bosh. The modern large size blast furnaces are expanding horizontally rather than vertically to provide larger volume for complete processing of the charge material. Hearth diameter and height are mandate to regulate for a predetermined burden distribution and coke rate, the coke throughput determines the iron production rate. An approximate relationship between coke burning rate and hearth diameter is derived,

$$D = C \cdot Q^{1/2} \tag{3.1}$$

$$Q = K \cdot P \tag{3.2}$$

where $D =$ hearth diameter (m), $Q =$ coke throughput (tonnes/24 h), $C =$ throughput coefficient, which varies between 0.2 and 0.3, for large size blast furnace, $C = 0.2$; $P =$ hot metal production rate (T/day), $K =$ coke consumed (tonnes/THM). THM—ton of hot metal.

Example 1 A blast furnace plans to produce 10,000 tons iron per day, with an optimum coke rate of 500 kg/THM, the hearth diameter would be 14.2 m (reduction of coke rate increases the productivity). In order to estimate the hearth diameter, the hearth height is being measured from the tap hole to the tuyere axis and volume of that area is referred as hearth volume (V_H). The height between the hearth bottom to the tap-hole axis is known as hot heel and the volume of that area is called sump volume (V_S). Usually, the ratio of V_H:V_s = 60:40. V_s can be calculated as follows:

$$V_s = \pi \cdot R^2 \cdot h_s$$

$$V_H = 1.5 \times (\pi \cdot R^2 \cdot h_s)$$

$$h_s = 20 - 30\% \text{of } D \text{ (for large size furnace it is 20\%as minimum)},$$

$D =$ diameter of hearth, $R =$ radius of hearth, $h =$ hot heel height.

In consideration of the above equation, the hearth volume of a blast furnace of hearth diameter 14.2 m is 647 m^3.

However, useful volume comprises of the volume of the furnace which is full of the charged material and smelting products. It is the volume measured from tap hole to the stock level. The useful volume can be approximately calculated based on total hearth area. The mean value is as follows:

$$V_u\ (\text{m}^3) = K \cdot \text{hearth area in m}^2\ (K \text{ varies from 24 to 26}).$$

Thus, a blast furnace of 10 m hearth diameter would have working volume (V_u) of about 2000 m^3, 12 m hearth diameter furnace, V_u of 2800 m^3 and 14.2 m hearth V_u of 4000 m^3.

3.2.1 *Refractory Practice in Stack*

The stack of the blast furnace is subdivided into three zones, upper stack, middle stack and lower stack. In the upper shaft, abrasion and impact shock by charge burden have the direct impact on refractory performances. Further, the refractory is also attacked by CO gas, which is generated for indirect reduction. The refractory is gradually damaged due to CO-disintegration. Indirect reduction by H_2, forming

Table 3.2 Causes of refractory deterioration in BF stack

Area	Attack phenomenon	Effect on refractory life
Upper stack	Burden movement, temperature fluctuation, impact, CO-attack	Hot abrasion, thermal shock, CO-disintegration
Middle stack	Burden movement, severe temperature fluctuation, attack of gas, alkali-attack	Hot abrasion, thermal shock, deterioration for alkali attack
Lower stack	Burden movement, severe temperature fluctuation, attack of gas, alkali-attack, thermal fatigue	Severe spalling, abrasive wear, alkali attack, zinc vapor attack, cracking due to stress

H_2O, which accelerates deterioration of refractory by the water gas formation. The refractory in the stack is severely damaged by all-coke operation and introduction of PCI (pulverized cole injection).

Refractory Deterioration in BF Stack [6]
In middle and lower stack, severe temperature fluctuations frequently occur in the vicinity of inner wall, which generates cracks in refractory and resulted in falling of bricks. The wear mechanism of stack refractory is summarised in Table 3.2.

Optimum cooling is required to maintain the temperature of refractory within 450–700 °C to minimise the effect of CO attack in upper stack and alkali attack as well as the attack of zinc.

In 1960–70, in cast iron stave design, no cooling was provided in upper stack, caused severe erosion. Frequent refractory gunning maintenance was required to extend the campaign. In 1980, the stave cooling was introduced in stack and cooling was extended to upper stack to prolong the life, as shown in Fig. 3.3. With the upgraded design of cooling, the wall thickness was significantly reduced from average 500–600 mm to 150–200 mm, which helps for uniform burden distribution and to improve productivity.

CO-disintegration is one of the very important cause for failure of blast furnace stack refractory. The fireclay refractory is disintegrated into pieces in the presence of CO gas, which is available into blast furnaces. Gas containing CO releases carbon at 400–800 °C in the presence of metallic iron or iron oxides in refractory, as per Boudouard equation ($2CO = CO_2 + C$). Any form of iron present in the refractory acts as a nucleation site for carbon deposition. In blast furnace lining disintegration is caused by deposition of reactive carbon. The crystallization pressure of carbon, based on precipitation of carbon and volume expansion, causes complete disintegration of refractory bricks. The reaction of CO with Fe and iron oxides present in refractory bricks and precipitation of carbon are as follows:

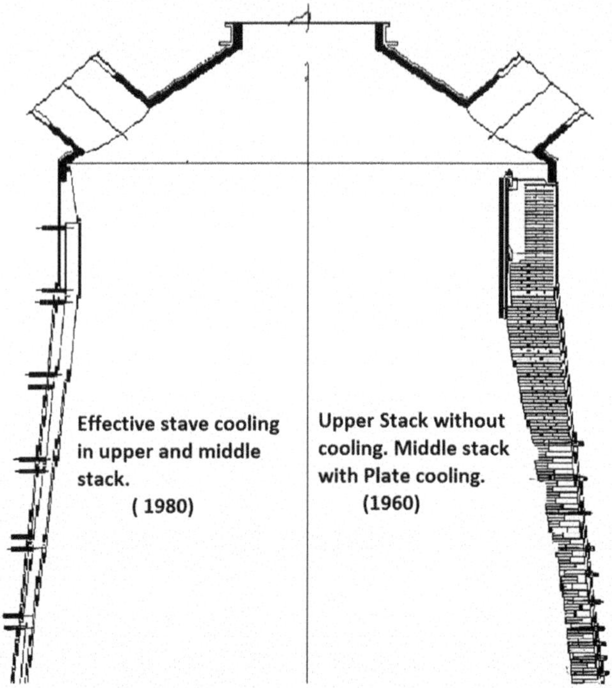

Fig. 3.3 Upgraded design of cooling in stack

$$3Fe + 2CO = Fe_3C + CO_2 \tag{3.3}$$

$$3Fe_2O_3 + 9CO = 6Fe + 9CO_2 \tag{3.4}$$

$$Fe_3C = 3Fe + C \text{ (precipitation)} \tag{3.5}$$

$$Fe_3C + 14CO = 3Fe_{20}C_9 + 7CO_2 \tag{3.6}$$

$$3Fe_{20}C_9 = 20Fe_3C + 7C \text{ (precipitation)} \tag{3.7}$$

Low Fe_2O_3 content (<1.0%) in fireclay and high alumina bricks fired at high temperature decreases the danger of CO disintegration in stack refractory.

Refractory practice in stack: Traditionally the three zones of stack are lined with 42–45% Al_2O_3 bricks in upper stack and 62% Al_2O_3 of dense quality bricks in middle and lower stack. Severe erosion was observed in lower stack due to alkali attack and thermal shock. Frequent repair was needed to extend the campaign. In modern large-size blast furnaces, SiC-based refractory is used in middle and lower stack. Refractory made of graphite and carbon is also in use in lower stack, resulted in very low erosion. This will be discussed in detail in Sect. 3.2.2.

3.2.2 Refractory Practice in Bosh and Belly

Bosh is the interface between hearth and belly/lower stack and is facing the slag attack generated below cohesive zone. It also experiences the attack by ascending gases and alkali deposits. Despite profiling of the large-size furnace as per theoretical and empirical formulas, the erosion of refractory in this area is very severe. The formation of primary slag containing high FeO at bosh is more serious than other part of the furnace shaft. However, if the intensity of cooling on furnace lining is strong enough to form a solidified slag layer over the refractory lining, it would protect the lining from severe erosion. Wear lines of a blast furnace and its original lining thickness are shown in Fig. 3.4, and it is a typical wear pattern in many blast furnaces as seen after only 3–4 years of campaign. In most of the cases, the shell became exposed and results in red-hot or cracking. This type of erosion is common in high-alumina refractory-lined furnaces as the corrosion of refractory with bosh slag contains high lime and FeO. The other factors that cause severe erosion are as follows:

- Attack of alkali and zinc vapour.
- Hot metal infiltration.
- High thermal shock.
- High abrasion.

Effect of slag composition of refractory used in lower stack, belly and bosh: In blast furnace, the slag generated at lower stack and bosh is called bosh slag, and the

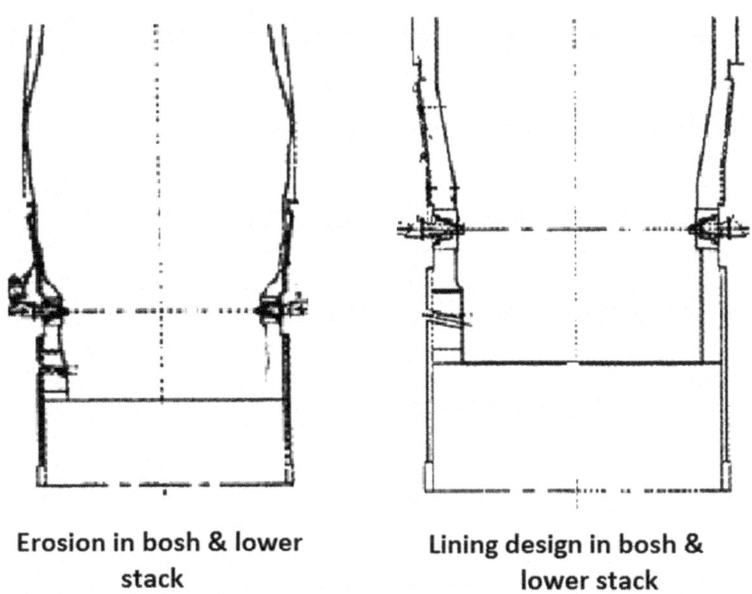

Fig. 3.4 Erosion of Bosh, belly and lower stack refractory lining

Table 3.3 A typical slag composition of a blast furnace

Composition (%)	CaO	MgO	FeO	Al_2O_3	SiO_2	$B = CaO/SiO_2$
Final slag	36.5	8	1.5–2.5	17	33	1.1
Bosh slag	40	10	10–12	13	27	1.48

final slag is generated in hearth. Though the final slag is acidic, the basicity of bosh slag is maintained with higher basicity to facilitate reduction of sulphur in hot metal. Further, the bosh slag is enriched with FeO, which helps to reduce the viscosity of slag. A typical composition of final slag and calculated bosh slag is shown in Table 3.3.

From the above analysis, it is found that the bosh slag is basic in nature and contains very high FeO, which are the causes for high corrosion of high alumina (62% Al_2O_3 Bricks) lined in belly and bosh areas.

Corrosion due to presence of high FeO in slag: Under the operating condition of a blast furnace, very low partial pressure of oxygen, under high pressure, has to be maintained in lower stack, belly and bosh. Thermodynamically, FeO is the stable phase in the operating condition of pO_2 of 10^{-6} to 10^{-9} under high top pressure of more than 2.5 bar; FeO reacts with high alumina refractory to form very low melting phases in lower stack.

Bosh and belly area, 60% Al_2O_3 bricks are used, which has more than 65% solubility under C/S ratio of more than 1.2 and FeO content of more than 8.0%. The 62% Al_2O_3 bricks contains corundum and mullite as major phases with the presence of SiO_2 as cristobalite. In the presence of FeO, the following reactions are taken place to form low melting compound:

1. FeO + Corundum (Al_2O_3) = $FeO.Al_2O_3$.
2. FeO + Mullite ($3Al_2O_3.2SiO_2$) = Iron – cordiorite ($2FeO.2Al_2O_3.5SiO_2$).
3. FeO + SiO_2 = Thyalite ($2FeO.SiO_2$).

All the above low melting phases are forming solid solution among each other resulted in further lowering of liquidous temperature as shown in the FeO–Al_2O_3–SiO_2 phase diagram (Fig. 3.5).

Corrosion of a fire clay refractory comprising mullite ($3Al_2O_3.SiO_2$) and SiO_2 by FeO occurs because of low melting formation at as low as 1200 °C, while in high alumina bricks containing more than 60% Al_2O_3, low melting forms at 1380 °C, which is much lower than the slag forming temperature. Use of more than 65% Al_2O_3 containing refractory in bosh area is not feasible due to high thermal shock due to high thermal expansion coefficient [1, 2, 26].

The refractory lining life is extended in modern blast furnaces by the following development in operation and refractory quality:

1. By maintaining a protective slag coating over alumina refractory as a freezing layer to protect further reaction.
2. Use of SiC-based refractory which is corrosion-resistant towards slag attack.
3. An increase in intensive cooling helps to create freezing layer in bosh and belly.

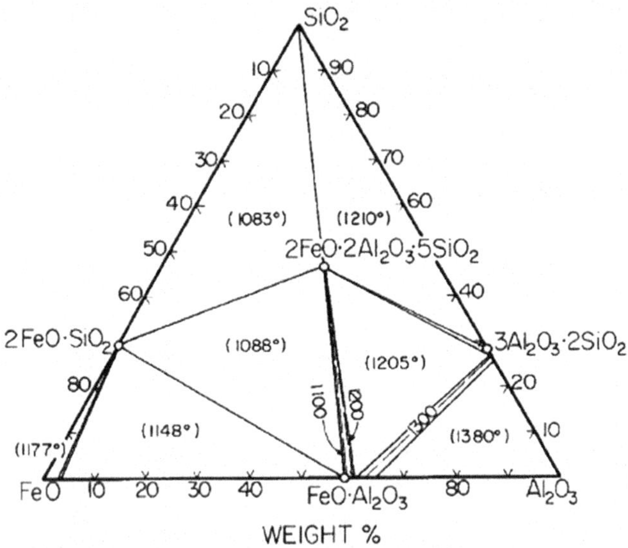

Fig. 3.5 Ternary phase diagram of $FeO–Al_2O_3–SiO_2$ system

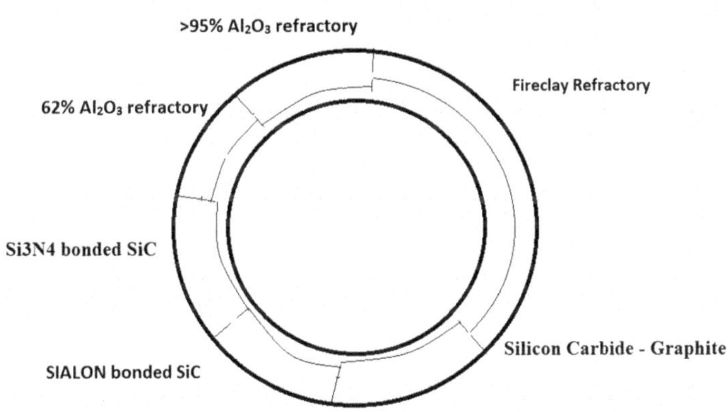

Fig. 3.6 Comparative remnant refractory thickness in Bosh

The different quality corrosion resistant refractories are tried in lower stack and bosh area, as shown in the Fig. 3.6. SiC-graphite combination is the best refractory used in these areas, but it should be protected from oxidation. Si_3N_4 bonded and SIALON bonded SiC are widely used for their corrosion resistance and thermal shock resistance. Corrosion resistance of >95% alumina bricks are better in corrosion resistance, but inferior to thermal shock resistance compared to fire clay and high alumina bricks. However, in small-size and medium-size furnaces, 62% Al_2O_3 dense-quality bricks are widely used.

3.2.3 *Refractory Practice in TJ Area*

Hot air is blown into the Blast Furnace through the nos. of tuyeres and all tuyeres are aligned in a horizontal axis, called tuyere axis. The tuyeres are made of copper with a cooling water circuit to avoid burning of copper body and also enable to cool the surrounding refractory. This area has the height heat load compared to other area of blast furnaces. The temperature generated by combustion of coke in front of tuyere exceeds more than 2200 °C under oxidizing atmosphere, and the area of combustion in front of the tuyere is called "Race way". Tuyeres are inserted into cooling jacket and the cooling jackets are surrounded by refractory lining as shown in Fig. 3.7. The lining refractory is getting cooled by plate coolers or stave coolers. Additionally, to protect the steel shell from overheating, a water cooled jacket was installed outside of the shell and hence the area of the blast furnace is called "Tuyere Jacket area".

Coke with other burden descend through the shaft enter into the oxidizing zone in front of tuyeres, where intensive combustion of coke takes with a jet of blowing air.

$$C + O_2 = CO_2 \tag{3.8}$$

As there is excess carbon in the hearth, CO_2 is reduced to CO, and the production CO (Boudouard reaction), is the source of reduction of iron oxide to metal iron in blast furnace. CO thus always evolved from the oxidation zone in front of tuyeres as a product of combustion.

$$C + CO_2 = 2CO \tag{3.9}$$

Blown air jet contains natural moisture and often mixed with moisture, and generates hydrogen as another source of reducing gas. N_2 is also generated in this high combustion zone. As it is a very high temperature zone adjacent to tuyere, the high alumina refractory is used as a conventional lining (Fig. 3.8). The essential properties of the refractory used in TJ area are as follows:

– Very high refractoriness.

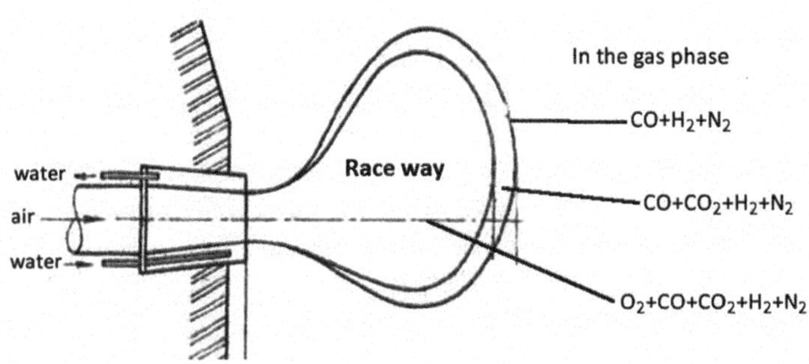

Fig. 3.7 Tuyere position and race way

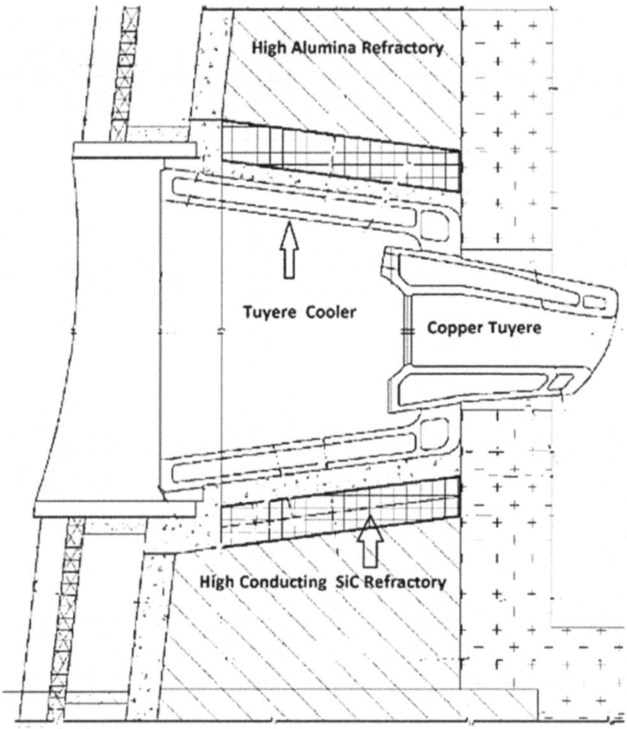

Fig. 3.8 Arrangement of tuyere

– High creep resistance.
– Resistance to alkali and zinc vapour attack.
– High hot abrasion resistance of coal and coke dust.
– Resistance to slag attack.

Modern Refractory Lining Practice

In small blast furnaces, 62% Al_2O_3 bricks were used and the life was 6–8 years due to poor resistance to alkali and slag attack. In large-size furnaces, above 90% Al_2O_3 refractory and Si3N4 bonded SiC refractory are in use and the life was extended to more than 12 years. In order to satisfy all the high-temperature properties, two types of refractories are developed and in use worldwide.

1. Carbon refractory: For many years, in large blast furnaces, tuyere belt was constructed with machined carbon blocks, each tuyere surround being composed of four symmetrical quadrants. But the main disadvantage was the problem of oxidation in the condition of water leakage. A more

(continued)

Fig. 3.9 Tuyere blocks assembly made of Alumina-chrome castable refractory

oxidation-resistant carbon product was developed by the process of impregnation of carbon into the pores and application of coating. The new product shows oxidation resistance up to 1000 °C, when ordinary carbon oxidizes above 600 °C; with the use of superior carbon refractory, the life was improved significantly.

2. Corundum-Chrome oxide castable: It is a special type castable with fused corundum as based material and a special binder, incorporating 8–10% chrome oxide fines and very little cement to prepare a low cement castable. The main characteristics of the castable are as follows:

   - It requires only 3.5–4.5% of water for casting.
   - Very high crushing strength, at 1000 °C fired castable: 90 MPa.
   - At 1600 °C, hot crushing strength is more than 45 MPa and hot modulus of rupture has a value of about 18 MPa.
   - When heated above 900 °C, chromic oxide starts to diffuse into corundum grains and forms a highly refractory solid solution.

   The refractory design in tuyere area of a modern blast furnace is shown in Fig. 3.9.

3.2.3.1 Use of Ramming Masses in Blast Furnace Lining

Between the gaps of tuyere coolers and SiC bricks, carbonaceous ramming mass is used. The critical function of ramming mass is to tight fill irregularly shaped spaces. The ramming mass helps to adjust the expansion gaps during operation. The similar quality ramming masses are used around the gaps in plate coolers and box coolers in bosh and stack area. The main advantages of using ramming mass are as follows:

- Reversible compressibility to adjust the expansion–contraction of other components of refractory.
- High thermal conductivity to ensure proper cooling of lining refractory.
- Easy to use as a gap filling material in a complicated area.

3.2.4 Refractory Practices in Hearth

The size and productivity of blast furnaces has grown steadily over the last few decades. For a long time, the evolution was largely dominated by the construction of giant blast furnaces with high production rate, and a continuous strive was made in upgradation of hearth refractory. The campaign life of a blast furnace depends on performance of hearth. Any damage in hearth refractory causes the stoppages of the furnaces for complete relining. Most modern blast furnaces have similar operational characteristics and the refractory problems encountered are alike. In fact, hearth erosion depends more on high production rate than size of the furnaces. Other causes of high erosion in hearth wall and bottom linked to the nature of burden and to the hearth inner profile [7, 8].

The wear in the refractory lining in this area is by alkali attack, dissolution of refractory in molten iron, erosion by iron and slag and liquid metal penetration. In early days, the complete hearth wall and hearth bottom were lined with high alumina bricks and campaign was only 6–8 years due to very high erosion. Use of carbon refractory in hearth side wall and later in bottom is a breakthrough in the area of blast furnace hearth lining. Usage of carbon as lining material has substantially prolonged the campaign to more than 20 years hearth life. The advantages of using carbon are as follows:

- Withstand very high temperature under reducing atmosphere,
- No softening or melting (sublimation),
- Non-wetting towards hot metal and slag,
- Excellent thermal shock resistance,
- Very high thermal conductivity for proper cooling,
- Volume stable in a wide range of temperature,
- High hot strength and abrasion resistance.

Wear mechanism of hearth refractory: Considerable experience has been gained by operations in the last few decades using carbon in bottom and hearth of blast furnaces. Although a good explanation of the factors which are responsible for the destruction of carbon refractory is not yet known, it is believed that its compatibility with blast furnace slag, hot metal chemistry and volume stability and refractoriness make it the most suitable alternative material for blast furnace hearth. The early trials of carbon in hearth were justified by the fact that high alumina refractory showed high shrinkage in service which often led to serious breakouts. With the carbon lining in hearth wall, the problem of frequent breakouts was eliminated. The high thermal conductivity of the carbon was found effective in efficient cooling and thus extended the life of the hearth wall and bottom. Though carbon refractory has many advantages over oxide refractory, it has few major disadvantages also and hence one has to study carefully the wear mechanisms of the lining material, especially predominantly use of carbon refractory. The following wear mechanisms are responsible for hearth erosion [9]:

1. Carbon dissolution in hot metal: It has a direct influence on hot metal carbon saturation. The hot metal carbon concentration at the tuyere level is around 2.5–3.5%. Tapped hot metal is carbon saturated and the concentration of carbon depends on temperature and silicon content in hot metal. Hot metal can be saturated in carbon in the presence of coke and pulverized coal injection (PCI).
2. Effect of chemical attack: The potassium oxide is absorbed in carbon blocks and migrated into the block to a temperature zone of 900 °C. K_2O reacts with crystalline phases of carbon blocks, such as mullite, corundum and silica and low melting phases, Kalsi lite ($K_2O{\cdot}Al_2O_3{\cdot}2SiO_2$) and leucite ($K_2O_3{\cdot}Al_2O_3{\cdot}4SiO_2$) are formed. These reactions are resulted in a volume expansion, which causes destruction of the brick texture. Further absorption of excess potassium forms potassium-carbide compounds of the formulas C8K, C24K and C60K, which causes swelling of the carbon blocks and forms a brittle layer.
3. Effect of metal penetration in hearth: Carbon hearth refractories come into direct contact with liquid iron. The density of hot metal is 6.9 gm/cc and of average density of carbon blocks is around 1.95–2.85 gm/cc, hence due to wide difference in density, the ferro-static load on carbon hearth lining is significantly high and causes penetration of hot metal through the pores of carbon blocks. The penetration of hot metal thus forms a dense brittle layer over carbon blocks and dislodged from the original blocks during turbulent flow of hot metal into hearth, causes thinning of carbon blocks.
   Metal penetration through open pores in carbon block is the major factor for hearth erosion. It depends on capillary force of hot metal over carbon blocks and on the radius of open pores. Amount of penetration reduces with reduction of pore diameter, and the optimum pore diameter can be calculated from the following empirical formula:

$$\Delta P = \frac{2\,\gamma\,\mathrm{Cos}\,\theta}{r} \tag{3.10}$$

   where, ΔP = capillary pressure in open pores of carbon blocks, γ = surface tension of liquid hot metal, θ = contact angle and r is the radius of pores. It was observed through mathematical modelling and experimental results that if the radius of open pores of carbon blocks is less than 0.5 μm, penetration of hot metal can be avoided.
4. Effect of peripheral flow in hearth and elephant foot formation [10, 11]: The effect of hot metal carbon concentration and brittle layer formation due to alkali deposition and hot metal penetration on wear rate of carbon refractory is more pronounced in the "stagnant" condition of hot metal bath. For the "turbulent" flow of hot metal during tapping through tap hole, the wear rate is mainly controlled by the hot mechanical abrasion due to liquid flow against the wall. The presence of "Dead Man" at the centre of the hearth and its nature governs the effect of mechanical abrasion. In hearth solid coke is accumulated and liquid iron and slag dripping from the cohesive zone accumulate and mixed with coke. A "Dead Man" is the accumulation of coke, mixed with slag, and due to high

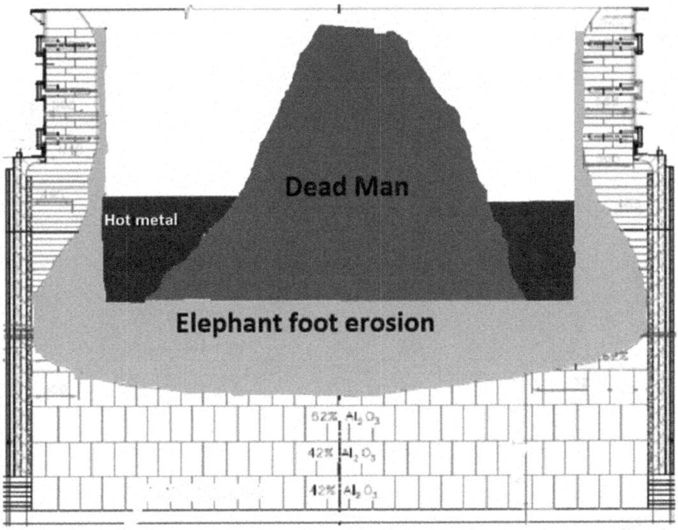

Fig. 3.10 Elephant foot formation

density, iron droplets accumulated in hearth. The porosity of the dead man is between 0.3 and 0.5, which indicates that only 30–50% of the total volume is available for the liquids if the dead man fully sits on the hearth. In fact, the dead man may either sit on the bottom of the hearth or float partly or completely in the liquid bath if the buoyancy force of the liquids is sufficient to lift the dead man.

A partly floating dead man sits on the central hearth bottom and gives rise to an annular coke free zone at the hearth edges, which offers little flow resistance when the hearth is drained. Therefore, a circumferential flow is formed, being the main reason for the "elephant foot" shaped wear in the hearth, as shown in Fig. 3.10. Due to sitting of Deadman, the direct flow of hot metal is restricted towards tap hole during tapping, and the circular flow along the periphery causes severe hot abrasion of side wall carbon blocks, which caused breakout in hearth in several occasions in many operating plants.

In consideration of above consequences, details of important **carbon and graphite** refractories are discussed systematically. Elemental carbon is found in nature in the form of diamond and graphite and as coal. For refractory application, natural and artificial graphite are widely used to manufacture carbon blocks. The raw materials for carbon source must have very low ash content with high yield. Graphite is a unique material with hexagonal platy structure in infinite two-dimensional array of carbon atoms arranged in hexagonal network.

The carbon–carbon co-valent bond in a hexagonal plane is strong indicated by interatomic spacing of 0.142 nm, where the van der Waal type bonding within the planes is weak, the interatomic space is 0.340 nm, as shown in Fig. 3.11. Graphite has a layered structure, occurs in flakes and shows anisotropic properties. Thermal

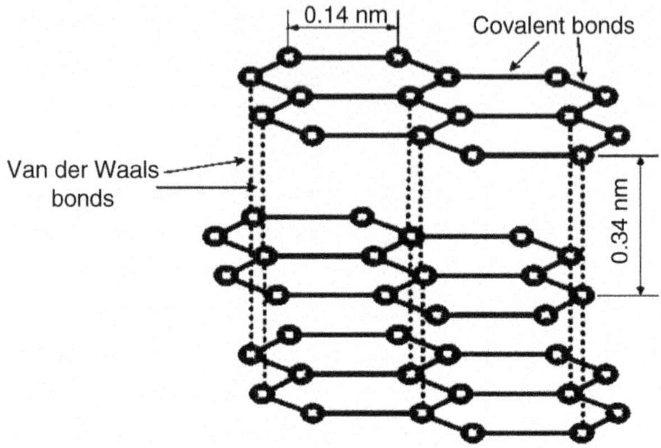

Fig. 3.11 Crystal structure of graphite

expansion perpendicular to the planes is 200 times than parallel to the planes. Thermal conductivity in parallel planes is 200 times that of perpendicular to the planes. Its compressibility is 10^4–10^5 times greater in perpendicular direction than that of parallel direction.

Refractory bricks made of carbon can be classified in three groups: amorphous, semi-graphite and graphite blocks. In recent years, another two different superior-quality carbon blocks are developed, micropore and super micropore. Those quality had been developed to increase corrosion resistance and hot metal penetration by reducing the average pore sizes with diameter (d) below 1 μm.

The raw material used for the manufacture of carbon and graphite bricks or blocks are coke, petroleum coke, calcined anthracite and natural graphite, pitch as dry ingredients and tar pitches, petroleum pitch and resin are used as binder. Other additives used are Si–metal powder, SiC, Al_2O_3, etc., to improve the strength and other corrosion properties. The dry mix was made by mixing all the dry granulated ingredients, the mix heated at around 160 °C depending on the type of bonds and mixed with the binder to prepare a green batch. Depending on the final product quality, the shapes are made in vacuum-pressed equipment, semi-dry pressing, isostatic pressing, ramming to make large-shaped blocks or through extrusion process. Large-size bottom block was made with an average size of 600 × 800 × 3800 mm and the weight is about 6–8 tons. Thus the green blocks and bricks were subjected in heat treatment up to approximately 1200 °C, under reducing atmosphere to develop carbon bond, or under nitrogen atmosphere to develop nitride bond. Further densification was done by impregnation of carbonaceous liquid through high vacuum treatment, dip into liquid bath and followed by applying pressure over liquid. The impregnation agent, entered into pores, is also converted into coke during further firing cycle.

For manufacturing graphite blocks, the main raw material is petroleum coke and natural graphite, which can be graphitized with a large yield. The additional production step to follow for manufacturing graphite blocks is "Graphitization". It is performed in Acheson resistance furnace or graphitizing through direct current flow in the graphitizing furnace.

Application of carbon and graphite bricks is limited due to oxidation by oxygen, hydrogen and CO_2 at above 400 °C. With a reducing atmosphere, the service limit is as high as 3000 °C. Above 1600 °C, the carbon blocks are partially graphitized and above 2000 °C, it is completely graphitized.

3.2.4.1 Source and Quality of Graphite

The graphite is available as natural flake graphite and synthetic graphite. The other form of carbon has been further heat-treated at a temperature between 2400 and 3000 °C, and the process is known as "Graphitization" by changing the crystallographic structure of carbon and also changes the physicochemical properties at elevated temperature [5, 12]. Purification of graphite refers to its ash content, which acts as the flux to form low melting compound in carbon blocks.

Carbon Refractory in Hearth Lining [13, 14]
Since 1960 carbon blocks and bricks are used worldwide in blast furnace hearth lining, and continuous development in quality had been brought to improve corrosion resistance and extend campaign life. Till 1965, the average hearth life was 8–10 years, by 1975 it was reached to 15 years and within 2001, it has improved to more than 20 years. The properties of carbon block and graphite blocks had been upgraded through continuous experiment and development work in Japan and other European countries, and the properties of different quality blocks used in a modern blast furnace are shown in Table 3.4.

Table 3.4 Properties of different qualities of carbon refractories used in hearth lining

Properties	Unit	Graphite	Carbon	Micropore carbon	Super micropore
Bulk density	gm/cc	1.62	1.56	1.62	1.72
Porosity	%	23	20	16	16
MOR	MPa	10	10	13	18
Crushing strength	Mpa	20	34	45	55
Thermal expansion at 1000 °C	%	3.1	3.5	3.5	3.5
Thermal conductivity at 400 °C	W/mK	160	18	16	20
Pore size >1 μm	%	NA	NA	6	2
Ash content	%	0.3	3	17	20
Fe_2O_3	%	0.1	0.6	<0.6	<0.6

Semi-graphite is a quality of material composed of artificial graphite particle mixed with carbonaceous materials such as hard pitch or tar and baked at a temperature of 800–1400 °C. The product is a carbon bonded graphite, exhibits higher thermal conductivity than carbon but not as high as graphite blocks. A true semi-graphite contains graphite bonded with carbonaceous materials. The semi-graphite type material is porous in nature due to low-temperature backing, and more often it is densified by re-backing, and the surface pore size can be reduced by impregnating into molten tar under vacuum treatment followed by applying high pressure.

3.2.4.2 Hot-Pressed Carbon

One American manufacturer utilizes the proprietary hot-pressing method to manufacture a true semi-graphite refractory. The product is comparatively less permeable and has a higher thermal conductivity than conventionally baked semi-graphite. Nowadays this product is produced by many Chinese manufacturers and widely used in China and other countries. Hot-pressed carbon composite refractory is also developed by adding corundum and formation of SiC in matrix.

3.2.4.3 Thermal Conductivity

Coefficient of thermal conductivity of the carbon composite bricks is around 13 W/mK, which can meet the requirement of blast furnace hearth. The temperature dependence of thermal conductivity of carbon composite refractory is shown in Fig. 3.12. The relation between temperature and thermal conductivity coefficient of carbon composite is expressed as $\lambda = 0.00539\ T + 18.159$ and the linear correlation

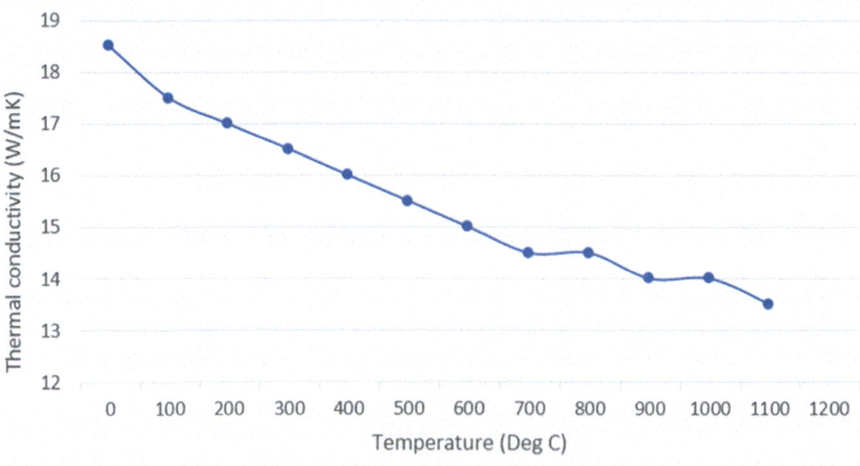

Fig. 3.12 Change in conductivity with temperature [15]

coefficient was found $R^2 = 0.987$. It is seen that the conductivity of the hot-pressed carbon is higher than conventional carbon due to its very dense structure, formation of SiC by reaction of Si with C. The thermal conductivity of carbon decreases when temperature increases, which is due to conduction of heat within inorganic materials, oxides and carbides, which is the result of a phonon collision.

Therefore, the reason for the decrease in thermal conductivity of carbon composite brick with increasing temperature is attributed to the deficiency in the amount of free electrons and thus has a poor ability of conducting heat.

3.2.4.4 Corrosion Resistance

The wettability of carbon compound bricks with slag is higher than micropore carbon and molten slag gradually penetrates into carbon blocks form brittle layer. However, the damage caused by the corrosion of the molten slag barely occurs in practical application. The main reason is that the increasing content of Al_2O_3 in the molten slag resulting from the reaction between the molten slag, and the carbon composite brick significantly increases the viscosity of the molten slag. At the same time, the cooling effect in the bottom and hearth of the blast furnace promotes the precipitation of magnesium aluminate spinel, which could also increase the viscosity of the molten slag. Therefore, the molten slag is difficult to infiltrate into hot-pressed compound carbon blocks. Hence, the composition of slag layer adjacent to hearth wall is different from bulk slag composition, and therefore it is confirmed that the source of the protective layer is not derived from blast furnace slag.

3.3 Refractory Maintenance Practice

Besides designing and operating a blast furnace for extended campaign life, there are various maintenance practices that can be implemented to extend the campaign life of a blast furnace and thus complete relining of the furnace has been differed for 3–5 years, resulted in the reduction of capital cost [16].

- Grouting.
- Gunning.
- Titania charging.
- Shell protection and welding.
- Control of water leakage inside the furnace.

The implementation of all those practices needs to be proactively planned as a part of a campaign extension strategy and not for an emergency repair. To make a schedule for all those practices, it is required to strengthen the furnace monitoring system, and by applying the state-of-the-art automation, the modern alarming parameters are able to show the actual health of the furnace.

The blast furnace campaign life depends on the condition of hearth refractory. There is no proven method for partial replacement of damaged hearth refractory without taking shutdown for long period and which is very close to complete hearth relining. Hotspot formation in below tuyere level and adjacent to tap hole is very common. It occurs due to the passage of hot air through the brick joints. Not paying attention to those problems may cause cracking of steel shell and breakout in hearth occurs. So in carbon lining, grouting in hearth is the most important maintenance practice to follow as per predetermined schedule. The following sections of this chapter will discuss the different maintenance practices in detail.

3.3.1 Robotic Stack Gunning

Since the 1990s, gunning in blast furnace stack by "Robotic method" has made a breakthrough in the area of blast furnace campaign extension. In this process, it is possible to repair very large affected area in stack within very short time. The success of the process depends on the skill of the executing agency, contractors and use of proper equipment. A plan needs to follow for a successful robotic gunning:

- Modify the shell design of upper stack for easy access and fixing remote machine into the furnace.
- To select an expert agency who can supply all equipment, gunning refractory material and has expertise to carry out the jobs within the optimum time.
- To improve blowdown process so that burden can be decent below tuyere level with almost clean refractory surface, so that gunned refractory can adhere to the old lining properly.
- Selection of the suitable gunning materials with the experience on local erosion mechanism.

3.3.1.1 Arrangements for Robotic Gunning

The first step was to instal opening doors at 180 °C. apart in upper stack of the furnace. One would be larger, through which the cantilever type "gantry" can be installed to hold the gunning robotic "pod" at the centre of the furnace and the other one to provide support.

The next step was to instal all of the gunning equipment. A gantry equipped with a trolley, pulleys and cable drum are clamped to a flange on the large gunning door. Then a gunning "pod" with light and a removable gunning pipe is hung from the gantry trolley. The gantry pod is controlled by remote from outside of the furnace. The remote box controlled the vertical movement of the gun, rotation direction and rotation speed, the feeding of powder refractory, compressed air, water and electrical power supplied into the gantry. The condition of stack before gunning and arrangement for gunning are shown in Fig. 3.13. Once the equipment was installed, the

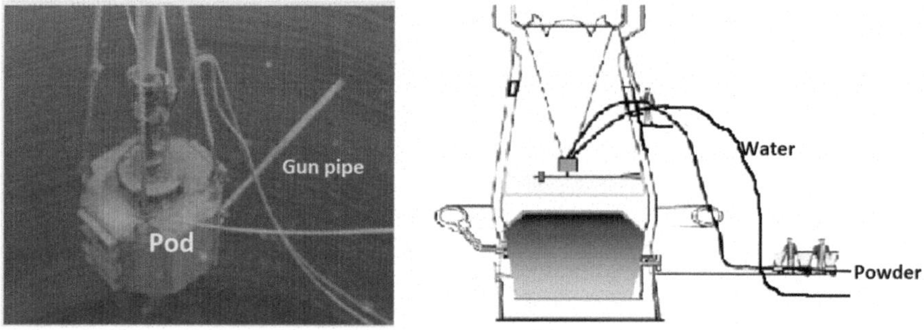

Fig. 3.13 Arrangement of robo-gunning in BF stack

refractory powder material is poured into a mixer at ground level and the mix is pumped up to the furnace top through hose and water is lifted through another hose.

Water mixed with castable at the tip of the gun and sprayed the wet mix on to the hot surface of old refractory. With the support of plate coolers, it is possible to develop more than 400 mm refractory thickness over old lining in lower stack.

3.3.1.2 Blast Furnace Wall Cleaning

The success of stack gunning depends on blowdown of the furnace and wall cleaning from scab, jams and dust. If the wall is not properly cleaned and dust-free, the new gunning castable would not adhere to the old lining and rebound losses would be very high. Short life after gunning repair had been reported when the refractory was applied over the slag layer, which melted during operation and refractory was dislodged. The remote gunning technique uses high-pressure water jet to clean the wall, and it also helps to dislodge big jams or scabs due to high shrinkage during cooling and cracking by thermal shock.

If it is required for gunning repair in bosh area, it is required to down the burden below tuyere level and in that case the following problems were faced in many steel plants:

1. Due to burning of dead man coke by air sucked from the tuyere opening, temperature was increased and pod was overheated and gunning operation was discontinued. To avoid the problem, it was recommended to close all the tuyere openings by temporary brick work at the tuyere coolers opening from outside, so that fresh air could not enter into hearth.
2. CO gas generation into the furnace is another problem to carry out the job within the stipulated time. The CO gas needs to be burned continuously and blue flame into the hearth burden has to be observed.

Development of Refractory for Stack Gunning

Since 1990, major initiative had been taken to develop the suitable gunning castable, suitable for robo-gunning. The following essential properties were considered to design the material:

1. Minimum rebound loss.
2. High durability.
3. CO-resistant.
4. Properties to prevent chemical attack.
5. Good adhesion capability with old refractory in hot surface condition.
6. No cracking of gunned castable in no-dryout and very fast heating.
7. To maintain high gunning rate.
8. Optimum cost.

50% Al_2O_3 Chamotte-based gunning castable with low iron was first developed for gunning in upper stack and application through robot was successful. Later the demand was accelerated to repair up to lower stack and presently down to the lower of bosh area. So, as per severity in application conditions, other improved quality castables were developed. SiC-based gunning castable was developed for gunning in lower stack and bosh area under high-temperature and excellent abrasion and corrosion resistance. Different types of castables with their properties are summarized in Table 3.5.

Table 3.5 Different qualities of gunning castable

Gunning materials	Unit	Type 1	Type II	Type III	Type IV	Type V
Area of application		Throat	Upper stack	Middle/ lower stack	Belly/ bosh	Belly/ bosh
Al_2O_3	%	50	58	70	40	25
SiO_2	%	40	40	26	5	7
Fe_2O_3	%	0.8	0.8	0.5	0.2	0.2
SiC	%	0	0	0	>50	>60
Bulk density	gm/cc	2.18	2.26	2.45	2.5	2.58
Linear change at 1100 °C	%	−0.1	−0.2	−0.11	−0.08	−0.06
Cold MOR	Mpa	10	9.6	9.4	10.5	11
Rebound loss	%	<10	<10	<8	<5	<5
Thermal shock resistance in water quenching	Nos. of cycles	8	10	>15	>20	>25
Thermal conductivity at 1000 °C	W/Mk	1.46	2.3	2.1	4.5	>5.5

3.3.2 *Grouting Refractory*

Grouting is a maintenance practice that is used to restore the worn-out refractory thickness with new refractory material that is pumped into the furnace through a series of pipe nipples placed on the shell from outside. Grouting has been successfully employed to extend furnace campaign life. In 1970s, the grouting was introduced in the stack of the furnace, and later this technique was extended to lower stack, bosh and belly also. Since past three decades, the grouting process was initiated in hearth side wall and around the tap holes. Grouting is a good campaign extension technology for a localized area to recover the critical condition of overhearing of shell as an immediate action to avoid further damage or cracking with a very short downtime. If large area requires grouting at a high frequency, the furnace downtime and grout cost need to be optimized, and if it becomes un-acceptable, another repair process has to be adopted.

3.3.2.1 Decisions on Grouting in Stack and Bosh

In an ideal condition, grouting maintenance has to be started before the "hotspot" or cracking of shell due to overheating to avoid serious injury and equipment damage. Decision for start grouting can be taken based on the following observations:

- Continuous monitoring of shell temperature by the process of thermal imaging, if the shell temperature exceeds more than 120 °C, steam sprays are installed, and when the temperature reaches 450 °C, grouting has to be started.
- Measurement of temperature difference of outlet and inlet temperature of plate or stave coolers. Action has to be taken if ΔT exceeds 7–8 °C.
- In small- and medium-size blast furnaces, core drilling is performed at various heights in the bosh and stack around the circumference of the furnace. The length of the core would indicate the remnant refractory thickness. Grouting may be started when the core thickness is less than 300 mm.

3.3.2.2 Preparation for Grouting

Grouting castable has to be pumped into the furnace brickwork to restore the refractory thickness. The grouting has to be done through nipple fixed on the steel shell. Hence for grouting operation, the following activities have to be carried out as preparation for grouting.

1. Grouting nipple has to be installed and shell has to be core-drilled with a minimum diameter of 35–50 mm, depending on the flowability of castable. If it is required to deposit the new refractory on the hot face of the bricks, it requires to core-drill the bricks, and grouting has to be done against the burden inside the furnace. Grouting nipples to be welded coinciding the drilled hole. Other end of

Fig. 3.14 Screw Pump for Grouting

the nipple would be threaded or fixed with quick changing type coupling so that the hose, connected with the pump, has to be coupled with the nipple.

2. The distance between the two nipples has to be determined on the grout castable flowability. It is experienced that the grout castable material usually spreads 3 ft from the nipple and then begins to harden due to heat inside the furnace. Sometimes, it spreads less if the heat inside the furnace is high. However, as a guideline, the grout nipple should have a maximum spacing of 6 ft and this needs to be confirmed after proper trial. Once the grout nipples are installed, they must be fixed with ball valves and the ball valves have to be closed.
3. Installation of pump: Reciprocals diaphragm-type or screw-type pumps are in use. The screw-type pump is shown in Fig. 3.14. The grout quantity and the pump pressure need to be adjusted on past experience. Excess pressure would damage the old refractory lining and increases the chances of back pressure and jamming of pumping hose. Quantity of grout per nipple must be optimized based on remnant refractory thickness measured. However, as per working experience, the average quantity grouted per nipple is 25 kg to obtain the smooth grouting operation and satisfactory performance.

Safety concerns arise during grouting the water-bonded castable generates hydrogen gas when it hits the hot coke, and this causes flame shooting out through the adjacent open nipples and even gas explosions in the furnace. Due to this problem grouting has to be done after blanking the tuyeres with clay or mud, so that coke into the furnace would not burn by fresh air. During grouting only working hole has to be opened and other holes will be kept closed by closing attached ball valves. It is also recommended that another one nipple and working nipple would be kept open and other all adjacent nipples have to be closed. Entrapped air into the gaps of the brick lining would come out from other opening when grouting is continued through another nipple. This results in free flow of grout castable and no possibility of back pressure.

Table 3.6 Properties of carbon-bonded grouting materials

Properties	Unit	Type I	Type II	Type III	Type IV
Al_2O_3	%	30	65–72	26–30	45–55
Fe_2O_3	%	10	2.5–5.5	24–28	3.5–5.5
SiO_2	%	45	30–35	30–34	24–28
LOI at 1000 °C	%	92	85–89	90–92	90–92
Setting temperature	°C	180–220	170–190	80–120	170–200
Cold crushing strength after heating at 600 °C	Mpa	15	16–18	13–15	16–18
Thermal conductivity	W/mK	10.5–12.5	5.5–8.5	11.5–13.5	10.5–12.5
Base material		Graphite	Bauxite—graphite	Graphite	Graphite
Service temperature		1800	1700	1800	1800
Grain size	mm	0.06	0.5	0.045	0.065
Type of bond		Organic	Organic	Organic	Organic

Refractory Castable for Grouting

The design of grout material to be compatible to the properties of original furnace lining material with respect to thermal expansion characteristics, adherence to the existing lining bricks and chemical durability. Other parameter is the flowability and setting characteristics, which depends on capacity of the grouting pump, effective height of the grouting point from the pump elevation and diameter of the hose and nipples. Since 1980s Castable slurry with more than 20% water addition was used to maintain smooth flow, which was pumped by low-pressure pump. The main disadvantages were high setting time, poor adherence with refractory and substantial steam formation into the furnace, Hence, after grouting the existing refractory was further damaged due to contact with a lot of water and steam pressure in evaporation of water. After introduction of high-pressure pump, now it is possible to inject low cement (LCC) self-flow castable with only 5–8% of water addition, which improved the durability of grout material and blast furnace operation started within a short shutdown. Another breakthrough in material development is the use of silicon carbide-based castable to grout in bosh and belly area. Use of silicon carbide improves thermal conductivity, thermal shock resistance and corrosion resistance and above all it becomes the best compatible material with lining bricks in those areas. Properties of different quality grout castables are as shown in Table 3.6.

Type of grout castable		55% Al_2O_3, CO-resistant	65% Al_2O_3	Al_2O_3—45%SiC	Al_2O_3—65% SiC
Area of application	Unit	Upper stack	Middle and lower stack	Belly	Bosh
Al_2O_3	%	55–60	65–72	35–45	20–30
SiO_2	%	25–35	20–26	<3.0	1.5–5.0
Fe_2O_3	%	<1.5	<1.0	<0.8	<0.8
CaO	%	4.5–7.0	1.5–3.0	2.5–35	<2.0
SiC	%	Nil	Nil	42–48	64–70
Water addition	%	5.5–6.0	5.5–6.5	55–7.5	6.5–8.5
Density	gm/cc	2.1–2.3	2.2–24	2.0–2.2	22–2.6
Cold crushing strength at 110 °C/24 h	kg/cm^2	450	550	350	300
After firing at 600 °C/3 h	kg/cm^2	480	600	480	380
After firing at 1500 °C/3 h	kg/cm^2	650	770	560	480
Cold mod of rupture	kg/cm^2	65–75	100–120	35–55	35–55
Hot mod of rupture at 1400 °C	kg/cm^2	35–40	55–65	45–55	35–45

3.3.2.3 Decision on Grouting in Hearth Side Wall

The carbon refractory in hearth wall used to be damaged in crack formation due to thermo-mechanical stress created by ferro-static pressure and its flow dynamics. Cracks inside the carbon blocks also take place due to alkali concentration near 800 °C isotherm inside the block and removal of ramming masses between the blocks and steel shell as shown in Fig. 3.15. The main vulnerable areas in hearth are, adjacent to tap holes, opposite to tap holes and around tuyere areas. It is frequently experienced that the ramming mass between the shell and carbon blocks dislocated from the original rammed area created voids and resulted in short circuit between tap holes and tuyeres, and the steel shell becomes red-hot for passage of hot air.

The phenomenon is shown in Fig. 3.16. Ramming mass between the steel shell and carbon blocks confirms the contact of shell with refractory and helps for effective cooling. It also absorbs the stress generated due to expansion of lining refractory and prevents from damaging the carbon blocks [17]. It is experienced in several cases that during operation, the ramming mass is displaced from its designed position and creates voids, resulting in the following problems:

- The void forms a barrier between the cooling members and carbon blocks.
- Carbon temperature increases and alkali attack increases.

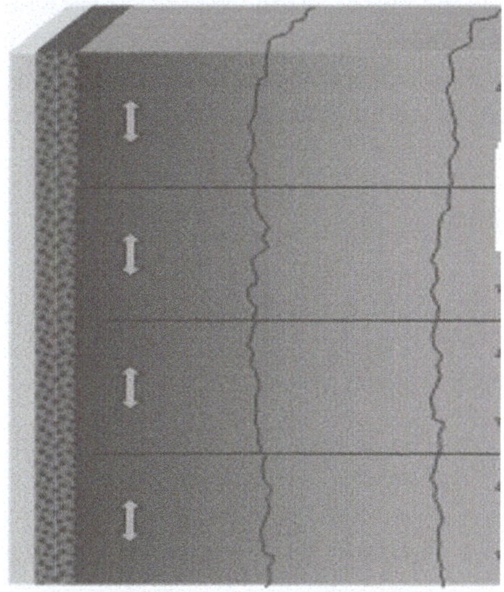

Fig. 3.15 Crack formation in carbon blocks

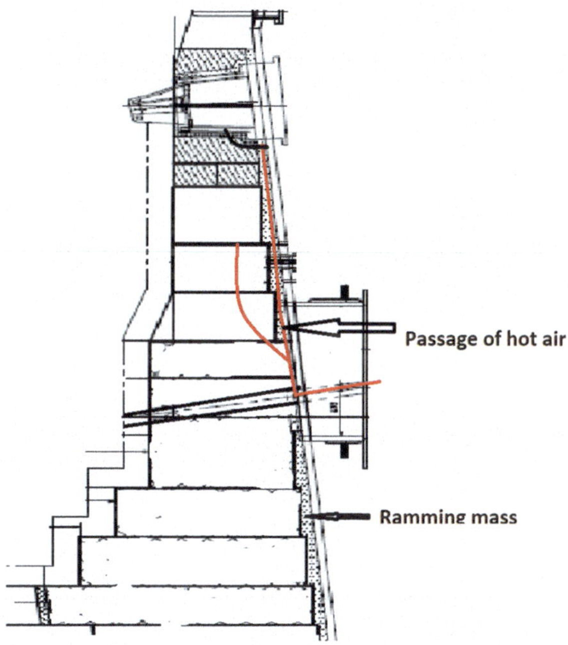

Fig. 3.16 Hot air passage from tap hole to tuyere

- Differential expansion causes stresses in each block.
- Cracks form in carbon due to stresses.
- Cracks form a barrier to the cooling and damage of carbon lining.

Grouting in hearth wall around tap holes and other area is very essential for safe and reliable operation of the furnace. As the modern blast furnaces are lining with carbon blocks in hearth, water bonded grout castable is not suitable to use. Non-aqueous, resin and tar-based grout material has been developed and in regular use to grout hearth wall.

The properties of such non-aqueous carbon-based grout material is shown in Table. 3.6.

Grouting Maintenance in Hearth

The main objective of such grouting or injection of carbon paste into hearth wall are as follows:

- To eliminate the shell overheating and control of warping of steel shell.
- To stop damage to the lining refractory.
- To seal the possible cracks and voids in the refractory lining.
- To prevent gas leakage.

The following procedure to follow in injecting the grout material:

- To install nipple of 2″ diameter fixed with ball valve,
- Connect the hose to the pump delivery side,
- To start the pump with grout material into the hopper,
- Trial with open discharge to ensure no clogging into the hose,
- Put off the pump and other end of the hose to be connected with the nipple by quick changing type couplers,
- Inject the grout paste, not more than 150–200 kgs through single nipple,
- To close the ball valve and remove the hose and connect to another nipple,
- After completion of injection, open the bypass line of the pump,
- After discharge all the material from hose dismantle the hose from pump and clean the hose and pump hopper,
- Two hours after completion of grouting release the ball valve and fill up the nipple with clay mud so that no gas leakage occurs during starting of the blast furnace.

Recommendation for Safety During Injection

- During the grouting process, use safety goggles, safety shoe, helmet and hand gloves.
- During shell cutting and drilling, to take care against gas leakage, use gas detector all the time and if the reading exceeds 50 ppm, remove the persons immediately.
- Before start the grouting, to ensure that the rotation of the pump is in right direction.
- Should not run the pump in off-load condition, to maintain the grout material into the hopper as filled by continuous pouring of the material. After completion of grouting, never remove the hose from nipple, before closing the ball valve, to avoid injury due to back pressure.

3.3.3 TiO_2 Injection [18]

Since the 1950s, one of the major initiative taken to inject Titania-bearing compound in blast furnace is based on the generation of high-temperature and high-wear-resistant Ti (C,N) compounds, which exhibits temperature-dependent solubility in hot metal. In the worn-out area of side wall, the heat loss through the wall is high and the hot metal temperature decreases, solubility of Ti(C,N) in hot metal also decreases and those compounds are precipitated out of the hot metal and deposited in the more severely eroded hearth wall, in the damaged area at "Elephant Foot". This practice includes the addition of titanium minerals, ilmenite or rutile with the burden or later the addition was started by injecting through tuyere along with PCI. The mechanism of addition of TiO_2 in blast furnace is shown in Fig. 3.17.

In the worn-out area, a solidified metal matrix in which loose Ti(C,N) particles were present, which is composed of a solid solution of titanium nitride and titanium carbide and the composition varies with TiN/TiC ratio. The Titania rich scaffold is likely to form and adhere to the wall of worn-out area to protect it from the subsequent. The addition of titanium ore through burden and added from top of the furnace increases the Ti content in hot metal, thus promoting deposition of Ti-rich scaffold. However, this process also increases the TiO_2 content in slag and increases slag viscosity. The deposit of TiO_2 in slag may lead to hearth drainage problem such as clogging and very slow discharge. To avoid the tapping difficulty, injection of TiO_2-bearing compound through tuyere is preferred, and it is done with coal injection (PCI). The advantages of TiO_2 injection through the tuyeres are summarized as follows:

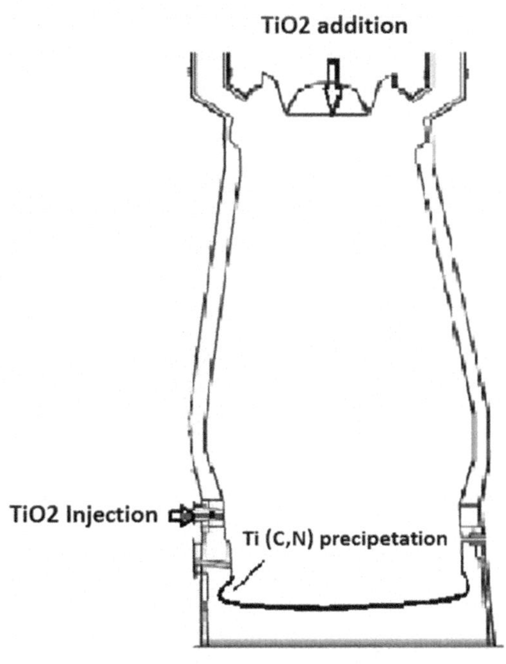

Fig. 3.17 Addition of TiO_2-containing minerals

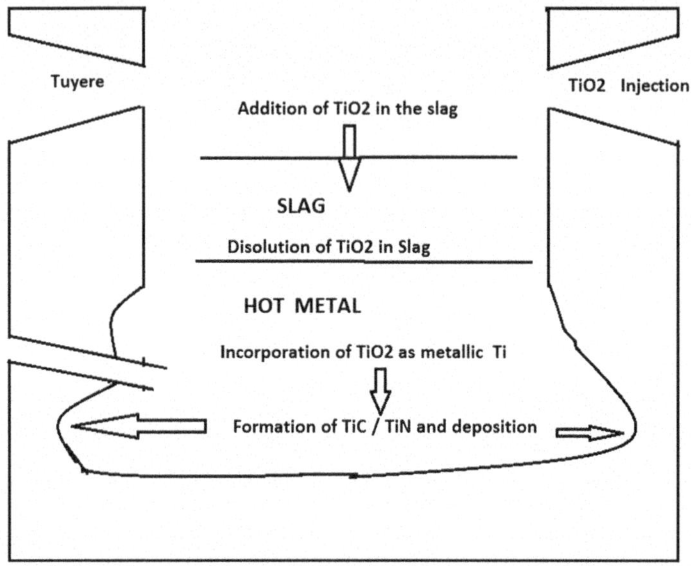

Fig. 3.18 Mechanism of Ti-compound precipitation

- TiO_2 load can be uniformly distributed throughout the hearth as the injection carried out through all tuyeres.
- TiO_2 can be injected through the selective tuyeres for addressing any localized erosion in hearth wall, hence the consumption of titania would be reduced.
- TiO_2 load is brought directly into the reactive zone of hearth wall, avoiding the way through the shaft.
- Higher mass transfer of Ti into the hot metal; kinetics of the reactions improved.

Deposition mechanism of titanium compound: Mechanism of titanium compound precipitation in worn-out area of hearth wall is shown in the schematic diagram in Fig. 3.18. It represents the protection mechanism of hearth wall by deposition of TiC/TiN. The mechanism can be explained in the following steps:

- **Step 1**: Addition of ilmenite or rutile by adding with burden or injecting powder through tuyere along with pulverized coal-injection. Dissolution of TiO_2 in slag.
- **Step 2**: Reduction of titanium oxide into metallic Ti and incorporated into hot metal: TiO_2 (dissolved in slag) + 2C (dissolved in hot metal) = 2 CO (g) + Ti (dissolved in hot metal).
- **Step 3**: Reaction of metallic Ti with dissolved carbon, formation of TiC and reaction with dissolved N_2 to form TiN and precipitation in the worn-out zone in hearth wall:

  Ti (dissolved in hot metal) + C (dissolved in hot metal) = TiC (s).
  Ti (dissolved in hot metal) + N2 (dissolved in hot metal) = TiN (s).

Table 3.7 Tentative properties of injection materials.

Chemical composition %	TiO_2	Fe_2O_3	SiO_2	Particle size
Rutile powder	95–97	0.8–1.5	1.5–2.2	<75 (μm): 97%
Ilminite powder	35–50	30–45	2.5–5.5	<200 (μm): 60%

Precipitation of those titanium compound is increased when temperature lowers. In eroded area, effective carbon refractory thickness reduces and hence heat loss through wall increases due to external cooling, resulted in temperature drop in those zones and precipitation of TiC/TiN increases forming erosion protection layer over old carbon blocks.

Addition of ilmenite: It is a natural source of TiO_2-containing mineral, contains about 30% of titania. It charges with the burden from the top of the furnace. The quantity of ilmenite addition depends on the mode of application, addition of 3–15 Kg/thm for preventive mode of operation and 10–30 Kgs/thm for remedial mode. The quantity is also kept limited due to the impact of Ti in hot metal and that of TiO_2 in slag and the change of slag viscosity. Maximum Ti to hot metal to maintain 0.3% and TiO_2 in slag is 4.5%.

Addition of rutile: Rutile is a beneficiated powder that contains approximately 95% to 97% TiO_2 and 0.8–2.5% Fe_2O_3. This powder can be injected into tuyeres through injection lance. Rutile injection is intended to address localized "hotspot" in hearth side wall and elephant foot area. Tentative properties of TiO_2-bearing injection materials are shown in Table 3.7.

3.4 Consideration to Prolong Blast Furnace Campaign

The cost of rebuilding or reline a blast furnace is very high, as well as the loss of production for long period till the furnace blown-in and becomes operational. So, it is a challenge to extend the campaign of a highly productive furnace, as long as possible. Higher coal injection rate, oxygen enrichment and high production levels, typical of today's demanding blast furnace operation, further challenge the designers with higher heat flux in the bosh and lower stack region [19, 20].

A long campaign life is the result of a focused and well-planned effort. It is required to establish a realistic campaign life target, which is related to past operational experiences and then a strategy can then be developed. Campaign life is based on the performance and life of lining refractory, refractory condition of the hearth in particular. The lining life depends on optimizing the following factors:

- Design,
- Selection of refractory materials,
- Constructions,
- Operational psychology,
- Maintenance.

As the choice of initial refractory and cooling system lays the foundations in any campaign extension programme, the following sections are focused on describing the system which experienced a life extension from 8 to 10 years to more than 20 years in medium-size and large-size blast furnaces.

3.4.1 Designing Features

Nowadays, the life of hearth mainly determines the blast furnace campaign. Penetration of hot metal, oxidation and chemical reactions are the main factors for hearth refractory erosion. Thermo-mechanical and chemical attack can be slowed down or eliminated if the lining refractory is under low temperature, below the threshold temperature of chemical reactions and hot metal penetration. Hence, on the basis of heat transfer calculation and heat load and heat flux removal, blast furnace designers are continuously striving to improve the structure of hearth, bosh and stack. Blast furnace lining against steel shell is a circular construction. The design involves the understanding of the thermo-mechanical behaviour of the lining with optimum allowable shell temperature and minimum stress on the shell and refractory lining. In most of the cases, the structural or pressure-vessel grade of steel is used in blast furnace shell, temperature has to be maintained below 450 °C.

3.4.1.1 Design on Thermal Gradient

The parameter of thermal conductivity of a material or assembly of multiple layers of different material is the coefficient of thermal conductivity, λ. The measured value depends on material property and temperature. The unit for λ is W/mK. If the conductivity of a refractory material is measured with the share of radiation and convection also, the conductivity coefficient would be expressed as follows:

$$\lambda = \lambda_{\text{conduction}} + \lambda_{\text{convection}} + \lambda_{\text{radiation}} \tag{3.11}$$

At lower temperature conduction is predominant, and in high temperature, radiation. In the case of fluid in direct contact of refractory, usually the convection heat flow is not significant. The thermal conductivity of different material depends on the following factors:

- Temperature.
- Bulk density.
- Porosity and particle size distribution.
- Pore structure and pore size distribution.
- Chemical composition.

The various types of refractory have different thermal conductivity, depending on temperature of application and atmosphere. The conductivity value may also change during operation due to gas absorption, reaction layer formation from chemical reactions and liquid metal penetration into refractory and hence it is necessary to consider correction factor based on laboratory experiments and actual operating data.

3.4.1.2 Heat Conduction in a Multilayer Refractory Lining in Straight Wall

In blast furnace, usually we need to consider the following three or four layers for thermal calculation:

- Hot face scaffold formed at different height and location,
- Original refractory lining (in hearth carbon blocks and in bosh, stack, Al_2O_3–SiC).
- Back fill ramming mass, installed between steel shell and permanent lining.
- Steel shell.

The thermal conductivity of three layers can be calculated as shown in Fig. 3.19.

The equation of heat flow for each of the three layers, which is same for all three layers in steady state, is correspondingly as follows:

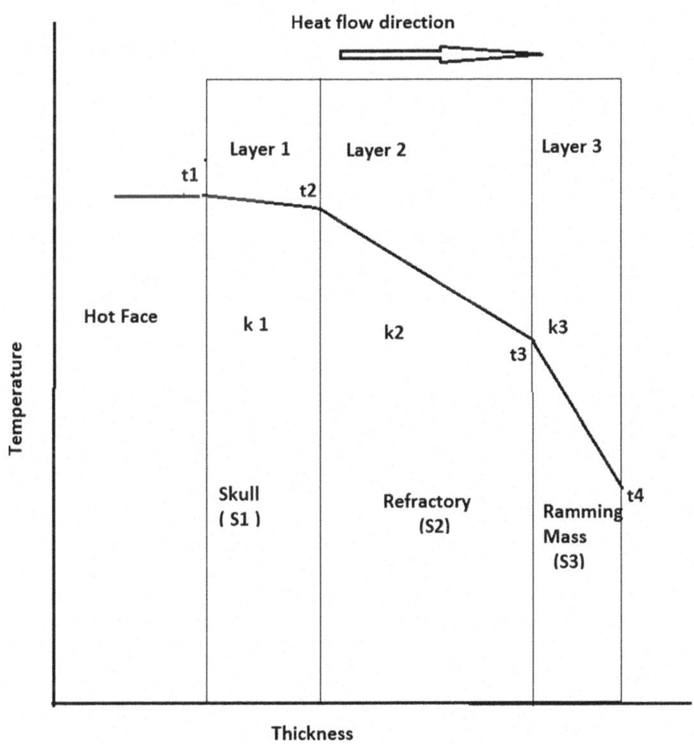

Fig. 3.19 Heat transfer through flat surface wall

$$Q = (\ t1 - t2), \quad t1 - t2 \frac{Q\ S1}{A\ k1} \tag{3.12}$$

$$Q = A\frac{k2}{s2}(t2 - t3\ ), \quad t2 - t3 = \frac{Q\ S2}{A\ k2} \tag{3.13}$$

$$Q = A\frac{k3}{s3}(t3 - t3\ )4, \quad t3 - t4 = \frac{Q\ S3}{A\ k3} \tag{3.14}$$

Adding of the above three equations provides the entire temperature gradient as shown:

$$(t1 - t2) + (t2 - t3) + (t3 - t4\ ) = \frac{Q}{A}\left(\frac{S1}{k1} + \frac{S2}{k2} + \frac{S3}{k3}\right) \tag{3.15}$$

$$\text{Or, } Q = A\frac{t1 - t4}{\frac{S1}{k1} + \frac{S2}{k2} + \frac{S3}{k3}} \tag{3.16}$$

where, Q = heat flux, A = Area, t_1, t_2, t_3 and t_4 are interface temperature, S_1, S_2 and S_3 are the thickness of different layers and k_1, k_2 and k_3 are the thermal conductivity of different layers. For large-size blast furnace, the side wall of hearth, bosh belly can be considered as a small section of flat surface wall and the above method of calculating heat flux can be applied. But, for small- and medium-size furnace with hearth diameter less than 8 m, the surface of heat flow may be considered as circular.

3.4.1.3 Heat Flux Calculation through Circular Wall

Heat flux through the circular multiple wall includes the following factors as shown in Fig. 3.20.

- Heat transfer from medium inside the furnace to inner brick surface.
- Heat conduction in various layers.
- Heat transfer from cold face shell to atmosphere.

For heat transfer inside furnace atmosphere,

$$Q = \frac{\pi D1}{a1}(t_a - t_1), t_a - t_1 = \frac{Q}{\pi\ D1\ a1} \tag{3.17}$$

For thermal conduction in the first layer from hot face,

$$t_1 - t_2 = \frac{Q\ \ln D2/D1}{\pi/2\ k1} \tag{3.18}$$

For thermal conduction in second layer,

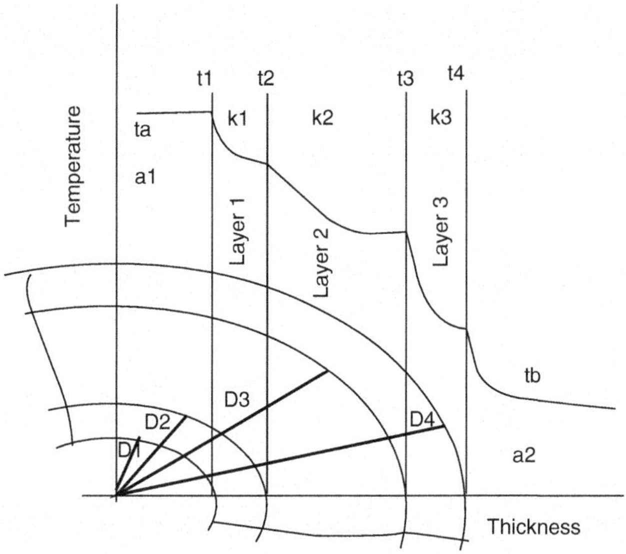

Fig. 3.20 Thermal gradient in a circular refractory construction

$$t_1 - t_3 = \frac{Q \ \ln D3/D2}{\pi/2 \ k2} \tag{3.19}$$

For thermal conduction in third layer,

$$t_3 - t_4 = \frac{Q \ \ln D4/D3}{\pi/2 \ k3} \tag{3.20}$$

The heat flow in medium 2:

$$Q = \frac{\pi \mathrm{D4}}{a2}(t_\mathrm{b} - t_4), t_\mathrm{b} - t_4 = \frac{Q}{\pi \ D4 \ a2} \tag{3.21}$$

$$t_\mathrm{a} - t_\mathrm{b} = \frac{Q/\pi}{R} \tag{3.22}$$

$$\text{R (Heat flux resistance)} = \frac{1}{a1D1} + \frac{\ln D1/D2}{2 \ k1} + \frac{\ln D2/D3}{2 \ k2} + \frac{\ln D3/D4}{2 \ k3} + \frac{1}{a2D4} \tag{3.23}$$

where, Q = heat flux; D_i, D_2, D_3, D_4 = outer diameter of different layers; a_1 and a_2 = heat transfer coefficients of the media; k_1, k_2, k_3 = heat transfer coefficient of different layers of refractory.

Table 3.8 Radiation coefficient of surfaces at temperature 0–200 °C

Type of surface material	Radiation coeff. (W/m^2 K^4)
Black body	5.77
Refractory brick surface	5.3
Construction material	5.3
Coloured layers	5.3
Steel with rolling or casting skin	4.5
Steel rusted	4.1
Copper oxidised	4.2
Aluminium coating	2.5
Steel, galvanised	1.5
Aluminium crude or oxidised	0.5

The heat transfer coefficient $a1$ on the hot side is usually given at >100 W/m^2 in refractory lining system. Its accurate value is not important for the precise results. For the cold side however, the value has to be evaluated and calculated precisely. It is based on the share of convection and radiation. The convection pert depends on the flow condition of hot air on the refractory surface or surface of steel shell. It is a complicated calculation method to determine the radiation share of media conductivity, $a2$, and without getting the value through calculation and other operational data the following data can be used to calculate radiation coefficient in W/(m^2·k^4) for temperature range of 0–200 °C, according to VDI 2055, as given in Table 3.8.

3.4.1.4 Design on Thermo-Mechanical Stress

A simplified numerical model is explained to show the behaviour of lining refractory and steel shell to develop stress at elevated temperature [21]. Several assumptions are made to reduce the numbers of variables to only those necessary to consider to explain the lining behaviour under stress. The assumptions are as follows:

1. The refractory material is linearly elastic.
2. The temperature gradient is linear in refractory thickness, therefore only steady-state heat flow is considered.
3. The mortar joints cannot support the tensile strength, this is true for mortar-less dry joints and use of mortar can support tensile stress in very small quantity.
4. Hot face of the refractory is under high compression and cold face attached to the steel shell is under tensile stress.
5. No expansion allowance is taken in consideration.

Typically as explained in Fig. 3.21, the refractory lining is experiencing a compressive thermal stress and the shell is experiencing tensile stress due to thermal expansion of refractory and shell. Redial crack develops at cold side of the lining as refractory weak under tensile stress. Behaviour of a circular lining with shell under thermo-mechanical stress is defined as follows:

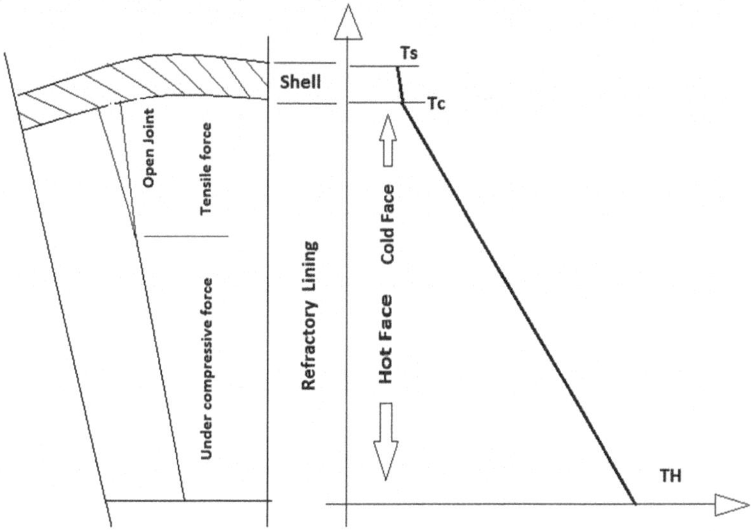

Fig. 3.21 Thermo-mechanical stress in circular construction

$$H_c = PR/t \tag{3.24}$$

(The ratio of shell radius to the shell thickness is more than 10, calculation is based on thin shell concept) where H_c = circumferential Stress (Hoop Stress), R = vessel radius, t = shell thickness and P = redial pressure on steel shell.

Dividing both side of the above Eq. (3.24), by modules of elasticity, (E).

$$H_c/E = € \text{ (hoop strain)} = PR/Et \tag{3.25}$$

The incremental change in peripheral length along the lining thickness ΔC is defined as follows:

$$2\pi R€ = \Delta C = 2\pi PR^2/tE \tag{3.26}$$

The radial displacement can be explained as follows:

$$\Delta R \text{ (radial displacement)} = \Delta C/2\pi = PR^2/Et \tag{3.27}$$

The above equation is the basis of evaluation of circular refractory wall lining design which consists of the refractory lining and steel shell. The shell expands radially and less than the lining so that opening of lining joints would be eliminated. The radial pressure (P) exerted by the thermal expansion of lining as well as the pressure of fluid inside the furnace is of equal magnitude on the shell and lining. However, this pressure causes inward radial displacement of the lining defined as follows:

$$\Delta R_L = PR_m{}^2/t'_L E_L \tag{3.28}$$

where, R_m = effective middle lining radius, t' = working thickness of lining and E_L = MOE of lining. Similarly, for steel shell,

$$\Delta R_s = PR_s{}^2/t_s E_s \tag{3.29}$$

where, R_s = radius of shell, t_s = thickness of shell and E_s = MOE of shell. The sum of the inward lining displacement and outward shell radial displacement ($\Delta\zeta$) is defined as follows:

$$\Delta\zeta_a = P\left(R_m{}^2/t'_L E_L + R_s{}^{2/} t_s E_s\right) \tag{3.30}$$

3.4.1.5 Thermal Expansion of the Lining and Shell

Radial thermal expansion of refractory lining (Δr_L) is defined as follows:

$$\Delta r_L = a_L\, R_m \cdot \Delta T_L \tag{3.31}$$

where, a_L = coefficient of thermal expansion of lining refractory, ΔT_L = temperature increase at the middle of the lining. Similarly, radial thermal expansion of the shell is expressed as follows:

$$\Delta r_s = a_s\, R_s \cdot \Delta T_s \tag{3.32}$$

where, a_s = coefficient of thermal expansion of shell at operating shell temperature (less than 350 °C) and ΔT_s = increase in temperature from room temperature. Since the expansion of lining would be more than the shell to keep the total lining thickness under compression, the resulting difference of the thermal growth in defined as follows:

$$\Delta\zeta_b = a_L\, R_m \cdot \Delta T_L - a_s\, R_s \cdot \Delta T_s \tag{3.33}$$

To satisfy both the force and displacement conditions, the internal thermal interference between the lining and shell is given as follows:

$$\Delta\zeta_b = \Delta\zeta_a \tag{3.34}$$

3.4.2 *Quality Upgradation*

In conventional refractory lining in blast furnace stack and bosh, the lining was made of 62–85% Al_2O_3 containing high alumina bricks. Since the twenty-first century, there was no brick work in those areas in newly built or rebuilt furnaces. The silicon carbide bricks with nitride bond or oxy-nitride bond embedded in stave coolers in stack and micropore carbon or graphite blocks are used in belly and bosh areas. TJ area was also lined with anthracite-carbon or micropore carbon blocks. Quality of carbon refractory for hearth side wall and furnace bottom and introduction of ceramic cup lining have increased the campaign of blast furnace to more than 20 years.

The new lining concept of "Permanent Lining" in hearth wall by formation of skull adherence to the hearth wall has prevented the refractory erosion due to brittle layer formation. The major development of refractory quality in blast furnace and its advantages is described in the following chapters.

3.4.2.1 Upgradation of Carbon Refractory in Hearth Lining

The corrosion resistance of carbon blocks has been improved based on the following understanding [20, 22]:

1. Penetration of liquid hot metal through pores,
2. Brittle layer formation in hot face of the blocks,
3. Decarburization in direct contact with molten iron.

The penetration was prevented through reduction of pore size by densifying the refractory structure introducing silicon-oxy-nitride whiskers grown in the pores of carbon blocks. To prevent decarburization, alumina is added to improve the high-temperature strength and resistance to mechanical wear. The thermal conductivity of the carbon blocks had also been improved. Types of carbon blocks used in hearth lining are summarized as follows:

- Micropore carbon.
- Super micropore carbon.
- Graphite.
- Hot-pressed carbon.

3.4.2.2 Micropore and Super Micropore Carbon

To prevent the penetration of molten iron, the micropore carbon has been developed. In those blocks, the pore size has been reduced from few mm to less than 1 μm by formation of Si-O-N whiskers within the pores. In addition, alumina is added in the block to reduce the interacting surface area of carbon with molten iron. Another fundamental development was brought by increasing thermal conductivity of carbon

Table 3.9 Development of hearth carbon (Ref: In-Service Performance of Micropore Carbon in Newcastle BF 3, by Barry Fraser, AISTech 2004 Proceedings – Vol 1)

Product	Carbon	Micor-pore	Super micro-pore	Ultra micro-pore
Year of introduction	1957	1978	1986	1994
Anthrasite carbon	80	80	40	0
Graphite (%)	20	10	40	76
Al_2O_3 (%)	0	5	6	11
Si (%)	0	5	9	12
Binder	Pitch	Pitch	Resin	Resin
Total porosity (%)	18	18	19	23
Thermal conductivity (W/mK)	17	13	23	33
Pore Dia <1 μm (% of total porosity)	14.5	85	94.7	99.1

blocks. In recent years, a new type of carbon blocks has been developed by mixing the high-conducting carbon with an optimum amount of oxides and silicon powder and using the resin as binder.

These developments were made to reduce the occurrence of embrittlement and to reduce localized erosion. Graphite was added to increase thermal conductivity and alumina to reduce dissolution of the matrix. Matrix dissolution was also reduced by developing resin bonded binders. Metallic silicon was added to reduce pore size. Most of the silicon forms silicon carbide; however, the remaining metallic silicon forms Si-O-N whiskers (Si_2ON_2) in the pores when post heat treated in air. Table 3.9 summarizes the development of hearth carbon. From this table, it is possible to note that the levels of silicon and alumina have been increased to produce super and ultra micropore materials.

The reaction of carbon with alkali is very complex. Ordinary grade of carbon blocks reacts easily with alkali and forms a "brittle" layer due to expansion, within the carbon block, where the temperature exceeds 800 °C. The expansion depends on calcined temperature of the blocks during manufacturing as shown in Fig. 3.22. Hence manufacturing parameters have a direct impact of properties of carbon. The better solution to oxidation and alkali attack is to use micropore carbon and super micropore carbon in critical areas of hearth lining.

Addition of artificial graphite in the presence of natural graphite to the mix recipe and SiC formation in material matrix during high-temperature backing in product processing. The oxidation of carbon lining by water and steam due to leakage of coolers is a problem in blast furnace operation, and the research work shows that the micropore carbon has better oxidation resistance than conventional carbon blocks.

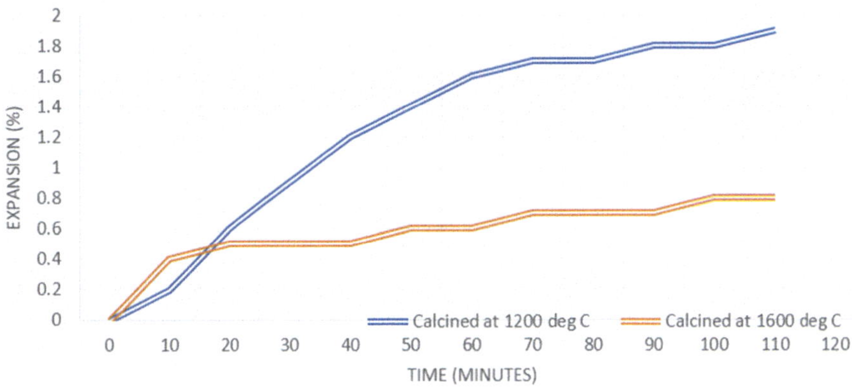

Fig. 3.22 Alkali expansion of anthracite carbon sample calcined at different temperatures

3.4.3 Monitoring of Refractory Condition

Blast furnace campaign life depends on the refractory condition of hearth, and visual inspection in bosh, belly and hearthside wall is not possible as those areas are filled up with burden, coke and liquids [23]. Condition of stack area can be estimated visually, after opening the top hatch covers. However, it is essential to understand the refractory condition in all critical areas to estimate the extended campaign and planning for future repair and relining. Hence, in-direct methods are adopted to estimate the remnant thickness of refractory in hearth and bosh area from outside of the furnace and even furnace in running condition. Nowadays an online monitoring system has been developed, where the hourly data can be generated to understand the refractory lining condition. The processes for monitoring condition are explained as follows:

3.4.3.1 Use of Duel Thermocouple

The process is based on the assumption of one-dimensional steady-state flow of heat. Using pairs of thermocouples with a known separation distance and heat flux can be calculated from the following equation:

$$Q = -\lambda \frac{\Delta T}{\Delta X} \tag{3.35}$$

where, Q = heat flux (W/m^2), λ = thermal conductivity (W/m°K or W/m°C), ΔT = temperature difference between pairs of thermocouples (°C or °K) and ΔX=distance between pairs of thermocouples. The refractory wear line is defined refractory/liquid of hot metal or the refractory/skull interface. This interface is defined as freezing point of hot metal at 1150 °C isotherm [24]. The liquid metal can penetrate up to the

Fig. 3.23 Schematic diagram of dual thermocouples

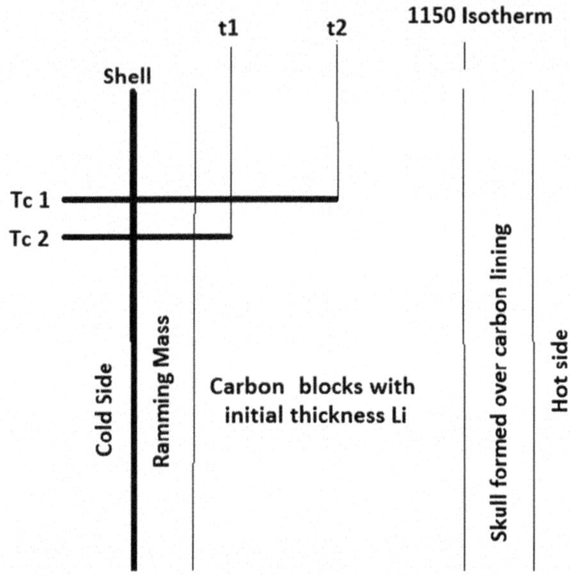

depth of refractory where this isotherm, and not beyond that line of isotherm. If the 1150 isotherm lies within the carbon block thickness, the new distance indicates the remnant thickness of the refractory. If the 1150 isotherm lies beyond the original thickness of hearth refractory, then the additional thickness denotes the thickness of skull, which protects the refractory lining.

Example 3
The schematic diagram shown in Fig. 3.23 shows the arrangement of duel thermocouples.

where Tc1 and Tc2 are the duel thermocouples with a temperature reading of $t1$ and $t2$. It is considered that 1150 isotherm is on the interface of refractory and skull.

Y = Additional distance from tip of Tc2 to 1150 isotherm;

Hence, under steady state,

$$Q = \lambda \frac{(t2 - t1)}{(x2 - x1)} = \lambda \frac{(1150 - t2)}{Y} \tag{3.36}$$

$$Y = \frac{(1150 - t2)\,(x2 - x1)}{(t2 - t1)} \tag{3.37}$$

Remnant thickness = $Y + x^2$; if the remnant refractory thickness ($Y + x^2$) is more than the original thickness of block, L1, it confirms the formation of protection skull layer. Hence,

Case 1: $Y + x2 < L1$: refractory is corroded, and no skull in front of refractory.

Case 2: $Y + x2 > L1$: Protective coating formed.

3.4.3.2 Acoustoultrasonicecho Technique

This unique non-destructive measurement technique has been developed on the fundamental theory on sound wave propagation, reflection from different media and resonance. It directly measures the residual thickness of the hearth wall refractory. The carbon block surface is acoustically excited by a sinusoidal force with frequency range from 1 to 5 kHz, applied by an electro-dynamic generator. The acoustic response of the echoes is received by an accelerometer. The residual thickness is determined by using the following relationship:

$$E = V/2t \tag{3.38}$$

where, V = sound velocity in carbon (2300 m/s); t = resonance frequency; E = residual thickness. This method has been extensively used in many blast furnaces worldwide.

A receiver placed beside the external surface of the shell and receives the signal caused by reflections from the internal and external interfaces of carbon and skull or carbon and liquid hot metal. The signals are transferred to a digital data acquisition system and are analysed both in the time and frequency domain. The critical factors to be considered are change in refractory thickness, temperature and sound wave speed. During propagation of sound wave through the carbon blocks, the partial or total reflection takes place when there is a drastic change in density and the elastic properties of material (Fig. 3.24). As the sound velocity in carbon blocks is higher than the skull of liquid hot metal, a total internal reflection takes place and the interface of carbon blocks and skull or liquid metal and the remnant thickness of carbon blocks can be calculated. In an operating furnace, refractories are subjected to severe thermal gradient across the thickness, which affect the wave speed inside refractory. As the temperature increases, the wave velocity decreases.

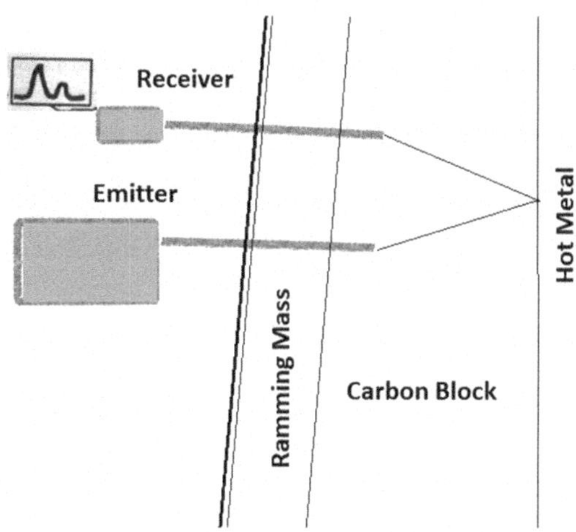

Fig. 3.24 Schematic diagram for acousticultrasonic test

Table 3.10 Properties of hearth refractory and hot metal

Properties	Density (gm/cc)	Thermal conductivity (W/mK)	Heat capacity (J/Kg/K)	Viscosity (kg/m/s)
Bottom refractory				
Carbon	1.56	12.5	700	NA
Graphite	1.35	80–800 °C	650	NA
Wall refractory				
Carbon	1.57	14.7	750	NA
Alumina	2.75	2.5	600	NA
Molten iron	6.9–7.0	46–49	910	−1.1

For measurement of remnant thickness of hearth bottom and side wall, the requirement of material's properties is shown in Table 3.10.

3.4.3.3 Anomalies in Defining Refractory Lining Thickness

The detection of remnant refractory thickness in hearth wall and bottom would be correct and reliable when the wave reflection takes place at the interface of refractory and skull or liquid metal, but in many cases the abnormal results are recorded, other than refractory thickness, received by the receiver. These reflections could result from several different phenomenon in the hearth lining. The anomalies in an operating furnace are as follows:

Cracking: when there are discontinuity or presence of cracks inside the refractory, the waves tend to reflect from the crack interface with higher frequencies. The pick frequency can then be used to detect the presence of crack and crack length.

Oxidation layer: In carbon hearth lining oxidation takes place when water is in contact with carbon lining due to water leakage in coolers. The degree of oxidation has a significant effect on the reflected signals. The reflected wave would show very low frequency when it is passing through oxidized layers.

Metal penetration: Metal penetration through the joints causes abnormal detection of lining thickness. The penetration of metal can be identified by the sonic detection method. However, the penetration cannot be detected if the metal fins is smaller than half a wave length of the acoustic wave.

Presence of brittle layer: Formation of a brittle zone into carbon and graphite refractory is commonly seen in a blast furnace and the block loses its bonding property, becomes soft. The reason for formation of brittle zone is due to reaction with alkali or liquid metal penetration. The presence of brittle layer can be detected by acoustic sonic modelling.

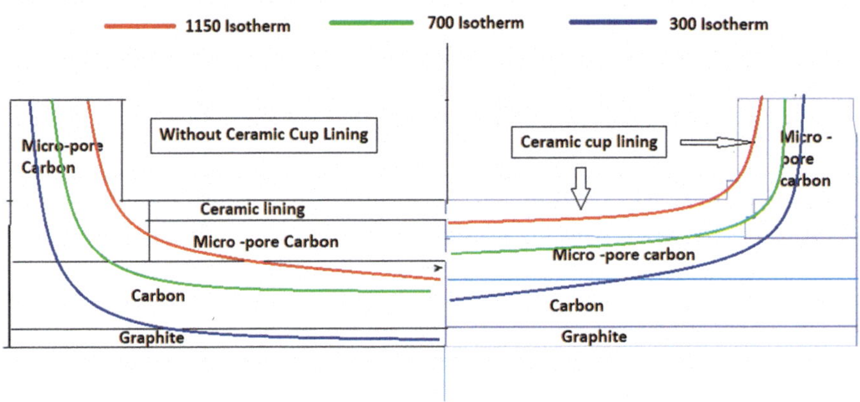

Fig. 3.25 Systematic diagram of hearth refractory with and without ceramic cup

Material Solution

Penetration of hot metal into the carbon block is one of the major problems for brittle layer formation, which results in hearth erosion and the hot metal penetration takes place when the temperature inside carbon blocks exceeds 1150 °C, which is the freezing point of pig iron. Above 1150 °C the pig iron containing 3–4% C remain liquid with very low viscosity, and it easily penetrates carbon blocks. Pig iron became solid at temperature below 1150 °C. So, it is a well-known technique to reduce the penetration by keeping the temperature of carbon blocks below 1150 °C, and it is possible by introducing sufficient forced cooling and an oxide ceramic layer over the carbon blocks so that carbon blocks would not come in direct contact with molten iron, and 1150 isotherm would be maintained within ceramic layer or this may remain within the first layer of carbon. The concept of lining ceramic layer is a material solution to obtain higher campaign life of blast furnace hearth. There are two types of lining concept, "ceramic pad" and "ceramic cup", as shown in Fig. 3.25.

A ceramic cup is an additional engineered lining approximately 300 mm thick in front of a modern carbon lining. This can be regarded as a "insulation oxide lining" to control the 1150 isotherm as well as uncontrolled heat extraction by external cooling, which may cause chilling of hearth. The use of a ceramic cup has met with considerable success in the last decade and has made a significant contribution to BF hearth design.

3.4.3.4 Design of Ceramic cup

The suggested refractory for ceramic cup lining is based on mullite, or corundum or corundum with chrome oxide for bricks or precast shapes. For last 50 years, the composite design of carbon hearth and combination with ceramic cup was standardized, and it is adopted worldwide. Lot of literatures tell about the experiences of using the composite lining as summarized below:

- The depth of wear in bottom and side wall is reduced, provided the oxide refractory withstand the corrosion by hot metal and slag and does not float.
- The ceramic cup lining over carbon blocks acts as an insulation, reduces heat flow through the wall and hence heat loss is reduced and temperature of liquid iron increased by 15–25 °C which results in reduction of coke rate as a cost-saving measure.
- A thermal calculation shows that as for ceramic lining in hearth the best ratio between ceramic cup thickness and carbon blocks lining thickness is 2 for medium size and 2.5–3 is for large-size blast furnace, for example:
  - 350 mm for ceramic cup lining
  - 800–860 mm for carbon lining.

3.4.3.5 Advantages of Using Ceramic Cup

- Improved resistance to hot metal erosion and oxidation resistance assured.
- Prevent hot metal penetration into carbon.
- Alkali resistance improved; brittle layer formation negated.
- Operating efficiency improved.
- A reduction in heat losses, which provide an increase in the temperature of hot metal.
- No need for ilmenite additions till ceramic lining exists.

Selection of Refractory Materials [6, 25]

Ceramic cup material made of oxide refractory has bulk density of 2.8–3.2 gm/cc, which is much lower that the density of hot metal, 6.9–7.0 gm/cc, so the refractory oxide layer may be dislodge and float by the upward thrust of hot metal during its turbulent flow. This problem can be solved by interlocking brick work in bottom and side wall and therefore the material properties recommended are as follows:

- Positive permanent linear change.
- Low thermal expansion.
- Adequate weight of the bricks and blocks, larger-size blocks are preferred.

Table 3.11 Properties of ceramic cup materials

Refractory materials	Mullite	Corundum-mullite	Corundum-SiAlON bonded	LCC (Precast)	Corundum-chrome oxide
Al_2O_3 (%)	74	85	70	90	87
SiO_2 (%)	20	12	Nil	6.5	0.5
Cr_2O_3 (%)	Nil	Nil	Nil	Nil	10
SiAlON (%)	Nil	Nil	15	Nil	Nil
Bulk density (gm/cc)	2.55	3.05	3.15	3.25	3.15
Permanent linear change at 1500 °C (%)	0.5	0.1	0.4	0.8	0.4
Thermal conductivity (W/mK) at 1000 °C	2.2	3.2	3.5	3.8	3.5
Thermal expansion ($\times 10^{-6}$ K^{-1})	5.5	7	5.2	7.8	8.6
Hot Mod. of rupture (N/mm^2)	5	6	18	8	12
MOE at 1000 °C (GPa)	15	16	20	30	35

Cracking Mechanism

Cracking of bricks may occur due to thermal gradient across the lining thickness and the shell induced hoop stress towards the hot face. Selection of refractory material depends on ability to absorb such stress of thermal gradient during preheating. The property may be explained by the following equation for crack initiation for calculating crack initiation factor, R' and higher the value is higher resistance to cracking.

$$R' = \lambda \frac{\delta\,(1-\gamma)}{E\,a} \tag{3.39}$$

where, R' = Crack initiation factor. λ= bending strength, δ = thermal conductivity, γ=poison ratio, E = Mod. of elastricity (MOE) and a = Thermal expansion coefficient. The metal temperature is usually above 1500 °C, hence material of high hot strength must be used. According to experience on material application, mullite bricks perform satisfactory in hearth bottom area. Therefore, its crack initiation factor, hot modulus of rupture, MOE and low thermal expansion may be taken as a reference for material selection. The material properties used in ceramic cup are shown in Table 3.11.

Example 4
The R' value of reference mullite bricks can be calculated as follows:

$$a = 5.5 \times 10^{-6}, \delta = 2.2\ \mathrm{w/mK}, E = 19\ \mathrm{Gpa, and}\ \gamma = 0.32, \lambda = 5\ \mathrm{N/mm^2}.$$

$$R' = \lambda \, \frac{\delta\,(1-\gamma)}{E\,a},$$

R' (W/m) = 5 × 2.2 (1–0.32) × 10,000/15 × 5.5 × 10.197; R′ = 88.9 W/m; [conversion detail, 1 N/mm^2 = 10 kg/cm^2, 1 GPa = 10,197 kg/cm^2]. Similarly, it can be calculated for other types of refractory and optimize the refractory selection.

Example 5
R′ for mullite-corundum bricks,

$$R'\ (\mathrm{W/m}) = 6 \times 3.2\ (1 - 0.32) \times 10{,}000/16 \times 7.2 \times 10.197; R' = 98\ \mathrm{W/m}.$$

3.4.4 Thermal Solution

There are two ways or approaches for influencing the chemical wear. These are referred to as the "thermal approach or solution" and the "materials approach or solutions". As it is described in Chap. 3.1, the processes of chemical wear are based on the temperature and the rate of every chemical reaction, and the rate decreases with falling of temperature and at a specific temperature level approaches zero for thermodynamic reasons. The temperature at which the rate of reaction tends to zero is known as lower reaction temperature. The objective must therefore be to cool down the brick temperature down to that temperature in order to bring the chemical reaction to minimum rate. The fundamental factors to be observed are shown below:

$$S = \lambda(T_u - T_a)/(T_i - T_u) \times (1/\alpha_i - 1/\alpha_a) \tag{3.40}$$

where, S = residual thickness of the refractory lining found under equilibrium condition, λ = thermal conductivity of refractory, T_u = lower reaction point, T_a = temperature of coolant media under equilibrium, T_i = furnace inside temperature, α_i = heat transfer coefficient of furnace inside, α_a = heat transfer coefficient of coolant.

For a predetermined operational condition, the value of T_a, T_i, α_a, α_i cannot be changed and under that condition, the value of λ and T_u can be increased to maximize the remnant refractory thickness S, at any point of time, and as those values depend on the quality of refractory, bricks of higher conductivity and high T_u would improve the refractory erosion resistance. It is also possible to reduce corrosion of refractory by reducing the temperature of coolant used in coolers. The T_u and thermal conductivity, λ of different types of refractories used in blast furnace lining is shown in Table 3.1.

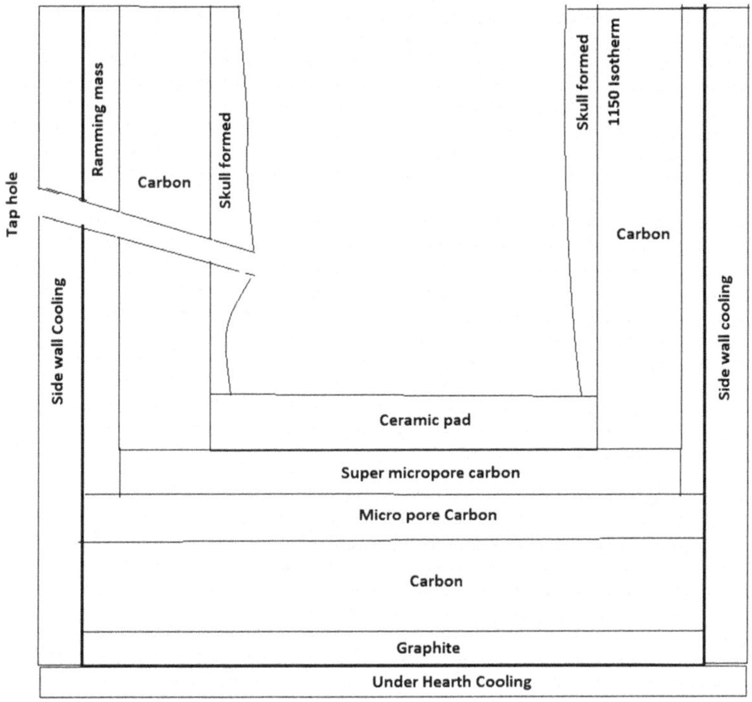

Fig. 3.26 Protective skull formation over hot face carbon

3.4.4.1 Formation of "Freezing Layer"

To protect the hearth refractory and approach to "end-less lining life" technique of freezing layer formation over hearth wall and bottom is an example of such thermal solution. The concept of freeze lining is successfully utilized for iron-making blast furnaces, direct reduction furnaces, electric arc furnaces and other metallurgical process lining applications. The theory is to combine effective cooling, with a thin, thermally efficient, conductive carbonaceous lining to provide a refractory hot face temperature that is significantly process metal solidification temperature. This in turn provide a chilled surface on which an insulating layer of process slag and metal is solidified, as shown in Fig. 3.26. When designing blast furnace hearth wall, the theoretical thinking predicts that the hearth wall would erode to the location of the 1150 isotherm. This philosophy is based on the fact that hot metal containing >2% carbon solidifies at 1150 °C, so cooling should be adequate to maintain this isotherm at the hot face of skull and the hot face of carbon blocks would be ideally less than 800 °C, which prevents alkali attack in carbon lining also [20, 23]. Hence, the main advantages of freeze lining by self-skull formation are the ability to reduce lining thickness, increase furnace inner volume without changing the diameter of shell, reduce capital cost and reduce refractory installation time. This is possible because instead of having to instal a thick mass of refractory to withstand wear, the freezing lining concept of a thermal equilibrium has to be achieved with protective slag and metal accretions on the lining hot face.

3.4.5 *Bottom Pad Cooling Layer*

The freeze lining concept for hearth bottom carbon lining includes the protecting ceramic pad along with bottom cooling. A layer of very high conducting graphite is utilized as the lowest carbon lining layer. This highly conducting layer helps to uniform distribution of heat and prevents localized over-heating in bottom shell. The average temperature at the graphite layer is maintained around 220–250 °C. This cooling layer of graphite also helps to cool the other refractories lined above it, thus the freezing line, 1150 °C isotherm, is maintained within the ceramic layer.

3.4.6 *Blast Furnace Repair Processes*

Building of a blast furnace involves high capital cost. Thus it becomes more popular to extend the campaign life, to reduce the iron-making cost. A longer campaign life can significantly lower the cost and increase productivity as downtime needed for repair is significantly low. It is difficult to prolong the campaign life by robot-gunning or shotcreting maintenance process, if the hearth is damaged or severe refractory erosion in stack, bosh and belly. Hence, campaign life has to be increased by relining or change of refractory in stack, if the predicted hearth condition is found in good condition to run the furnace for another equivalent campaign. Usually the life of stack refractory is less than the hearth refractory lining due to continuous improvement of refractory quality and design of hearth, so it is essential to change or repair the stack refractory.

3.4.6.1 Relining Process

The relining process involves the change of complete refractory in a blast furnace in "Like to Like" design. However, the relining is not an important activity to extend the furnace campaign life, also it is an opportunity to modernize the facilities of the iron-making process. The main objective is summarized below:

- To prolong campaign life,
- Upgrading the function and efficiency of the existing equipment by overhauling,
- To use upgraded quality of refractory,
- To improve environment and operation control.

For a successful relining of a furnace, it requires proper planning and consideration of modernization on existing operating parameters and utilization of downtime, to keep it as short as possible to improve productivity. Past experience is the key actor for such planning as summarized below:

- To identify what proved to be beneficial during the campaign can again be applied in the same or modified way.
- To identify the changes required to improve the next campaign and increase productivity.
- To prepare the pert chart indicating the critical activities and parallel activities. Relining period can be shortened by taking more activities in parallel.
- Safety measures for all activities.

3.4.6.2 Expansion of Blast Furnace Inner Volume

Blast furnace inner volume can be increased by increasing the inner diameter of hearth with the furnace shell to be kept no change in relining. From the operational experience, it was established that blast furnace with large in diameter and small in height is better than one large in height and small in diameter. Hence, it is possible to increase the hearth diameter by selecting conductive refractory with higher strength. The proportional increase in diameter is feasible in stack, belly and bosh areas by improving the cooling efficiency and selection of high-conducting refractory in silicon carbide or silicon nitride grade. Refractory thickness in stack can be reduced by changing the types of coolers for plate type to stave cooling. Thus, it is possible to improve the production of hot metal in the process of relining of the furnace.

3.4.6.3 Relining Activity

The main activities in refractory relining involve dismantling of old and relining of new refractory. During blowout of the furnace, it is required to drain out all metal and slag including major pert of dead man by melting. The process is called "Salamander Tapping". Satisfactory salamander tapping saves lot of time during dismantling process. This process usually started before blow-out of the furnace, when slag, hot metal and partially melted dead man are in liquid condition with low melt viscosity. Usually the tapping position is in second or third row of carbon block, and the carbon block is drilled with 10–150 up-word angle, and it is chosen according to the estimation of wearing line by computer aid, as shown in Fig. 3.27. Drilling has to be done with 80–120 mm diameter drill-bit, and the core drilling is done till the depth of inside hole temperature reaches 900 °C and then operation was changed to oxygen lancing till the salamander starts coming and collected in a refractory lined vessel through trough.

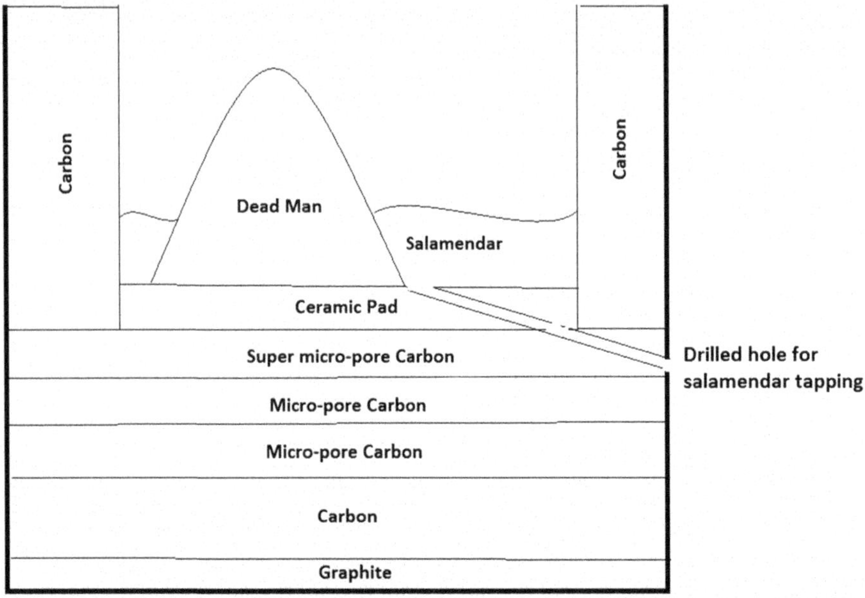

Fig. 3.27 Schematic diagram for salamander tapping

3.4.6.4 Dismantling of Old Refractory

A big opening is made around tap hole at cast house by cutting shell plate and removing refractory from outside of the furnace and racking of hot coke started from the opening. As a parallel activity, top dismantling of charging equipment has to be completed and after cooling of the furnace by water quenching, hanging scaffolding has to be inserted from throat of the furnace. Man can start dislodging of old refractory standing on the hanging scaffolding, and digging would be continued by descending and adjusting the position of scaffolding. After dig-out of old refractory up to below tuyere level, the handing scaffolding would be removed from throat a to contract a permanent platform in bosh level, as shown in Fig. 3.28.

Removal of old refractory in hearth is the most critical job, which requires experience and skills. Usually solid salamander inside the hearth is very hard, and it may require blasting of salamander and cleaning of big-size lumps and jams after blasting. Several numbers of holes have to be drilled into solid salamander, about 1500 mm depth, and optimum amount of explosive has to be filled into the holes for blasting.

Utmost safety precautions have to be followed for blasting operation. After removal of salamander, the carbon layers have to be removed and disposed periodically till the dismantling of refractory is completed.

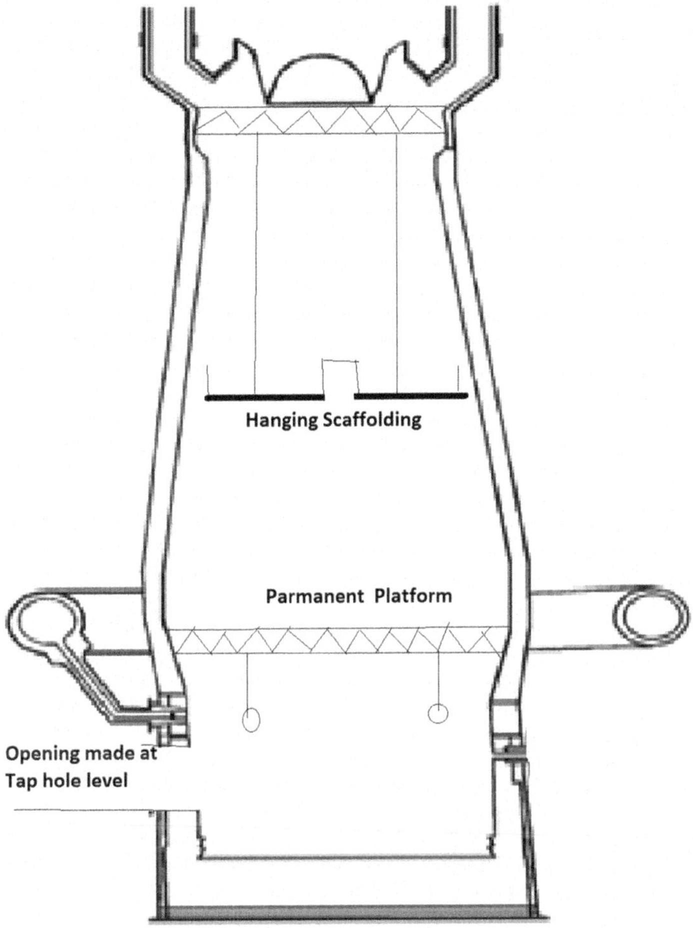

Fig. 3.28 Arrangement of platform and hanging scaffolding

3.4.6.5 Lining of New Refractory

The lining job in blast furnace has to be carried out in two stages. The primary job is to lay carbon blocks in hearth bottom and side wall, up to the tuyere level and the parallel job would start lining bricks from bosh, above permanent platform and to be continued up to throat. Jobs in both the stages would continue simultaneously and would be completed within same time. Hence, lining blocks in hearth area is critical activity and brick laying in stack is a parallel activity. For an example in a medium-size furnace with 7–11-meter hearth diameter, lining jobs in both the stages would be completed within 40–45 days.

Bottom blocks lining in hearth starts from the centre of the bottom plate and hence to find out the centre point in hearth pad is a very critical job. The centre point

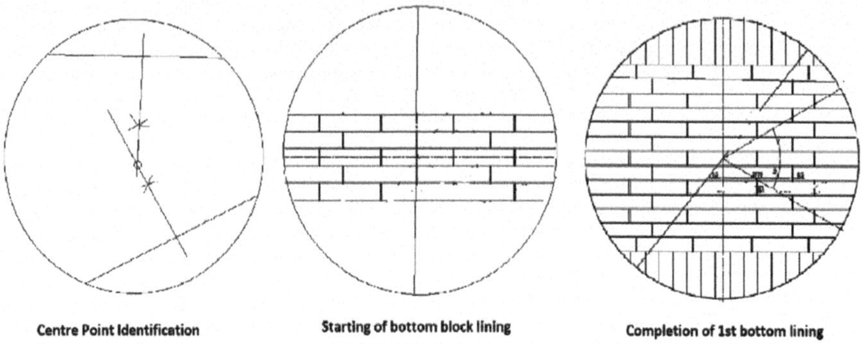

Fig. 3.29 Lining of hearth bottom blocks

Fig. 3.30 Arrangement of stack refractory

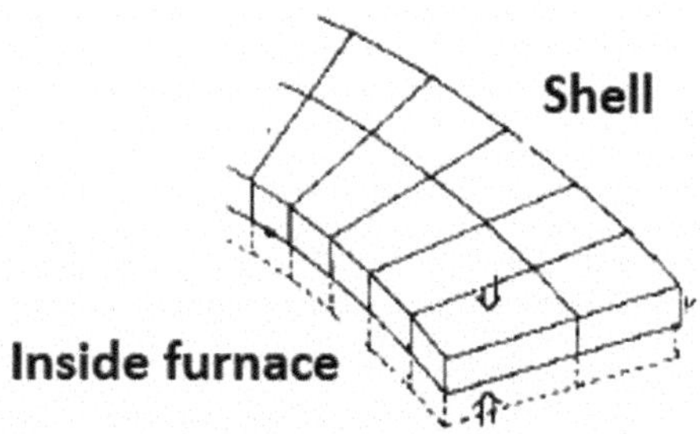

has been identified from three reference points: geometric calculation, reference from tuyere level and reference from stock line level at throat of the furnace. Geometric calculation is made by drawing two arcs in hearth bottom plate, to draw perpendicular and intersection of the perpendiculars refers to the centre point, as shown in Fig. 3.29.

After completion of first bottom, level has to be checked and second bottom would start by rotating the lining axis by 45^0 from the axis of first bottom and in the same design the lining would be continued up to the tuyere level. Tuyere blocks have to be assembled and lined with reference to the centre of each tuyere opening.

3.4.6.6 Brick Lining in Stack

Side arch bricks are used in stack, bosh and belly. They are joined by refractory mortar. The refractory mortar has a good deformability and an elastic modulus corresponding to the stress level of bricks. The deformability also depends on temperature accompanies large permanent deformation, which causes opening of the joints under stress in elevated temperature. Hence selection of mortar quality is of prime importance. Thickness of the joints also plays an important role on refractory performance. Proper thickness of the joints can reduce the thermal stress, especially compressive stress to less than the fracture strength. The thermal expansion of the brick is compensated by the deformability of the mortar in joint, then chances of joint opening or cracking of the bricks under compressive stress can be eliminated. In the stack lining such an arrangement is commonly adopted that the press forming direction of the brick corresponds to the vertical, as shown in Fig. 3.30.

3.4.7 Change of Stack Refractory

The blast furnace in small and medium size with carbon hearth lining has higher potential of hearth refractory compared to that of stack, bosh and belly. So, to extend the campaign life and improve productivity in economical way are to change the stack refractory without changing the carbon refractory. On taking such decision, it is required to evaluate the condition of hearth and if found that it has higher potential, decision for changing stack refractory can be taken. The major challenge to carry out this job is to protect the carbon hearth and to create a CO gas-free atmosphere so that men can enter into the furnace to work. In order to protect the carbon from physical damages and oxidation, the following measures have to be taken:

1. No force cooling by water spray. Natural cooling has to be followed or cooling by nitrogen in leak proof condition.
2. Carbon refractory has to be covered by coke bed.
3. Refractory capping over coke bed to prevent escaping of CO gas.

3.4.7.1 Gas-Free Atmosphere

To change the refractory in belly, bosh and lower stack area, the brick lining job has to be carried out inside the furnace and man has to work continuously. One of the main constrain is the presence of CO gas in hearth. It is a challenge to create a gas-free atmosphere by capping the hearth. It is a process to cover the coke bed in hearth by granulated slag and refractory concrete. The process involves the following steps:

1. Emptying the furnace below tuyere level,
2. Allow to burn the remaining coke to minimize CO gas formation,
3. Charging granulated slag from furnace top and later from tuyere opening,
4. Pumping the high strength castable to cover the hearth and form an impervious concrete layer of 600 and 800 mm thick,
5. Curing the castable for 24 h,
6. Checking the CO gas level inside the furnace and confirm it below 30 ppm.

Fig. 3.31 After capping in hearth (Tata Steel, Jamshedpur)

3.4.7.2 Refractory Concrete for Capping

It is a medium cement high-strength castable with high strength at intermediate temperature, 800–1200 °C, and it has a special property of adhering to the surface of hot coke, granulated slag or sand. It does not crack after setting over the hot coke. The material is having the following properties:

Chemical composition: Al_2O_3 (%): 55–62; Fe_2O_3 (%):1.6–1.8; CaO (%):1.5–2.5.
Physical properties: Bulk density (gm/cc): 2.25; cold crushing strength (kg/cm^2): 650 at 110 °C/24 h, 550 at 600 °C/3 h, 700 at 800 °C/3 h, 850 at 1200 °C/3 h. Hot MOR (kg/cm^2) 55–60 at 1400 °C.

After capping, the concrete has to be allowed to cure for 24 h, gas level has to be checked, and when it comes down to less than 50 ppm, man can enter the furnace through tap hole opening and hanging scaffolding has to be inserted into the furnace for dismantling of old refractory. Relining of new brick would start from lower bosh area and gradually completed and locked the lining below stock shield plate. The capping process is shown in Fig. 3.31.

3.5 Cast House Refractory

Cast house refractory is consist of tap hole mixes and trough and runner lining materials. In the past, inexpensive low-grade materials were used as cast house refractories, but with increase in blast furnace size, high top pressure and productivity, they are being replaced by high-performance materials. The consumption of those refractories is also high, and it becomes an important factor in total refractory cost. The performance of those refractories depends on design of tap hole, taping of hot metal philosophy and hot metal temperature.

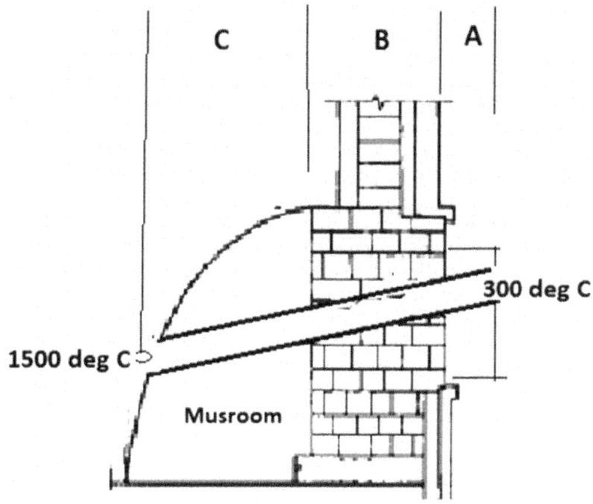

Fig. 3.32 Schematic representation of optimum tap hole length

3.5.1 Tap Hole Design

Tap hole angle plays an important role in tap hole mix performances. Earlier, in small-size and medium-size furnaces. The angle is kept between 10 and 15 deg., but with increase in furnace size the angle is reduced to 3–5 deg., which increases the taping rate, which has a direct impact on erosion and corrosion of tap hole mix. Corrosion resistance and maintaining tap hole length for smooth drainage with consistent liquid flow is the challenge for the day. The design of tap hole is shown in Fig. 3.30. After plugging the hole, the tap hole mix would experience the wide temperature range from 300 °C in outer side to 1500 °C of hot metal temperature, inside the furnace, and it has to possess equivalent thermal properties, i.e. all the temperature within 300–1500 °C. As shown in Fig. 3.32, the optimum tap hole length of a blast furnace can be calculated:

$$X = A + B + C; L = X/\mathrm{Cos}\ \theta.$$

where, A = thickness of tap hole spool box in m; B = refractory thickness at tap hole in m.

C = thickness of mushroom at tap hole elevation; Θ = tap hole angle; L = optimum length of tap hole has to be obtained after drilling the hole.

Example 6
In a 4600m^3 blast furnace, refractory thickness at tap hole is 1800 mm with 5^0 of tap hole angle, width of tap hole spool box is 800 mm and mushroom to maintain 600 mm as minimum for smooth furnace operation. What is the optimum tap hole length to maintain?

A = 600 mm, 0.8 m, B = 1800 mm, 1.8 m and mushroom thickness = 600 mm, 0.6 m and Tap hole angle = 5^0; Hence, L = (0.8 + 1.8 + 0.6)/Cos 5 = 3.212 m.

Fig. 3.33 Adherence test of Tap hole clay

3.5.2 *Tap Hole Clay and its Performances*

One of the basic requirements for stable tap hole operation is to maintain optimum tap hole length by forming stable mushroom inside the furnace hearth. In order to maintain stable mushroom, it is essential that the new clay mud pushed into tap hole must have good adhesion to existing old clay mud to form good mushroom around inside of tap hole opening. Innovative characterization technique was developed to compare the adhesion property between old and new clay mud of respective clay mud.

The cylindrical clay sample after preheated at 1450 °C under reducing atmosphere for 2 h, a 25-mm-diameter hole was drilled through the centre of sample. The central hole was filled with the new tap hole clay of same quality, preheated at around 70 °C. The whole arrangement was kept inside an electrically heated furnace, to fire at 1400 °C for 2 h under reducing atmosphere. The sample was then allowed to cool inside the furnace at room temperature, and then the cold crushing test was carried out.

After taking out the above sample from furnace, the sample was kept on cold crushing strength machine as shown in Fig. 3.33. Then the load is applied with the help of CCS machine through plunger, only on clay sample filled in drilled hole. Maximum load at which new clay mud starts to extrude out from cylindrically shaped old clay mud gives adhesion strength of the clay. The test arrangement is shown in Fig. 3.33.

Thermal expansion is a very important characterization of tap hole clay. The tap hole clay inside the tap hole is experiencing a variation in the temperature from 300 °C, at outside of the furnace shell, to more than 1500 °C, inside hearth. In general, for good tapping practice, the clay should have expansion characteristics in the abovementioned temperature range, and at above 1000 °C in particular. Due to unwanted shrinkage, normally spalling, bleeding and problem of self-opening, etc.

are experienced. For revealing the expansion/shrinkage behaviour, the reversible thermal expansion determination is recommended.

3.5.2.1 Wear Mechanism of Tap Hole Clay

Factors that affect the wear rate of tap hole clay are various and are closely related to the blast furnace operation practice, such as flow rate, temperature, chemical composition and other factors of hot metal and slag. The wear rate increases two to three times with increase in high top pressure. Of various wear mechanism, the influential one is corrosion by hot metal and slag. Porosity of the filled tap hole mix and degree of crack formation are related to wear. The wear rate of tap hole of large blast furnaces is generally estimated at 0.1 mm/min to 0.5 mm/min, but the enlargement of tap hole is not always linearly proportional to time. To make it wear-resistant, the setting and sintering properties need to improve. The tap hole mix to be capable for setting and sintering within the shortest possible time after pushing into the furnace.

The corrosion of tap hole mix depends on the following factors:

- Basicity of slag, basicity varies from 0.67 to 1.1, with changing the basicity, the melting point of slag and viscosity of slag also vary, which have a direct impact on erosion of tap hole.
- Consistant tap hole length and stable mushroom.
- Hot metal temperature.
- Casting rate, normally maintained from 4 to 7 tons per minute.
- Si content in hot metal.

3.5.2.2 Moisture and Volatile Matter

The durability of tap hole clay is affected by the moisture and volatile matter content. In the presence of high volatile matter, the mix becomes porous after pushing into the furnace and the presence of moisture oxidizes carbon and added silicon carbide, loosen strength. The greatest pert of the volatile matter content is attributed to coal tar. It is desirable that the low-temperature fractional distillation has to be as low as possible in binders. When subjected to heating, it increases the tap hole mix porosity and permeability and makes the structure brittle.

3.5.2.3 Checking the Properties of Tap Hole Clay to Ensure Performance

- Workability of the mix at different temperature.
- Pushability of the mix by determining Morgan Marshall Index (MI).
- Strength at high temperature range.
- Linear change at different temperature.
- Slag corrosion test.

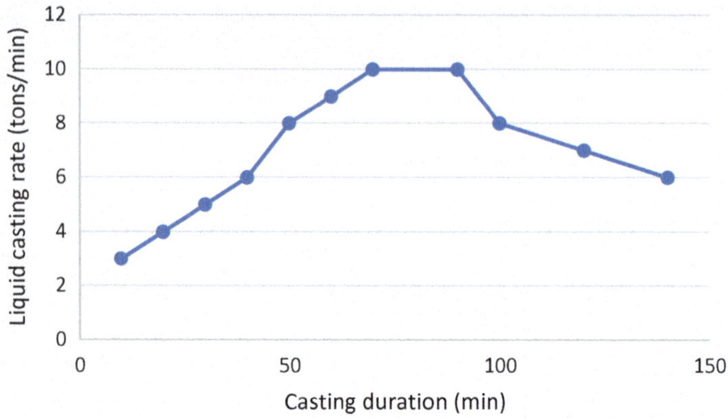

Fig. 3.34 Change of casting rate with cast duration

3.5.3 Hot Metal Trough and its Design

Iron troughs include the main trough, iron runner and slag runner, of which performance of main trough plays the most important role on cast house refractory performances [27]. The performance of main trough depends on the following factors:

- Design of main trough.
- Material selection.
- Installation processes.

The design of a trough depends on hot metal production per day, nos. of cast per day, slag and metal separation time, slag volume and hot metal temperature. For a purpose of illustration, a blast furnace is producing 9000 tons of hot metal per day; the furnace is operating two tap holes in alternate casting and nos. of cast per day is 10. Hence average casting rate would be 6.25 tons/min. Throughout the casting, the rate would vary, at the beginning the rate would be lower than average and after around 120 min, the rate would be as high as 10 tons/min and at the end of the cast, it is around 6 tons/min, as shown in Fig. 3.34.

As illustrated in Fig. 3.32, the iron trough would cope to drain out the hot metal with a casting rate of 10 tons/min. For designing a trough, the following factors have been considered:

- Casting rate: 10 tons/min.
- Residence time: it is experienced that the within 8 min of liquid flow, slag–metal separation takes place.
- Working capacity: Tonnage of iron content in the trough = Residence time × maximum flow rate = (8 × 10), 80 tons of iron.
- Working volume = working capacity/density of liquid iron. (80/6.9 tons m^{-3}) = 11.6 m^3.

- Slope of the trough: 1.5% (for non-drainable trough).

It may be assumed that for complete separation of slag–metal, the liquid would travel 18 m and hence length of the trough is 18 m. The average cross-section area of the trough would be 0.65 m^2.

Dimension of trough at skimmer: Depth of hot metal = % Slope × length of the trough = 1.5 × 18,000/100 or 270 mm.

It is required to maintain around 250 mm iron pool at tap hole side and hence at skimmer side, the metal depth would be 270 + 250 = 520 mm.

The average width at the slag–metal intersection $= \frac{\text{Working Volume}}{\text{Length} \times \text{average depth}}$, or $\frac{11.6}{18 \ x \ 0.52} = 1240$ mm.

Depth of slag: It is assumed that the slag generated is around 320 kg/THM

The total tonnage of slag in the trough at any point of time: 320 × 80/ 1000 tons = 25.6 tons

Slag volume: $\frac{\text{Weight of slag}}{\text{density of slag}} = \frac{25.6 \text{ tons}}{6.9 \text{ tons/m}^3} = 3.71 \, \text{m}^3$

Height of slag layer: $\frac{\text{Slag volume}}{\text{Length of trough} \times \text{width of trough at interface}} = \frac{3.71 \text{ m}^3}{18\text{m} \times 1.24 \text{ m}} = 167$ mm

To avoid over-flow of liquid during casting, it is recommended to maintain a free board of 350 mm over slag line.

Hence the total height of trough at skimmer = metal height + slag height + free-board = 520 + 167 + 350 = 1037 mm, and the all relevant dimensions would be as follows:

- Length of the trough up to skimmer block: 18 m.
- Width of the trough at slag–metal interface: 1240 mm.
- Total height: 1037 mm.

Other important dimensions are as follows:

- The inner wall of the trough lining to be maintained 4–8° with vertical line for smooth descending of liquid without adhering the liquid with metal and slag.
- Position of slag dam would be 150–250 mm above the metal dam.
- Refractory lining profile has to be designed properly such that shell temperature to be maintained below 150 °C during operating the trough.

Cooling of trough: It is a known fact that if the hot face of refractory can be kept at comparatively lower temperature, by means of external cooling, the reaction of hot face would be less. Cooling helps in bringing down the hot face refractory temperature and wear by way of chemical attack, infiltration, and thermal stress would be substantially decreased. It also reduces oxidation and de-carburization of carbon and SiC-containing refractory. The cooling may be designed as forced air cooled or water cooled. Erosion in the bottom of trough is very minimum, and it is severe in slag–metal and slag–air interface, so cooling panels have to be designed to protect the side walls.

Types of trough design: The type of construction of main trough plays an important role in deciding the trough life, selection of lining materials and minimize

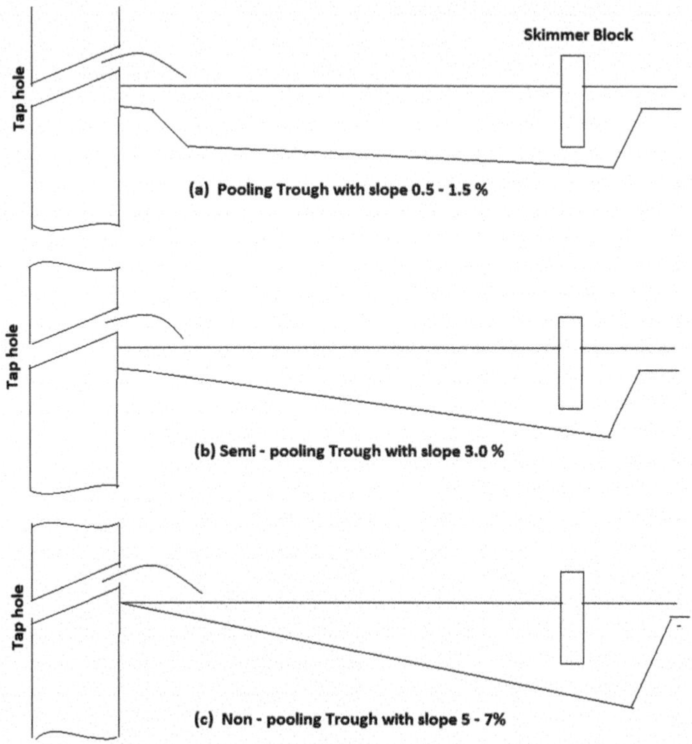

Fig. 3.35 Different types of trough

delay in cast house activities. Particularly depends on slope of the bottom, which determines the flow rate. Based on slope, the following three types of trough designs are available.

3.5.3.1 Pooling Trough

This is also known as non-drainable trough. This trough holds liquid hot metal for days together, till the trough is drained out for inspection of refractory lining condition. All large-size blast furnaces have opted this design for efficient slag–metal separation, faster evacuation of liquid, which leads to increase in productivity. The slope of such a trough is 1–1.5% and so the stream velocity on the bottom is very low, thereby the refractory erosion in bottom is not significant. However due to high velocity and turbulence, the erosion inside the wall is very severe. As the liquid pool is always maintained into the trough, the refractory lining is protected from severe thermal shock. High-performance, corrosion-resistant low cement castable performs satisfactory in this type of trough hot face lining. Shell of the trough is exposed in all sides and arrangements for forced cooling are feasible for long life. This design is very popular for double tap hole operation. The schematic diagram of the design is shown in Fig. 3.35a. The life of this type of trough is as high as more than 6–8 lakhs tons of throughput in a single campaign with one or two major repair and hence consumption is very low, 0.8–1.2 kg/THM.

3.5.3.2 Semi-Pooling Trough

Such a trough has high slope of 3.0% and so the wear is predominant in wide walls as well as bottom. Height of the pool is low at the impact zone and hence refractory wear at that area is also high. It requires frequent refractory repair at bottom and side walls. This design is suitable for medium-size furnaces for single tap hole operation. The design is shown in Fig. 3.35b.

3.5.3.3 Non-pooling Trough

The slope of this type of trough is maintained as high as 5–7%, depending on tap hole elevation and length of the trough and the stream of hot metal directly strikes on the refractory and hence the bottom erosion is severe. Side walls are not affected much. Such trough requires very frequent inspection and patching repair after almost every cast. This type of trough is suitable for small-size blast furnace with single tap hole operation. Severe thermal shock is experienced by working lining and use of castable is thus not suitable, it is lined with carbonaceous ramming mass, for easy to dig out the affected infiltrated area and immediate repair by patching with new ramming mass is used as working lining. Usually this type of trough is embedded into the cast house and arrangement for forced cooling is not feasible. Durability of working lining is very low and refractory consumption is very high. The design is shown in Fig. 3.35c.

3.5.4 Refractory for Hot Metal Trough and Iron Runners

In 1950s all non-pooling or semi-pooling type troughs were used in small-size and medium-size blast furnaces which were made of low-grade and low-cost materials and tar used as binder. Manual ramming was the predominant method of installation. The campaign life was very low and most of the blast furnaces were in single tap hole operation [28]. Later, the corrosion resistance and thermal shock resistance of ramming mass mix were improved by adding SiC, and resin is used as a binder. The hot strength of tar-bonded and resin-bonded ramming mass mix is shown in Fig. 3.36.

In the 1970s, with the introduction of larger-size blast furnaces, development of modern trough lining material started to use in pooling type trough. The new technology aimed at improved installation efficiency, extended trough lining life

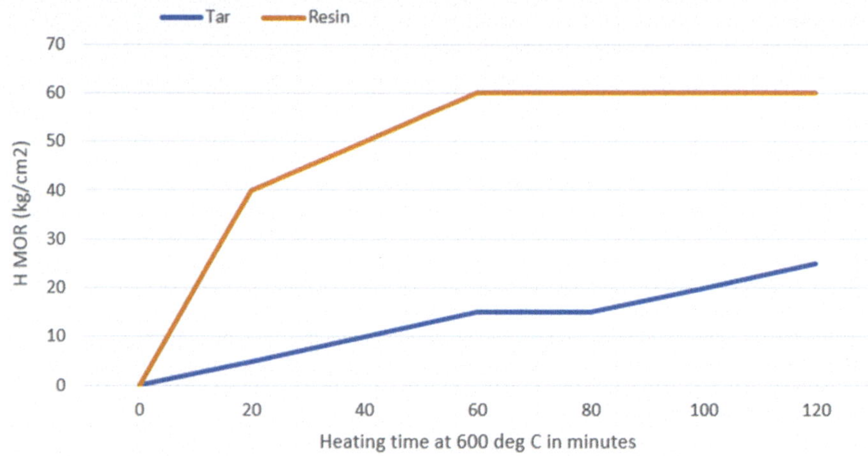

Fig. 3.36 High-temperature strength of tar- and resin-bonded ramming mass

Table 3.12 Tentative properties of trough lining materials

		Ramming mass	Castable		
Properties	Unit	Tar bonded	Resin bonded	Low grade	High grade
Physical					
Bulk density	gm/cc	1.70–1.90	1.70–2.0	2.4–2.6	2,5–2.65
Porosity	%	28–30	25–27	2.5–4	2.5–4
Compressive strength	kg/cm^2	90–120	220–240	NA	NA
At 110 °C/24 h	kg/cm^2	50–70	60–80	400–450	500–550
At 800 °C/3 h	kg/cm^2	60–68	160–180	300–420	480–500
At 1200 °C/3 h	kg/cm^2	120–180	300–320	500–650	450–500
At 1400 °C/3 h	kg/cm^2	150–220ss	350–400	600–650	600–700
Cold MOR	kg/cm^2	28–35	40–45	80–120	80–120
Hot MOR at 1200 °C	kg/cm^2	15–22	30–35	25–35	40–50
Chemical					
Al_2O_3	%	50–65	60–65	65–70	74–85
SiO_2	%	30–45	30–35	8.5–12	4.5–6
Fe_2O_3	%	3	2,5	1.5–2	0.8–1.2
CaO	%	0.5	0.5	2–2.5	1.5–2
SIC	%	Nil	4.5–6.5	4.5–8	12.5–18
C	%	8.5–12.5	10.5–12	3.5–4	5.5–6
Volatiles	%	13–18	6.5–9	NA	NA

and to improve cast house safety and cleanness. Carbon-containing low-cement castable had been developed with addition of metal powder and SiC to improve hot strength, thermal shock resistance and corrosion resistance. The tentative properties of different qualities of trough lining materials are shown in Table 3.12.

3.5.5 Wear Mechanism

Nowadays the non-drainable-type troughs are used predominantly in large size blast furnaces and are lined with high-grade castable. In backup, lining bricks or low-density lower conductive castables are used to keep the shell temperature below 120 °C during trough operation. No significant wear is observed in iron runners or in slag runners. The main wear is observed in metal–slag interface and in slag–air interface. The influence of major factors to the trough wear is explained below:

- Profile and critical dimensions of main trough,
- Drying and preheating conditions of castable lining,
- Fluctuation of temperature during operation,
- Temperature, chemical composition and casting rate of hot metal and slag.

The apparent wear mechanism involves mechanical wear by flow of liquids and dissolution of working lining material in hot metal and slag. The main wear mechanism are as follows:

- Infiltration of liquid through decarborized porous layers.
- High-temperature reaction with slag, presence of FeO and MnO influences slag corrosion and oxidation.
- Position of 1150 isotherm.
- Oxidation.

Non-oxide components in trough lining castable are carbon and SiC, and they oxidizes in the presence of oxygen, FeO and MnO according to the following reactions:

- $2C + O_2 = 2CO$
- $2CO + O_2 = 2CO_2$
- $SiC + 2CO = SiO_2 + 2C$
- $2FeO + SiC = SiO_2 + 2Fe + C$
- $2MnO + SiC = SiO_2 + 2Mn + C$

3.5.6 Modern Refractory Practices

Properties required of trough materials include the followings:

- Good workability.
- Uniform high-density structure and high strength at operating temperature,
- Resistance to hot metal and slag.
- High temperature volume stability.
- Oxidation resistance and thermal shock resistance.

The most trough material nowadays used are C–Al_2O_3–SiC–(Si_3N_4)-type materials. The type of castable used are low-cement castable or ultra-low-cement castable. The recent development is gel-bonded cement-free castable. Conventionally one type of castable is used in both metal zone and slag zone. One of the significant advantages of this method is the easy installation. However, the corrosion mechanism of metal zone and slag zone is different, hence zonal lining concept had proven the escalating improvement of refractory performance and enhancement of campaign life.

3.5.6.1 Use of Antioxidants

One of the major problems for carbon-containing refractories is oxidation of non-oxide addition, such as carbon. To protect the castable from oxidation, few additives are added as anti-oxidant. Anti-oxidants get oxidized and reduce the partial pressure of oxygen available to oxidize the carbon bonds. Metallic silicon, aluminium powder or non-metallic carbides such as SiC and B4C were used as antioxidants. Aluminium (Al) is one of the most extensively used antioxidants in refractory compositions. Aluminium reacts with graphite flakes to form Al4C3 at a temperature of 650 °C:

$$4\mathrm{Al}(s) + 3\mathrm{C}(s) \rightarrow Al4C_3\ (s)$$
$$6\mathrm{Al(s)} + 3\mathrm{CO(g)} = \mathrm{Al_4C_3} + \mathrm{Al_2O_3}$$

The silicon metal reacts with the carbon to form silicon carbide and reacts with CO or directly comes in contact of air, it oxidises to SiO(g) and later SiO is further oxidised to SiO_2 and forms a layer of fine silica over castable surface, prevents further oxidation. The reactions are shown below:

$$\mathrm{Si(s)} + \mathrm{C} = \mathrm{SiC}$$
$$\mathrm{SiC(s)} + \mathrm{CO(g)} = \mathrm{SiO(g)} + 2\mathrm{C}$$
$$2\mathrm{SiO(g)} + \mathrm{O_2} = 2\mathrm{SiO_2(s)}$$

3.5.6.2 Effect of Spinel Content on Refractory Castables

Spinel improves the corrosion resistance of the material. Spinel does not dissolve into the penetrated slag as spinel is stable in the presence of blast furnace slag. In situ spinel formation in the castable matrix, by adding magnesia to the high-alumina composition, could result in finer spinel and a better distribution throughout the refractory microstructure, increasing slag penetration and corrosion resistance. Pre-formed spinel castables present high volumetric stability and do not have the problems like MgO hydration and expansive behaviour at high temperatures like in situ spinel. Preformed spinel is added to alumina castables for two major purposes, to increase the slag resistance and to improve the thermomechanical properties.

3.5.6.3 Effect of Silicon Carbide

Silicon carbide is introduced into the trough castable to enhance the thermo-mechanical properties of the composite under operating conditions. Slag resistance of the castable significantly increases with the increase in the amount of silicon carbide due to the low wettability and low reaction of silicon carbide. Silicon carbide possesses the advantage of inhibiting sintering shrinkage and increases the thermal conductivity of the refractory material. In metal line, the temperature varies in a small range during service condition, for which no spalling damage occurs.

Addition of higher amount of SiC is not effective as it oxidizes in the presence of FeO and MnO,

$$SiC + 2FeO \rightarrow SiO_2 + 2Fe + C$$

Slag resistance of the castable increases with increase in silicon carbide content. Increase in non-oxide components in the batch reduces the chances of direct bonding among alumina particles, thereby reducing the sintering and densification, consequent upon which strength will decrease. Instead of using one kind of mix for the trough, it is always better to use two kinds of the mix and low SiC-containing castable for the iron line, and high SiC-containing castable in the slag line. Usually 8–12% SiC is added in metal zone castable and 18–30% SiC is added in slag zone.

However, the high oxidation of SiC by FeO is the main cause of refractory corrosion in metal zone. Thus, one alternative to reduce this effect consists of adding MgO or $MgAl_2O_4$ as a component of trough castables, which could provide a higher FeO corrosion resistance. On the other hand, spinel accelerates the oxidation of SiC.

3.5.6.4 Use of SiO_2-Sol as a Binder

The use of cement-free castable has brought in significant development to improve productivity in terms of campaign life improvement and quicker installation. Presence of cement causes formation of low melting compound with alumina and silica at operating temperature and limits the refractory life, which requires minor and major repair of the trough. Use of gel-bonded castable replaces cement bond system by SiO_2-sol. The SiO_2-sol is the colloidal form of silica; colloidal SiO_2 particles gel around the refractory particles of the castable. Colloidal silica particles can be linked together using different setting mechanisms, such as gelling and coagulation, providing initial strength when applied to castable systems. Concerning gelling, the interaction of silica particles converted to the siloxane bonding (Si–O–Si) and the build-up of a three-dimensional network, which provide strength at intermediate temperature. Hence, a gel-bonded castable is basically a cement-less castable that develops initial strength through gelation of the sol, thus the absence of CaO allows a better ceramic bond in the castable without formation of low-melting phases. A gel-bonded castable forms the final fired matrix through reaction between gel

Table 3.13 Properties of trough lining refractory materials

Properties	Unit	Working lining castables Metal zone	Backup lining Slag zone	Gel bonded	Bricks	PCPF blocks
Chemical						
Al_2O_3	(%)	60–65	60–65	65–70	55–62	65–74
Fe_2O_3	(%)	0.8–1	0.8–1	1.0–1.8	1.5–2.5	0.8–1.2
SiO_2	(%)	3.5–5	3.5–5	4.5–5.5	30–35	25–30
CaO	(%)	1.5–2	1.5–2	0.6–0.8	0.4–0.8	
SiC	(%)	8.5–12	25–30	12.5–18	Nil	20–30
Physical						
BD	gm/cc	2.6	2.8	2.6	2.2–2.4	2.8–2.85
CCS at RT	kg/cm^2	NA	NA	NA	450–500	750
At 110 °C/24 h	kg/cm^2	350–400	320–350	300–350		
At 1000 °C/3 h	kg/cm^2	450–500	400–450	300–350		
At 1600degC/3 h	kg/cm^2	400–450	400–450	600–750		700
PLC at 1000 °C/3 h	%	−0.1	−0.2	0.5		
PLC at 1600 °C/3 h	%	0.8	1.4	−0.2	−1.5	0.6
Thermal conductivity at 600 °C	W/mK	3.5–4.5	6.5–7.5	4.5–6.5	1.8–2.2	3.1–3.3

particles and reactive/finer components of the castable at elevated temperature. As gelation is a very faster process without any hydration reaction like cement–water bond system, no time required for curing as well as faster heating is possible, which helps for shorter downtime for repair.

3.5.6.5 High Performance Backup Lining

Usually the backup lining in trough is lined with high alumina bricks or castable. The backup lining was suffering from low strength, poor corrosion resistance and thermal shock resistance. Hence, for any failure of hot face Al_2O_3–C–SiC castable, break out in trough took place, as those back up lining materials were unable to protect metal penetration or slag infiltration. In order to improve the situation a new type of backup lining material is introduced, called PCPF (pre-cast and pre-fired) blocks of larger size. The composition of the blocks is very close to hot face castable material with lower conductivity. The main advantages of using those blocks are summarised below:

- Backup lining contain >15% SiC to improve corrosion resistance,
- Use of large-size blocks reduced nos. of joints,
- High hot strength,
- Excellent thermal shock resistance,
- Over-lapping construction prevents leakage of hot metal and slag,
- Helps to maintain 1150 isotherm within working lining under forced cooling, prevents metal infiltration.

The properties of high-performance lining materials are shown in Table 3.13.

References

1. Interceram, Special Issue 1983. Vol. 32.
2. Proceedings XXVth International Colloquium on Refractories, Aachen, Germany 1982.
3. A.J. Dzermejko, D F Baret et al., Chapter 4: Iron Making Refractory systems, in "The making shaping and Treating of Steel", 11th Edition, Iron Making volume, AISE, Pittsburgh, 1998.
4. Monitoring and control of hearth refractory wear to improve blast furnace operation, by R.M. Duarte, I. Ruiz-Bustinza, D. Carrascal, L.F . Verdeja, J. Mochon and A.cores. Published in Iron making & Steel making, July 2013.
5. R. Hood, The theory and practice of blast furnace hearth management at Corus Scunthorpe, AISTech, USA 2007.
6. Miyamoto M, Onoye T, Anarita K, Effect of Alkalies and Zinc on the wear of blast furnace refractories and tuyere displacement, Transaction ISIJ, Vol 21 pp 839–845 (1981).
7. P. Sylven, The Truths and Myths of Freeze lining Technology for blast furnace, the Moscow Institute of Steel and Alloys, Moscow. Russia 2004.
8. M.J Venturini, M Nogues et al. "New Developments for extension of Blast Furnace campaign life: monitoring the refractory wear and controlling the liquid flow pattern in the hearth" 3rd European Inro Making conference, Belgium, September 1996.
9. Masami K, Nobuyuki Nagai, Teruhisa U, et al. The On-line monitoring method of lining Erosion in the blast furnace, Iron Making Conference Proceedings (511–515).
10. Heshun Zhang, Research on Dead man state during tapping process, Iron Making & Steel making, Vol. 45, no 7 2018.
11. Shibata K, Kimura Y, Shimizu M. et al. Dynamics of Dead man coke and hot metal flow in a blast furnace hearth. ISIJ Int. 1990 Vol 30, no 3.
12. R. Van Leer, K. Andreev, Blast Furnace Bottom and hearth expansion Consideration, AISTech, USA 2007.
13. Zhao Hong-bo, Cheng Shu-sen et al. Analysis of all carbon brick Bottom and Ceramic cup systhetic hearth bottom, Journal of Iron and Steel research, International, Vol 14, No 2 2007.
14. Kenji Takeda, et al. Prevention of Hearth brick wear by forming a stable Solidified Layer, Iron Making Conference Proceeding, Chicago, USA, 1999.
15. Emad Mohamed M. Ewais, "Carbon Based Refractories, J of Ceram Soc, Japan 112 [10] 517–532 (2004).
16. R. Gronebaum, J. Pischke, "Investigation of Shot-cast mixes for blast furnace relining" 51th International Colloquium on Refractories.
17. Kasai K, Recent Advances in refractory technology for steel making, Nippon Steel Technical Report, Vol 61, pp 83–89 (1994).

18. Bao-Yu Guo, Paul Zulli, et al., "A Model to Simulate Titanium Behavior in the Iron Blast Furnace Hearth", Metallurgical and Materials Transactions B, August 2010.
19. Itotagawa S, Hashimoto S, et al. Taikabutsu Overseas, vol 45 no 2, pp 38–41 (1995).
20. Janusz T, Stefan Basista, SGL Carbon Group, Technical report.
21. Assessment of Refractory Condition in Blast Furnace Hearth using Computational Fluid Dynamics by Bryan WRIGHT, Paul ZULLI et al., Blue Scope Steel Research, published in 3rd International Conference on CFD, CSIRO, Melbourne, Australia, December 2003.
22. Michio Nitta, Hitoshi Nakamura et al. "Development of Carbon blocks for Blast Furnace Hearths", Nippon Steel Technical Report No 98, July 2008.
23. A. Sadri, R. Timmer, "Blast Furnace non-destructive testing (NDT) for defect detection and refractory thickness measurement ", Association for Iron & Steel technology, AISTech Annual conference, Vol 2, May, 2006.
24. M. Zagaria, V. Dimastromatteo, V. Colla, "Monitoring Erosion ang skull profile in Blast Furnace Hearth", Iron Making & Steel Making, April 2010.
25. D. A. Gunn, "A theoretical evaluation of the stability of Sialon – bonded Silicon Carbide in the Blast Furnace Environment", Journal of European Ceramic Society 11 (1993) 35–41.
26. Chester J.H, Refractories production and properties, 1973
27. Devid P, Joseph S, Trevor S, Sheffield Refractories Ltd, England, "Basic Principles of Blast Furnace Main Iron Runner design", Proceedings on International Conference on Refractories for Iron and Steel Industry, ICRJ, India, Nov 2008.
28. Cheng W, Zhongyang He, et al. Puyang Refractories Group Co Ltd, AISTech 2011 Proceedings, Vol 1, pp 767–772.

Chapter 4
Hot Stove and Hot Air Carrying System

4.1 Introduction

The hot stove is a thermal regenerator with the purpose to provide hot air, called hot blast, to the blast furnace at high temperature with constant flow. Regeneration is the term employed to describe the practice of using the waste heat normally available in the waste products of gaseous combustion. It is used for raising the temperature of fresh incoming combustible gas or air, and thus increasing the temperature of fresh air, at which combustible takes place. The efficiency of increasing the temperature depends upon the degree to which the constituents are regenerated.

Operation of stove is cyclic, divided into two periods, on-gas (or firing, when checkers are heated) and on-blast. During on-gas fuel gas is combusted to heat up the check work and increases the temperature of the checkers. The fuel gas used in hot stove is BF (blast furnace) gas, with an enrichment gas. The enrichment gas may be coke oven gas, liquefied petroleum gas (LPG) or natural gas. After on-gas the stove is switched to on-blast. On-blast is the period when pressurized air called cold blast is flown through the hot checkers and got heated. The switch between on-blast to on-gas takes time and this period is referred to as change over time.

The conception of the idea of using preheated air was due to Neilson in 1824. The "Mechanics Calculator," by Wm. Grier, fourth edition, published in Glasgow, 1836, contains details of the work. The inventor was the engineer and manager of the gas works of Glasgow and his attention was directed to the blast furnace work in consequence to develop a devise any means to purify the blast before it reaches to blast furnace. The experience of Neilson, however, led him to conclude that the air will not support combustion until heated to a temperature of 550 °C (1000 °F) and hence preheating of air was found necessary. For preheating of air, Neilson used cast iron pipes for air flow and the pipes were heated by means of coal. It soon became apparent that the fuel used in heating the blast could be very well replaced by utilizing the large amount of heat which escaped out at the top of the blast furnace. Later the pipe stove was developed in Great Britain and Austria to preheat the air by

S. Biswas, D. Sarkar, *Introduction to Refractories for Iron- and Steelmaking*,
https://doi.org/10.1007/978-3-030-43807-4_4

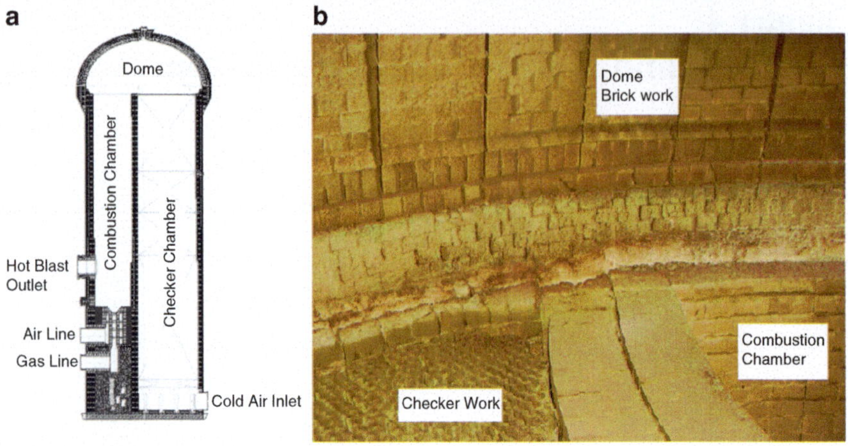

Fig. 4.1 (**a**) Hot blast stove and (**b**) refractory arrangement

utilizing blast furnace waste heat. The stoves were capable of dealing with 1700 cubic ft. of blast per minute at a pressure of 1.5 psi and maintaining any temperature up to 400 °C. The design of stove had undergone a continuous strive to increase the hot blast temperature by introducing modern hot stove with refractory checker works as regenerating system, and a modern stove can thus produce hot air of more than 1200 °C.

A modern stove consists of a mild steel shell, built to withstand the internal pressure and lined with refractory bricks. In plan, the interior is divided into a larger chamber and small chamber, by a cross separating wall, made of refractory bricks. The larger chamber is called "checker chamber", where refractory checkers are packed and the smaller chamber is called "combustion chamber", where combustion takes place. The blast is introduced into the combustion chamber, along with air for its complete combustion. The combustion is completed as the gases rise the combustion chamber, and they pass over into the large chamber at a high temperature. The chamber is filled with refractory checkers, arranged to form a series of channels, so devised that the maximum heating surface is exposed to the hot gases.

When the checker work is brought to the maximum high temperature, the cold blast is introduced below the hot checker work, and its passage through it, the cold blast is brought up to a high temperature by heat transfer through the checkers. The hot air then passes down to the combustion chamber and out of the stove through hot blast main to the furnace. The arrangement of a stove is shown in Fig. 4.1.

Hot Blast Carrying System

Hot blast carrying system transports hot air, typically at 1100–1250 °C or above, from stoves to the blast furnace. The system includes the following:

- Hot blast main,
- Mixer,
- Bustle pipe.

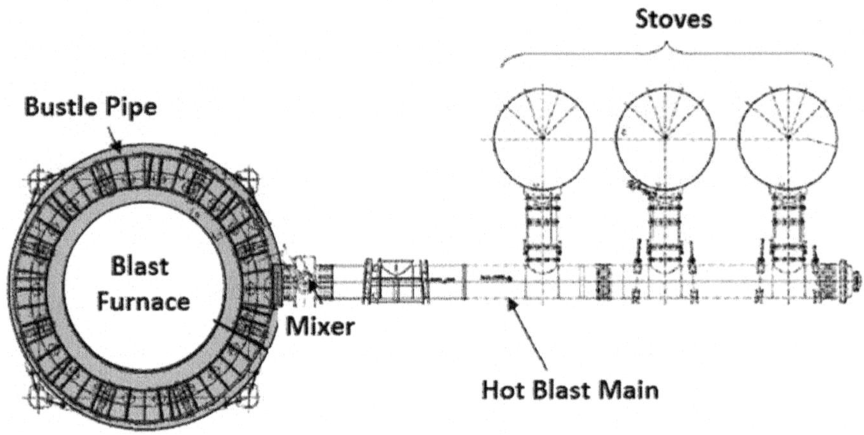

Fig. 4.2 Schematic diagram of hot blast carrying system

The hot blast main and bustle pipe are made of circular steel pipe, lined with refractory lining inside the pipe. The refractory lining is in multiple layers of different quality bricks and shapes, which protects the steel shell from overheating. The schematic diagram of hot air carrying system is shown in Fig. 4.2.

4.2 Design of Hot Blast Stove

With the improvement of blast furnace operation, the demand for larger hot blast stoves, higher hot blast temperature and higher blast pressure led to development of new design.

Basically, there are three different systems of hot blast stoves available:

- Hot blast stoves with internal combustion chamber,
- Hot blast stoves with external combustion chamber,
- Hot blast stoves with top combustion.

Within the three systems, there are several types further developed from different suppliers all over the world [1].

Hot Blast Stove with Internal Combustion Chamber

The internal combustion chamber is the conventional design of hot blast stove. This stove emerged around 150 years ago. In this design, both the checker chamber as well as combustion chamber are inside a circular shell, separated by a refractory partition wall, as shown in Fig. 4.1, and the combustion chamber occupies 10% of the total cross-sectional area of the stove [2]. The checker chamber is filled up by refractory checkers. There is an insulating lining just inside the steel shell. The combustion chamber is completely surrounded by a refractory wall, in old design, this wall was lined with super duty refractory bricks containing 50–60% of alumina

[2]. In modern design, for very high hot blast temperatures, in excess of 1100 °C, the entire combustion chamber and the dome are lined with silica refractory. Also, the top 8 m to 10 m of checkers are normally of super duty silica bricks.

Within the recent years, hot blast stoves with internal combustion chamber have been drastically improved in capacity and life campaign. The improvement is in the design and quality of refractory used in separation wall, dome and burners. In early design, the metallic burners were used, and the flame was directly heating on the combustion chamber separation wall, and caused for several refractory failures. In modern design, the ceramic burner is introduced, which improved the heating efficiency and campaign life of the stoves. However, the campaign life of this stove shortens due to the following disadvantages:

- "Short Circuit" or direct leakage of gases and hot air between the combustion chamber and checker chamber through cracks and brick joints in the partition wall. This may lead to a drop of hot blast temperature.
- Buckling of the combustion chamber wall in the direction of checker chamber, resulting from the temperature difference between the two chambers. It is also called "Banana Effect."
- Deformation and subsequent collapses of the combustion chamber lining, resulted frequent stoppage of stove operation and short campaign life.

Hot Blast Stove with External Combustion Chamber

European stove designers have found that the internal combustion chamber stove design is not suitable for dome temperature operation more than 1450 °C, even with the use of exotic refractory. The differential temperature between checker chamber and combustion chamber causes thermos-mechanical stress due to differential expansion of refractory on separation wall of the combustion chamber and failure takes place, which limits the life of the stoves. To overcome the difficulty, the stoves with external combustion chamber have adopted. Figure 4.3 shows such design in use today. The checker chamber and combustion chamber are separated into two different shells, connected at the top. In 1963, the Steel Company of Canada put into service an external combustion chamber stove, and at present the design is adopted worldwide. The stoves are performing satisfactorily with dome temperature more than 1450 °C and operating hot blast temperature of 1250 °C. The main advantages of this design are as follows:

- Enlarged heating surface of checkers,
- Dome temperature can be maintained more than 1450 °C.
- Complete combustion,
- High blast volume and pressure,
- Less refractory failure,
- High flexibility in operation,
- Long campaign life (more than 30 years).

Hot Blast Stove with Top Combustion

Compared to internal combustion and external combustion chamber stove design, top combustion hot blast stove has an advantage of obtaining very high hot blast

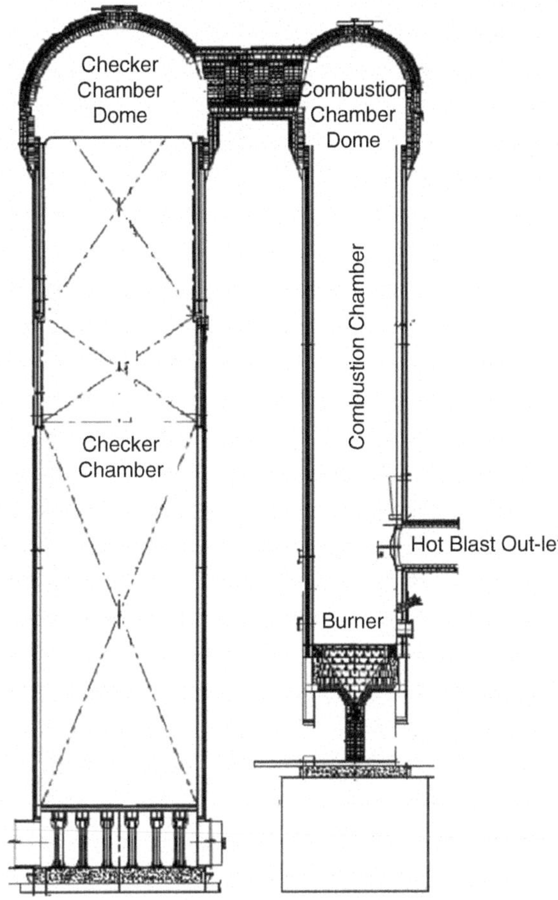

Fig. 4.3 Stove with external combustion chamber

temperature of more than 1300 °C [3]. Other features are high productivity, compact and symmetrical structure. Kalugin top combustion stove is a kind of top combustion stove, developed in the 1970s, which is widely used in the steel plants nowadays. The design of Kalugin stoves had undergone several changes for stable operation, robust refractory lining and prolonged life. The main feature of this kind of stove design is the placement of burners in the area of dome, and the outlet of hot blast is shifted at top of the stove and division wall in internal combustion chamber stove had been completely eliminated.

In this stove, gas is burnt in a small annular chamber located at the dome and having several zones of small ceramic burners installed at the lower portion of the dome. This stove provides complete combustion of gas and generates hot blast of temperature more than 1300 °C. The stoves of this design provide saving of capital expenditure and increased hot blast temperatures resulted in reduction of fuel consumption in large-size blast furnaces of working volume more than 5000 m^3

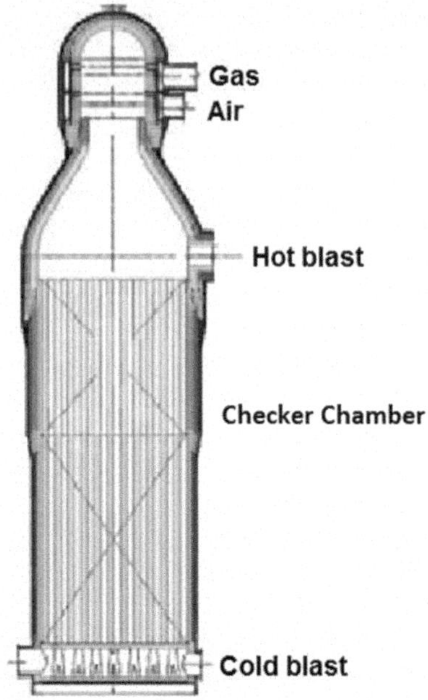

Fig. 4.4 Top combustion stove

[3, 4]. This design also improves environmental performance by reducing CO emissions by 50 mg/m^3. A schematic diagram is shown in Fig. 4.4.

Operator's experience of the top combustion stove shows the following advantages:

- A complete gas combustion is achieved.
- "Short circuiting" in such a stove is eliminated as there is no existence of combustion chamber.
- Hot blast temperature of 1400 °C can be achieved with commonly available refractory bricks and shapes in silica, fireclay, mullite and corundum refractory.

Types of stoves	Features	Advantages and disadvantages
Internal combustion chamber	• Dome refractory is supported on shell • Brick lined division wall between checker chamber and combustion chamber • Stove can be constructed in limited space • For small- and medium-size blast furnaces, alumina refractory can be used in checker and dome construction • In large-size furnaces, silica checkers and dome are used	• Cost-effective construction • For complete alumina-lined stoves with alumina checkers, repair is easy within short shutdown for faster cooling and heat-up of the stoves • Frequent refractory failure in division wall • Dome temperature can be obtained up to 1400 °C with silica refractory • Hot blast temperature cannot be obtained above 1200 °C

(continued)

Types of stoves	Features	Advantages and disadvantages
External combustion chamber	• Separate combustion chamber to avoid the brick lined division wall construction and permits a larger diameter of checker chamber • Silica refractory is used in dome, combustion chamber and checkers. During shutdown, the temperature of silica refractory has to be maintained above 700 °C • High heating surface area generates high volume of hot air	• No chances for failure of combustion chamber wall in the absence of division wall. • Very stable construction • Larger air heating surface provides larger volume of hot air • High hot blast temperature of more than 1200 °C can be achieved. • Larger space required to install external combustion chamber stoves • High cost of installation • Frequent refractory failure experienced in connection area of two chambers
Top-combustion	• In this design, the combustion chamber is eliminated • Ceramic burner is arranged at dome position • Circular ceramic burner is used in order to fully utilize the circular dome space and for complete combustion • Heating surface is effectively increased • Re-generative chamber uses high-efficiency checkers • Combustion air is preheated above 520 °C • The position of hot blast main and other machineries is in higher elevation compared to stoves of other design	• Cost-effective construction • Hot blast temperature of more than 1300 °C can be achieved • High combustion efficiency and long service life • Damages in dividing wall is fully eliminated as there is non-existence of combustion chamber • Hexahedral checker works with heating surface as high as $48 m^2/m^3$ are used • Performing maintenance job is difficult as the position of hot blast main and other machineries are in higher elevation

4.3 Refractory Lining Design

In order to achieve maximum performance in hot blast stove, the selection and design of refractory are the most important factors to consider [4, 5].

Refractory Materials

An increase in hot blast temperature is one of the basic directions in reducing coke consumption and to increase blast furnace productivity. It is known that each 100 °C increase in blast temperature reduces the specific coke consumption by 12–15 kg/ton of iron produced [4, 6].

With the increase in hot blast temperature, dome temperature is also increased. It became evident that high-quality refractory is in use to withstand the adverse operating conditions in stove operation. The following properties are considered for refractory selection at different areas of a hot blast stove [5]:

- High-temperature creep resistance.
- Refractoriness under load (RUL).
- Permanent linear change (PLC).
- Reversible thermal expansion (RTE).
- Thermal shock resistance.
- Resistance to alkali and iron oxide attack.

Most important property is the high-temperature creep, whereby the maximum deformation after 20–50 h under load of 0.2 N/mm^2 at the applicable temperature should not exceed 0.2% [5].

The next important property is thermal expansion. The expansion behaviour of different quality bricks should be known in order to calculate the required expansion allowance. In this respect, to protect the steel shell from overheating, in side wall and dome, multiple layers of bricks are in use. Insulation refractory of different qualities used against the shell, followed by dense bricks in inner wall to withstand the temperature and pressure. The refractory usage experience has shown that the most effective lining design is zonal lining concept, use of different quality refractory in different zones of stove. The following quality of refractories are used in lining stoves:

- Silica refractory.
- High alumina refractory.
- Semi-silica insulation.
- High alumina insulation bricks.
- Insulating boards and blankets.

In highest temperature zones, based on the technical consideration, silica refractory is preferred over high alumina due to very high creep resistance, high RUL, very closed to its melting temperature and superior thermal shock resistance above 600 °C. Another advantage of silica is that it is less sensitive to dust and alkali attack.

Design of Brick Shape

Design of brick shape and its interlocking arrangements plays an important role on refractory lining performances and trouble-free operation. In early days, several failures took place due to brick falling from side walls, collapsing of separation wall, displacement of checkers. Interlocking design within the bricks, shapes and checkers have made a significant improvement on structural stability of refractory work. The design considered in the following critical areas:

- Dome.
- Separation wall.
- Checker.
- Ceramic burner.
- Junction of hot blast outlet and other arch constructions.

In old design of stoves, traditional hemispherical design was used, where the dome was supported on the ring wall of combustion chamber and checker chamber. This causes failure of dome bricks due to differential expansion of both chamber

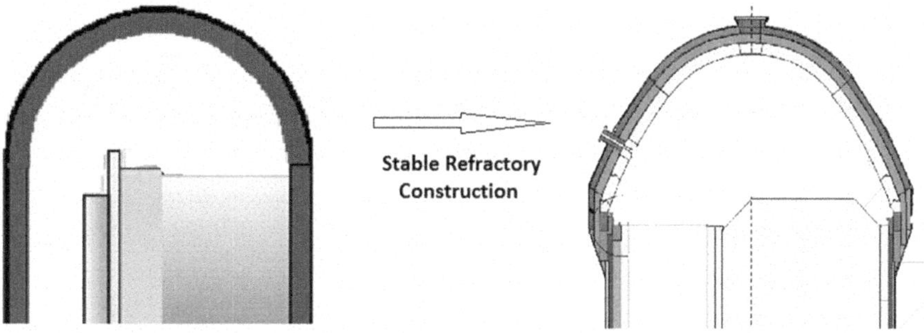

Fig. 4.5 Modification of dome for stable structure

refractory lining, and it also causes stresses due to difference in temperature between combustion chamber and checker chamber [4, 5]. To address the failure, an addition protection wall is constructed in the combustion chamber. Furthermore, a protection ring made of refractory shape was constructed at the base of the dome. The design was further developed to change the support of entire dome refractory from side wall to the shell, by extending the shell diameter at the base of the dome, it also helps to adjust the expansion of dome bricks without creating additional stress on side wall refractory. The modification is illustrated in Fig. 4.5.

Combustion Chamber

In old design of internal combustion chamber stove, the shape of combustion chamber wall was circular made of single layer of bricks, and there were many problems, such as inter-brick stress closed to the stove side wall, stress at the back side of the bricks due to thermal expansion, resulted in opening of joints and created "short circuit" and failure at checker chamber side, stress in the burner and hot blast outlet. The circular dome had been modified to elliptical shape with double layer refractory lining. In the new design, there is inter locking of structure with the radial side wall, which transfers the stress generated in the combustion chamber wall to the radial side wall.

Further, with the new design, the cross-sectional area of checker chamber had been increased, resulted in higher heating surface created, improved heating efficiency. Also, introduction of tongue and groove bricks has improved stability of inner combustion wall. The modern design of combustion chamber and its inter locking structure is shown in Fig. 4.6.

Checker Design

When hot blast stove was first constructed for generating hot blast for blast furnaces, split checkers were in use with a hole size of 40 mm and very thick wall thickness. This, thus used more refractory masses and fewer number of holes, resulted in lesser opening for air flow and hence lesser volume of heat exchange. During last four decades, the design of checker undergone many changes. The main factors considered for design modification are as follows:

– To increase air flow rate and air volume.

Fig. 4.6 Elliptical combustion chamber

- Excel heat exchange.
- To increase heating surface area.
- To increase heat capacity.
- Pressure drop across the checkers.

Given a constant cross-sectional area, A, for the checker work system, the average air velocity, V, per flue is expressed as follows:

$$V = \frac{Q}{A.\ R}, \tag{4.1}$$

Q = Total air volume flow rate,
R = Ratio of free opening in checker and total cross-sectional area of checker.
Pressure drop across a checker work system is expressed by the formula:

$$\Delta P = \frac{\gamma.\ F.\ LV2}{D.\ 2g}, \tag{4.2}$$

where,

γ = Specific weight of flowing gas, lb./ft^3.
F = friction factor.
L = Height of checker work, ft.
V = Flue velocity ft./s.
D = Hydraulic diameter of checker work flue, ft.

Keeping all operational parameters constant, the pressure drop is directly proportional to $V2/D$. Referring to Eq. (4.1), the pressure drop is proportional to the following:

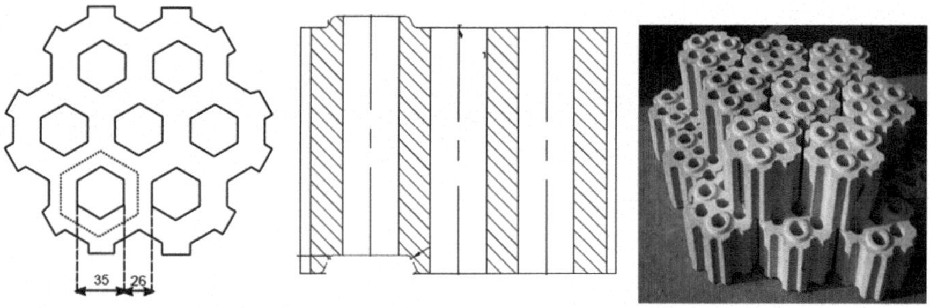

Fig. 4.7 High-efficiency checker

$$\frac{Q^2}{A^2.\ R^2.\ D} \tag{4.3}$$

Q is usually fixed by operating conditions, A is dependent on checker space availability and the pressure drop is inversely proportional to DR^2, where,

D = Checker flue hydraulic diameter,

R = Ratio of free checker work to total cross-sectional area of checker.

Hence, to improve the efficiency of heat transfer through checkers, the number of holes per checker helps to increase and to reduce the refractory wall thickness of checkers.

Considering the above conditions, the high-efficiency checkers had been developed. Hexagonal checkers with 7 holes, 12 holes and 19 holes are widely used in many steel plants. The schematic diagram of a 12 holes checker with 35 mm hole diameter and 20 mm wall thickness is shown in Fig. 4.7.

High Efficiency Checker Shapes

The thermal efficiency of the stove checker works system can be expressed by the following equation:

$$MC\left(\frac{\partial T}{\partial t}\right) = h\,A(\theta - T)$$

where M = mass of the checkers in stove, C = specific heat of checker refractory, T = checker work temperature, t = time, h = heat transfer coefficient, A = heating surface area and θ = heating or cooling fluid temperature.

Heat capacity, MC of the checker increases with an increase in heating surface area, and the heating surface area increases with an increase in the no. of holes, ratio of free checker area and reduction of flue wall thickness and hydraulic diameter.

The different dimensional parameters of the commonly used checkers of improved design are shown below:

Types of Checkers	Unit	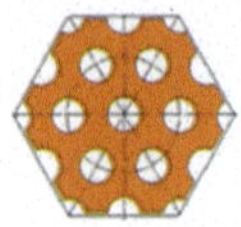	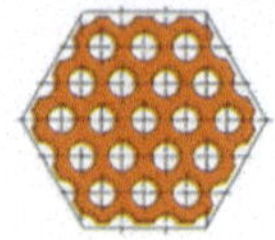	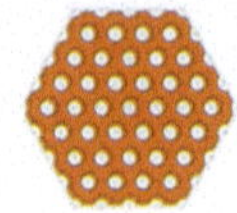
Nos of holes	Nos.	11	19	37
Heating surface	m2/m3	25	38	45
Weight	Kg/m3	1450	1390	1330
Free checker area / Total checker work area	Ratio	0.33	0.36	0.39
Wall thickness	mm	30	25	20
Hydralic Diameter	mm	50	38	32
Pressure loss Factor	mm	54	50	45

Ceramic Burner

Mechanical burners had been used in the past, where the air and gas are injected into the combustion chamber through the burner sleeve, with the help of separators. The air takes the outer annular path and the gas in the inner path. Air and gas then mixed into the combustion chamber at the mouth of the burner sleeve, as shown in the schematic diagram Fig. 4.8. The air–gas mix impinges on the refractory wall and takes a helical path to move up the chamber. The main disadvantages of this design are as follows:

- High heat loss,
- Low heating efficiency,
- Damage of refractory in combustion chamber wall due to flame impingement.

High-efficiency ceramic burner has gradually replaced the mechanical burners, and new hot blast stoves always include ceramic burners to secure complete combustion and long campaign life; heating efficiency has been increased in manifold. The ceramic burner is made of refractory bricks of different shapes, depending on different designs of ceramic burners. Andalusite and mullite-based refractories are used for the following essential properties:

- High volume stability,
- Very high softening temperature, above 1800 °C,
- Excellent thermal shock resistance,
- High creep resistance,
- Inert towards reaction with S and other components present in gas.

A schematic diagram of mechanical burner and high-efficiency ceramic burner is shown in Fig. 4.8.

A ceramic burner ensures no impingement of flame direct to the combustion chamber wall, smooth mixing of both air and gas and reduces heat loss. This also avoids chances of explosion inside the combustion chamber. As the flame is progressing to top through the combustion chamber, fuel is required to heat up the

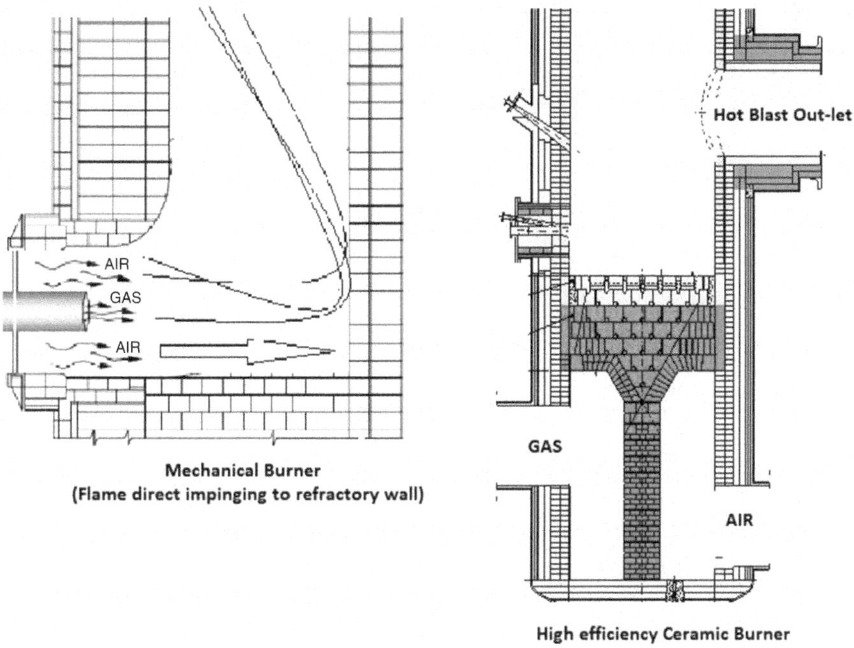

Fig. 4.8 Schematic diagram of hot blast stove burners

dome and checkers is low, unlike the mechanical burner, where the fuel is wasted in heating the lower part of the chamber. Tentative properties for ceramic burner refractory are given in Table 4.1.

4.3.1 High Alumina Refractory in Hot Blast Stove

For many years hot blast stoves have been built with internal combustion chambers with the use of chamotte refractories, which operated successfully with a hot blast temperature up to 1000 °C. With an increase in blast temperature to 1100 °C, high alumina refractory with 62–74% Al_2O_3 content is in use.

However, use of this refractory does not provide sufficient life of stoves, which required further development with respect to thermal shock resistance, creep resistance and resistance to alkali attack. In recent years considerable steps have been undertaken to improve and optimize the refractories for hot stoves. Depending on temperature and operating conditions, 62–74% Al_2O_3 bricks and checkers are used in the highest temperature zone, 50–55% Al_2O_3 bricks and checkers are used in intermediate temperature zone (600–900 °C) and 39–45% Al_2O_3 bricks are used in low-temperature zone, as shown in Fig. 4.9.

Table 4.1 Properties of ceramic burner refractories

Properties	Unit	Andalusite	Mullite
Chemical			
Al_2O_3	%	62–65	70–75
Fe_2O_3	%	0.8–1.1	0.6–0.7
Alkalies	%	0.7–0.9	0.7–0.8
Physical			
Bulk density	gm/cc	2.55	2.65
Porosity	%	17–19	17–19
Cold crushing strength	Mpa	60–65	70–75
RUL (ta)	°C	1650	> 1700
Creep at 1450 °C, under 0.2 Mpa load, 5–50 h	%	<0.22	< 0.2
Thermal shock resistance (water quenching)	Nos. of	30	30

The Al_2O_3-SiO_2 system is characterized by three phases, silica, mullite and corundum. They are thermodynamically stable at high temperature with congruent melting points of 1726, 1853 and 2054 °C, respectively. Due to very high liquids temperature, the alumino-silicate refractories are generally considered as having good refractory properties. For high-temperature application, more than 1600 °C, andalusite, kyanite, sillimanite and mullite-based high-alumina refractories are in use. In intermediate- and low-temperature zones, andalusite, kyanite and low iron bauxite or high-temperature calcined clay grog-based refractories are used.

Mullite, which is a mineral existing with the form of $3Al_2O_3$, $2SiO_2$ or $2Al_2O_3$. SiO_2, provides superior properties of low thermal expansion, low thermal conductivity, excellent chemical stability in stove operating atmosphere and high hot strength. Mullite rarely occurs in natural raw material, and synthetic sintered or fused mullite are used, which escalates the cost of those bricks [7].

Among the alumino-silicate raw materials, andalusite is the best natural source of mullite with a very high degree of purity. On heating andalusite, it transforms to mullite, above 1400 °C, with a volume expansion of +4.5%, as shown in the following equation:

$3(Al_2O_3.SiO_2) = 3Al_2O_3.2SiO_2 + SiO_2$
(Andalusite) (Mullite) (Cristabalite)
$2SiO_2 + 3\ Al_2O_3 = 3Al_2O_3.2SiO_2$ (Mullite).

Hence, the advantages of using andalusite-based high-alumina refractory in hot stove are summarized below:

- Can withstand temperature more than 1450 °C, suitable to use in dome, side wall of combustion chamber and checkers in high-temperature zone,
- High hot strength,
- Excellent thermal shock resistance,
- High creep resistance,
- Low linear change on reheating,
- Good CO-gas-resistant,

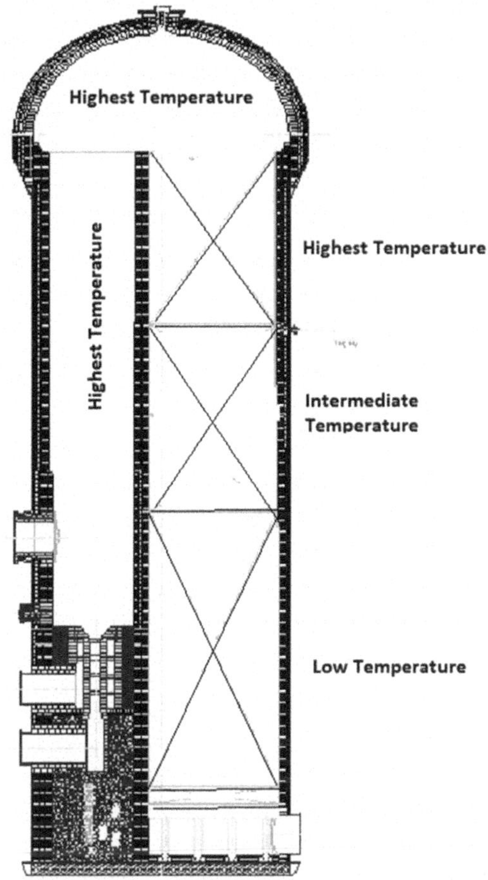

Fig. 4.9 A typical hot blast stove consists of different temperature zone profiles

– Resistance to alkali attack.

On the bottom of the stove where the service temperature is relatively low (<800 °C), standard chamotte bricks without andalusite and alumina content of 39–45% are used in checkers and side walls. Due to low reversible thermal expansion and high thermos-mechanical stability, andalusite-–chamotte type refractory are in use in checker chamber walls and checkers in medium-temperature zone. High-alumina refractory in different zones are given in Table 4.2 [8].

4.3.2 Silica Refractory in Hot Blast Stoves

With the increasing demand of high hot blast temperature of more than 1100 °C and prolonged life, the silica refractory became popular to use in high-temperature areas,

Table 4.2 Raw materials used in stove refractories

Areas	Zones in stove	Mullite (%)	Andalusite (%)	BFA (%)	Calcined alumina (%)	Chamotte (%)
Combustion chamber	Dome	20–25	30–35	25–30	15–18	Nil
	Upper and middle wall	0–10	65–72	Nil	15–18	Nil
	Lower wall lining	Nil	72–76	Nil	20–25	15–20
	Ceramic burner	30–60	40–72	Nil	15–18	Nil
Checker chamber	Upper zone wall lining and checker works	35–40	30–35	15–20	13–16	Nil
	Middle zone wall lining and checker works	0–10	65–72	20–25	15–18	Nil
	Lower zone wall lining and checker works	Nil	20–35	Nil	Nil	50–60

such as dome and upper highest temperature zone of checker chamber and combustion chamber. Silica refractory is considered for its very high softening temperature, excellent creep resistance and thermal shock resistance above 600 °C. For silica refractory, and almost quartz-free type with almost percentages of tridamite and cristabalite are preferred. The ratio of cristobalite and tridamite is determined by the used raw materials, firing temperature and soaking time in the kiln during manufacturing, where α-quartz is transferred into β-quartz, showing a volume change of 1.0%. With continues firing, β-quartz will be transformed to α-tridamite, with a volume increase of 14.4% and at higher temperature to α-cristobalite. In the operating condition of hot blast stove, cristobalite is the most stable phase at the temperature above 1400 °C. A study [9] on tridamite to cristobalite formation in used silica bricks of hot blast stoves confirmed that cristobalite has been often found to appear in the surface layer of silica refractory used in hot stoves due to the presence of impurities, and it also confirmed that cristobalite appeared in the stable region of tridamite, below 1470 °C. The results of chemical and mineral composition analysis of the original and used silica refractory are given in Table 4.3. Tridamite and cristobalite, except quartz, were included in original silica brick. It was observed that the surface layer, 10–15 mm thick, was white, and the inner layer was brown on the cut surface of used silica bricks. The mineral composition shows cristobalite on the surface layer and tridamite in the inner layer.

As it is thought that the surface layer is influenced by the penetration of impurities, the effect of impurities on the transformation of tridamite to cristobalite below 1470 °C was investigated and was found that in presence of Al_2O_3 as impurity, the transformation occurred even below 1470 °C.

Table 4.3 Chemical and mineralogical analysis of silica refractory (%)

	Original	Surface layer (used brick)	Inner layer (used brick)
SiO_2	95.6	92.8	94.8
Al_2O_3	0.8	1.6	1.1
Fe_2O_3	0.9	1.1	0.8
K_2O	0.2	0.3	0.1
CaO	1.6	2.2	2.1
Tridamite	53	0	80
Cristobalite	47	100	20

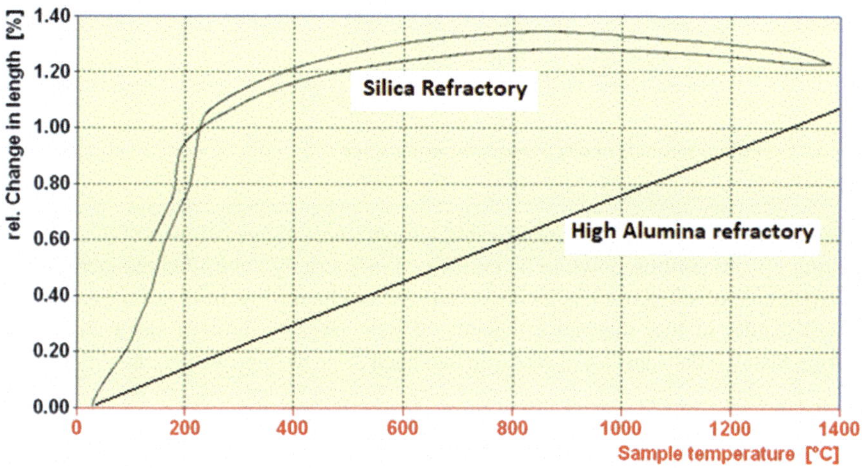

Fig. 4.10 Relative change in length of silica and alumina refractories with respect to temperature

Thermal Expansion of Silica Refractory

Expansion behaviour of the different quality refractories should be known in order to calculate the required expansion allowance. To calculate the allowance for expansion, the influence of permanent linear change (PLC) at different temperatures and reversible thermal expansion has to be considered. Design of expansion allowance in case of silica stove refractory is much critical compared to alumina refractory due to its non-linear expansion and change of expansion coefficient of silica refractory with temperature. This non-linear expansion is due to phase transformation during heating and cooling, as shown in Fig. 4.10.

Silica has a great advantage because of its expansion characteristics as shown in Fig. 4.11. Above 600 °C it has very small volume change with an increase in temperature. This means that during stove operation, the refractory construction remains stable in spite of the normal dome and checker temperature fluctuations during "on-blast" and "on-gas" conditions.

However, below 600 °C, the thermal shock resistance is very poor because of its very high thermal expansion during heating and contraction during cooling, which

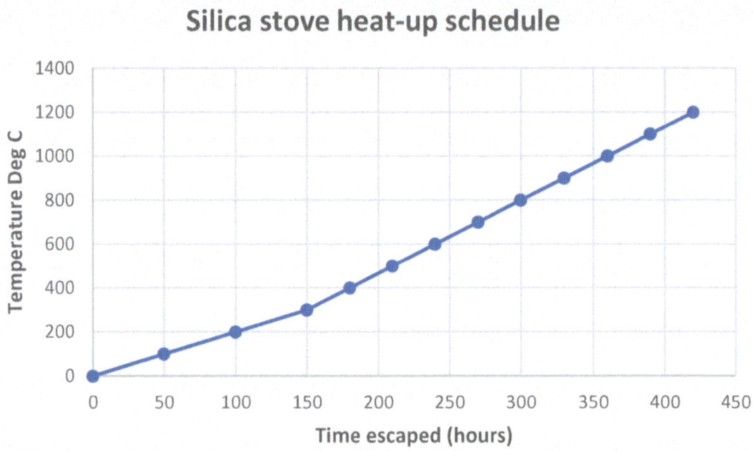

Fig. 4.11 Heat-up schedule of silica refractory in hot stove

causes cracking of refractory during rapid temperature change below 600 °C. Hence, heating up and cooling down to be performed in a strictly controlled manner. Other disadvantage of silica brick is its lower density and hence low heat storage capacity, which results in low thermal efficiency of heat transfer through silica checkers. During last two decades, lot of research had been done to improve heating efficiency.

Heat-Up and Cool-Down of Silica Stove Dome Refractory

Heating and cooling of silica refractory in hot stoves require adequate attention and expertise, it is required to take into consideration the nature and behaviour of the material. Not only the temperature changes at hot face are important but also the heat penetration into the lining as a function of time, explained by Frank T [5]. With FEM model, it is possible to determine the maximum developed stresses and also the location where the highest stresses occur inside the lining structure. From the calculation, it became evident that the maximum stresses do not occur on the hot face of the lining. The measured strength in laboratory was obtained 17 MPa, and a safety limit was set at a maximum stress of 7.5 MPa. During heat-up the stress limit would be much below when the recommended heating rate is 2 °C per hour up to 600 °C. With this heating up rate, the maximum peaks in stresses occur at approximately 200–350 °C. As the thermal characteristics of tridamite and cristabalite are reversible, the same philosophy is applicable during cooling of silica stoves. Prior to the actual cooling down, it is recommended to operate the stove at a reduced dome temperature, so that at the end of the last blast cycle the dome temperature remain at 1000 °C, which helps to reduce the total cooling time.

4.3.3 High-Temperature Corrosion Mechanism

The stoves and hot blast stoves are suffering from three types of corrosion:

High Efficiency Silica Bricks
In order to improve the efficiency of silica checker for escalating heat transfer, the heat capacity and thermal conductivity of silica refractory need to improve. Many experts are presently engaged to improve those properties. The processes adopted are by adding SiC in silica refractory and to manufacture dense refractory. Addition of TiO_2 in silica refractory and optimization of grain size distribution have reduced pore size as well as improved the density, which results in improvement in thermal conductivity.

Addition of SiC is a well-known process to improve thermal conductivity. It also improves creep resistance and hot strength. The tentative change of conductivity with addition of SiC is shown in Fig. 4.12.

- Corrosion of refractories,
- Reduction of shell plate thickness,
- Stress corrosion of shell.

The air and gas are contaminated with iron dust, iron oxides and alkali, which react with refractory to form low melting compounds and causes the failure of checkers and side wall refractories exposed in high temperature. The fused mass

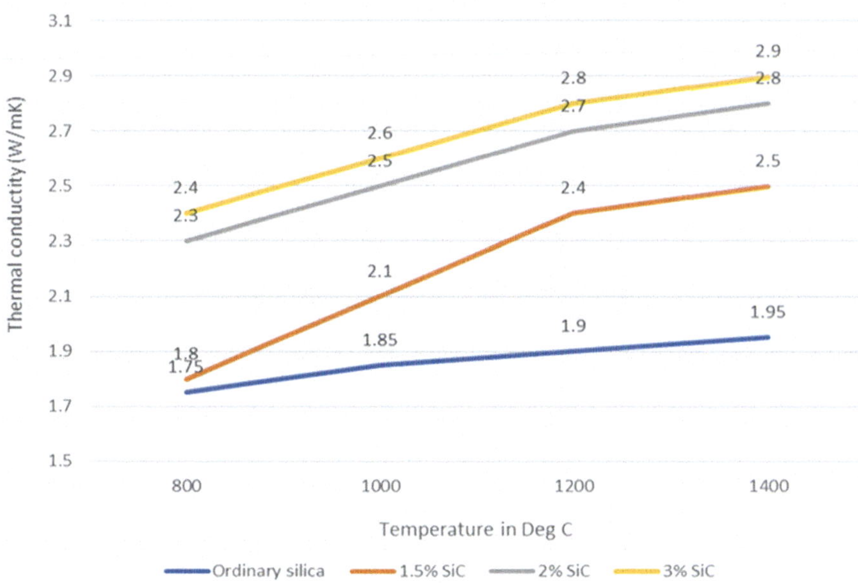

Fig. 4.12 Thermal conductivity of silica refractory

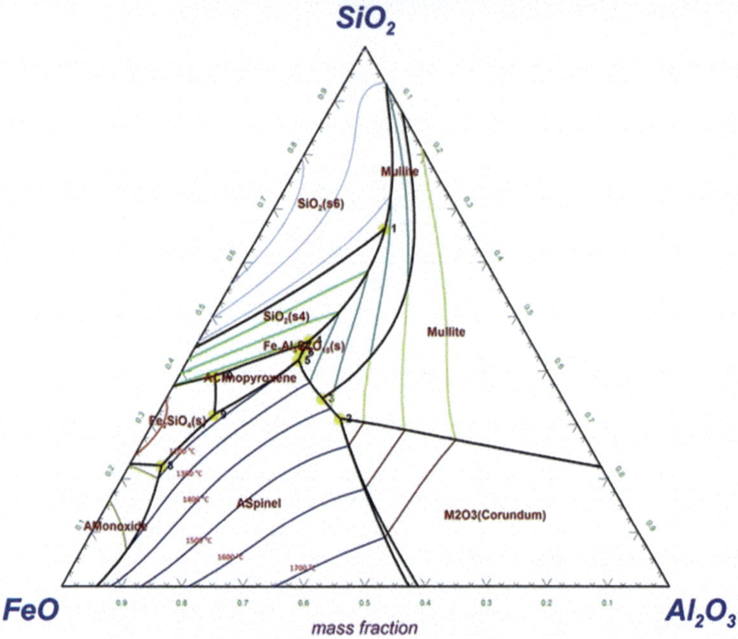

Fig. 4.13 FeO-Al_2O_3-SiO_2 phase diagram (source: Factsage)

deposits in the holes of checkers and clogging the gas and air passage through the holes. The top few layers of checkers are usually found completely fused and permeability of air had been reduced, which results in poor stove performance. Above 1400 °C, FeO reacts with high-alumina refractory to form the low melting phases, fayalite ($2FeO.SiO_2$), hercynite ($FeO.Al_2O_3$) and iron cordierite ($2FeO.2Al_2O_3.5SiO_2$) as shown in Fig. 4.13. With silica refractory, the effect of corrosion of both the FeO and alkali is much lesser than high-alumina refractory and hence the use of silica refractory extends the campaign of hot stoves.

Reduction of shell plate thickness is a common corrosion effect in steel shell. This is due to long exposure to heat and weathering. In order to identify such a reduction in thickness, the stove shell is mapped in a matrix and ultrasonic thickness measurement of the shell plate is carried out. The other type of shell corrosion is stress corrosion cracking (SCC) which takes place under certain conditions in high stress-level areas, such as heat-affected zones alongside welding seams. During cyclic operation of a hot stove, temperature of various interior portions of the stove vary from ambient to above 1400 °C, and this causes severe thermal stress on refractory and shell. In addition to thermal stress, the interior of stove shell is subjected to corrosive action of acids derived from nitrogen-containing oxides and sulphur-containing oxides produced during the combustion of gases. Such gases include N2O, NO, NO2 and N2O4, all of them are generally designed as NOx gases. Similarly, sulphur-containing gases are SO2 and SO3. The shell is susceptible to

SCC attack from these NOx-derived condensed corrosive acids. The formation of NOx rapidly increases when the dome temperature exceeds 1400 °C. The refractory lining of the stove is not typically gas tight, and NOx gases are passing through the pores and joints of the bricks and condensed on the inner surface of the shell, which causes stress corrosion cracking by nitrogen- and sulphur-containing acid concentrates. The concentration of acids takes place below 120 °C, hence one of the processes to avoid stress corrosion is to maintain the shell temperature above 120 °C by covering the steel shell by additional external insulation. The other way to prevent the SCC is to provide a protective epoxy paint coating in the inner part of the dome shell or lining an acid-resistance castable lining over the inner part of the dome shell so that NOx cannot come in contact with steel shell [3, 10].

4.4 Hot Blast Carrying System

Hot blast main and bustle pipe of blast furnace have a refractory structure comprising a multiple course of refractory bricks. In order to reduce heat loss through the exposed area, the mains carrying the hot blast from the stove to tuyeres should be as short as possible. At the same time, to protect the steel shell from overheating and heat loss to a minimum, it is necessary to line the inside pipe with refractory material with adequate thickness. Velocity of air inside through the mains is very high, and there is considerable attrition in consequence. The hot face refractory must satisfy the following properties:

- To withstand the operating temperature of more than 1200 °C,
- Resistance to thermal shock,
- Resistance to hot abrasion for high air velocity,
- High creep resistance,
- Volume stability at high temperature to avoid gap formation within the bricks.

It is usual that the hot face lining is 55–65% Al_2O_3 andalusite-based bricks, depending on the hot blast temperature and followed by another 2–3 layers of insulation bricks so that shell temperature to be maintained around 70–90 °C during operation. The hot blast arch lining, ceramic burner and the hot blast main, bustle pipes of a blast furnace are a highly thermally stressed zone. Use of good-quality refractory material for the manufacture of high-alumina refractory of those zones is a crucial factor, which decides the campaign of refractory lining and trouble-free operation.

Kyanite, sillimanite and andalusite are the three polymorphs of $Al_2O_3.SiO_2$. Crystal structure of kyanite is triclinic, and other two have a stronger orthorhombic structure. In Sillimanite, the tetrahedral Al and Si polyhedral are irregular with very short Al-O_3 and Si-O_2 bond distances, but in andalusite, Si-tetrahedra is not disordered, with similar Si–O bond distances. This crystal structure makes the andalusite to emerge as the strongest crystal structure for an aluminosilicate of composition

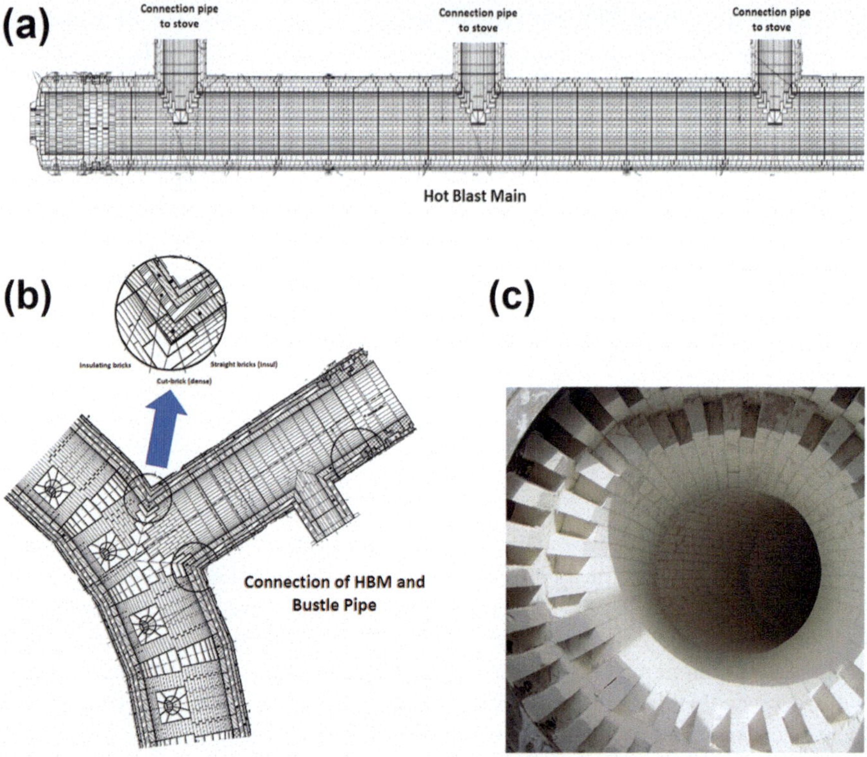

Fig. 4.14 Schematic refractory installation pattern in (**a**) hot blast main, (**b**) bustle pipe, (**c**) refractory brick lining in bustle pipe

$Al_2O_3.SiO_2$. This also gives andalusite a higher thermal shock resistance compared to others [4].

Refractory Construction Designs

The main considerations for the critical design in hot blast main and bustle pipe and connecting pipes with stoves are as follows:

- Joints of connecting pipes with hot blast main,
- Design of cold air line connected in hot blast main,
- Expansion compensators in hot blast main,
- Joint of hot blast main and bustle pipe,
- Joints of tuyeres connecting pipes in bustle pipes,
- Design of expansion joints in refractory lining in hot blast main and bustle pipes (Figs. 4.14a,b,c and 4.15).

Failure Mechanism in Hot Blast Carrying System

Cracking of the steel shell in hot blast main and bustle pipes is a major problem faced by most of the steel plants [11]. Interruption in the furnace operation is associated

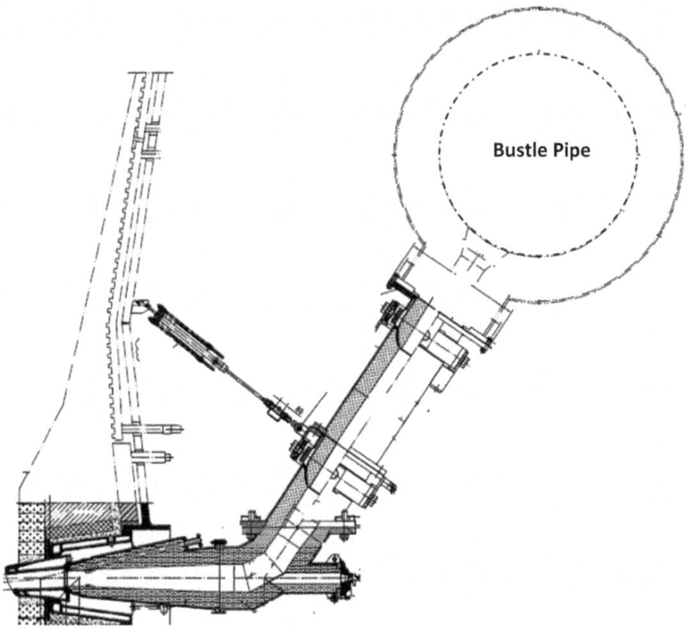

Fig. 4.15 Hot blast carrying system from bustle pipe to blast furnace tuyeres

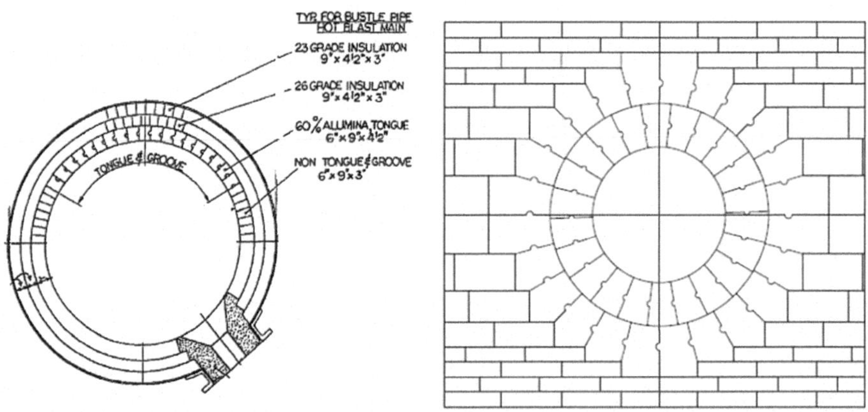

Fig. 4.16 Design of "Tongue & Groove" bricks of interlocking structure

with the time required to perform the welding of the cracked area of the shell and preventive repair of refractory by grouting from outside of the shell. The schematic diagram of refractory lining inside the pipes is shown in Fig. 4.16.

In circular construction of the pipes, the hot face lining in upper portion is designed with "Tongue & Groove" bricks of interlocking structure to prevent falling

Table 4.4 Properties of bricks used in hot blast main and bustle pipe

Properties	Unit	62% Al_2O_3	42% Al_2O_3	ASTM 26 (insulation)	ASTM 23 (insulation)	Ceramic fibre board
Al_2O_3	%	59–64	42–45	58	37–40	44–46
SiO_2	%	40–42	52–54	39	42–44	45–47
Fe_2O_3	%	1.2	1.5–2.5	0.8	0.7	1.1
TiO_2	%	0.3	1.6–1.8	0.1	1.2	1.1
Alkali	%	0.5	0.6	1.7	1.5	4.5
Application temperature	°C	1650	1450	1430	1260	1260
Refractoriness	Ceger cone	> 37	>34	NA	NA	NA
Bulk density	gm/cc	> 2.55	2.35	0.78	0.55	NA
App. porosity	%	14–16	16–18	NA	NA	NA
Cold crushing strength	Mpa	70–80	40–50	1.6	1.1	NA
Mod. of rupture	Mpa	13–15	6.5–10.5	1.4	0.9	0.6
Refractoriness under load (T 0.5%)	°C	1600	1450	800 (load of 0.07 Mpa)	550 (load of 0.07 Mpa)	NA
Reversible expansion at 1000 °C	%	0.65	0.6	NA	NA	NA
PLC at 1000 °C/5 h	%	0.5	0.5	0.22 shrinkage	0.2 shrinkage	NA
Thermal conductivity at 1400 °C	W/mK	1.22	1.18	0.33 at 1000 °C	0.2 at 1000 °C	0.07 at 800 °C

or dislodging of bricks. The backup lining is made of insulation bricks of different grades. Properties of the bricks are shown in Table. 4.4.

Hotspot, thinning and bulging of shell are also observed by operators. Thinning of the shell due to corrosion combined with high shell temperature has contributed to the cracking and bulging of shell plates [11, 12]. The yield stress of the shell plates decreases with increase in shell temperature, and shell temperature increases due to passage of hot air through refractory lining, gaps within the bricks and opening of mortar joints. The failure in bustle pipes takes place for the following major problems:

- Thermal shock,
- Change of direction of refractory expansion at all bends and joints causes non-linear expansion,
- Stress generation within the brick work due to inadequate expansion joints provided,

– Passage of hot air through the brick lining and travels up to the shell causes irregular expansion of shell due to overheating which causes loosening of brick lining inside hot blast main and bustle pipes.

Another factor that was recently investigated was the effect of build-up or accumulation of debris on the top of the hot blast main and bustle pipe. Dirt accumulated on the top of the bustle pipe retains moisture and leads to severe corrosion from the outside surface of the pipe. It is also observed that the temperature of the steel pipe and refractory is higher beneath a thick layer of debris. Heat transfer analysis shows that the shell temperature was found to increase in proportion to the thickness of the accumulated debris layer. For instance, a 4-inch-thick debris layer was predicted to raise the shell temperature to over 400 °C. Depending on the wall thickness of the pipe, this increase in temperature could result in permanent plastic deformation or rupture of the pipe. Additionally, the debris accumulation will result in a situation where the shell plate temperature is greater on the upper portion of the pipe than on the bottom. This top-to-bottom thermal gradient can cause significant thermal stress in the shell plate that will add to the mechanical stresses. Additionally, the presence of any debris on the shell would prevent proper measurements of shell temperature. Infrared temperature scans would show the temperature of the dust surface rather than the desired shell temperature (which could be much higher). These factors show the importance of routinely scheduling maintenance time for removing debris from the bustle pipe.

Thermal Stress in the Refractory Lining

Gaps that develop between the refractory layers and any interior cracks in the refractory lining can allow the hot blast to reach the pipe shell, resulting in a localized increase in the shell temperature. Initially, a heat transfer analysis was conducted to determine the temperature distribution through the different layers of bricks. The predicted temperatures were then used as input to a stress model in which the thermal expansion and load displacement characteristics of the bricks and steel shell are defined. This model provides an estimate of the thermal stresses that develop in the refractory and steel shell. In general, there will be a tendency for compressive stress to develop in the refractory components and tensile stress in the steel shell.

This occurs as the cooler, stronger shell plate restrains the outward expansion of the higher temperature refractory. The model was used to predict how the amount of thermal expansion allowance would affect the stresses in the brick. It is desirable to choose the expansion allowance such that the degree of compression that develops between brick is sufficient to keep the brick joints closed and the brick rings tight without exceeding the compressive strength of the brick. The results of the analysis show that, for each case, the radial and circumferential stress in the 60% alumina working lining brick is less than the compressive strength of the brick, which is about 8000 psi. Thus, it would not

(continued)

be expected that these bricks would fail in service due to the stresses associated with the expansion of the lining. The compressive strength of 2600- and 2300-grade insulation bricks, however, is only 275 psi. With no expansion allowance provided in the brickwork, both the radial and circumferential thermal stresses were predicted to exceed the strength of the insulation lining. Failure of the backup lining would likely result in an unstable working lining that could allow the hot blast to penetrate the lining and reach the steel shell. When the expansion allowance is provided, the stress in the refractory can be reduced significantly.

Selection of Grouting Materials
Grouting is a procedure that is done to fill any voids that may have developed between the bustle pipe shell and the refractory material. Typically, shell temperature readings are taken routinely, and the presence of hotspots indicates that voids may have developed in the lining allowing hot blast to penetrate the lining and possibly reach the steel shell. Once a hotspot is found, the shell plate is drilled and a grout pipe nipple is installed. Pumpable grout is then fed into the bustle pipe through the nipple. Grout will flow into any void that may be present. There are several grouting materials available. Selection of grout material is often made based on thermal conductivity of the material with the idea that a low conductivity grout will reduce the heat flux through the wall and keep the shell cooler. To evaluate this, calculations were performed to determine how the thermal conductivity of the grout and the size of the gap being filled would affect the temperature of the pipe shell. It is also preferable to select grout materials that adhere better to the steel pipe and lining instead of using thermal conductivity as a deciding factor.

4.5 Failure of Stove Refractory and Repair Methodology

Although various measures have been taken to upgrade refractory quality and design of refractory lining, to protect the hot blast stoves from typical damages, wear and tear to their various parts during operation for long life is unavoidable. However, the types of damages depend on the type of stoves, size and age and operation methodology. In addition to that, hot blast stove is a closed type of equipment, and hence usual monitoring of refractory condition is not possible from outside, even in shutdown. Hence, in most of the cases, damage to the stove is undetected till the failures occur.

Mode of Damages Take Place in Hot Stoves
The parts that are especially susceptible to damage are listed in Table 4.5.

Table 4.5 Typical damages in stoves and their specific regions

Type of stoves	Area	Type of damages
Internal combustion chamber	Dome	Vertical cracks and drop out of bricks
	Combustion chamber	Gap formation in partition wall, falling of bricks leading to shot-circuiting
		Bending of partition wall, Banana effect, falling of bricks in high-temperature zone
	Checker chamber	Mis-alignment of checkers, hole chocking in checkers due to alkali fused refractory leads to high pressure drop, reduce air flow
	Ceramic burner	Spalling and cracking with rapid temperature change, breaking of burner refractory due to falling of bricks from side wall
External combustion chamber	Connection pipe within domes	Spalling and cracking of bricks due to temperature change. Falling of bricks. Failure of compensator in cross pipes

Fig. 4.17 Cracking in dome refractory

Vertical Cracks in Dome

The vertical cracks are commonly observed in dome brick work, due to thermal stress generated in heating and cooling. The cracks sometimes propagate to the backup insulation layer, and due to passage of hot gas at more than 1300 °C, the insulation layer damages, which resulted in overheating of steel shell. The red spot is observed from outside, and it is usually repaired by grouting refractory mass from outside so that the passage of hot air is arrested. The cracking of refractory inside stove dome is shown in Fig. 4.17.

Failure of Separation Wall of Combustion Chamber

It is a very common failure in internal combustion chamber type stove. The partition wall separates the combustion chamber and checker chamber. Lower part of the wall in combustion chamber close to the burner zone is experiencing the highest temperature of more than 1200 °C, when the other side of the wall facing to checker

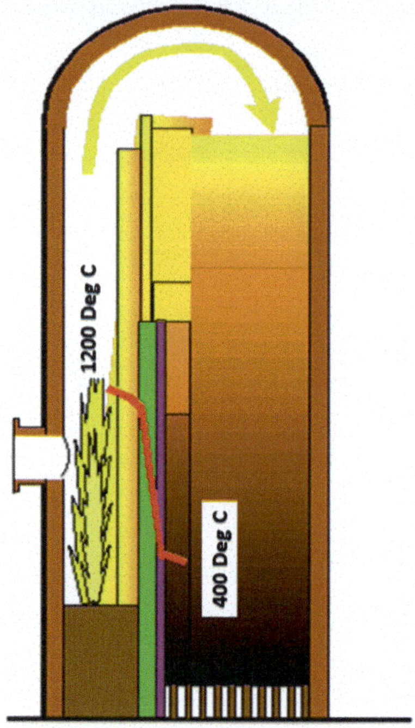

Fig. 4.18 Damage of separation wall

chamber is in low-temperature zone of 400 °C. So, the partition wall is facing a wide temperature gradient, causes thermal stress and crack forms, as shown in Fig. 4.18.

Due to cracking and brick falling in the partition wall, short circuit of hot air takes place and the lower side of the checker chamber becomes overheated and the efficiency of the stoves significantly lower down to minimum level. Other adverse effect of this type of failure is, during the gas cycle, hot gases would destroy the checker works and checker supports. During the blast cycle, cold air would reduce hot blast temperature. In this situation, the stove cannot be operated for long campaign, and repair is done by cooling down the stove, causes stoppage of operation for more than 20 days.

Damages in Ceramic Burner and its Repair

During long campaign of hot blast stove for more than 20 years of operation, the ceramic burners are damaged and the stove cannot be operating in normal mode. The causes of damages are dropout and pitting of plug bricks due to acid corrosion from fuel. Upper grating bricks are getting damaged due to repeated thermal load at 400 to 950 °C. Since the upper grating bricks are holding the plate bricks, as the damages of grating bricks propagated, combustion of ceramic burner would be disabled. The condition of a damaged ceramic burner and the burner after repair is shown in Fig. 4.19.

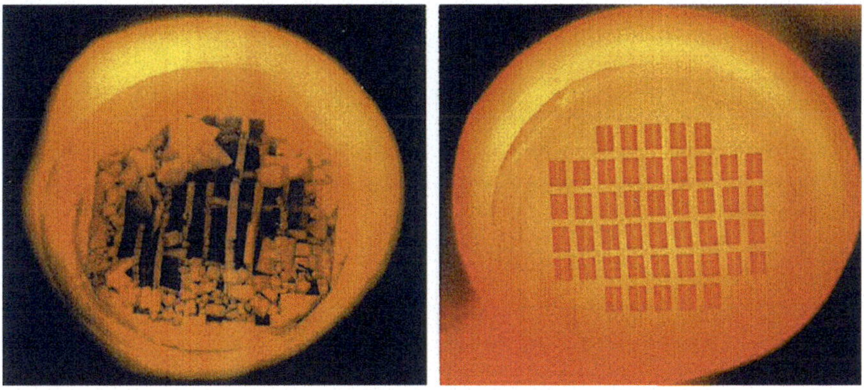

Fig. 4.19 Condition of damaged and after repair ceramic burners [12]

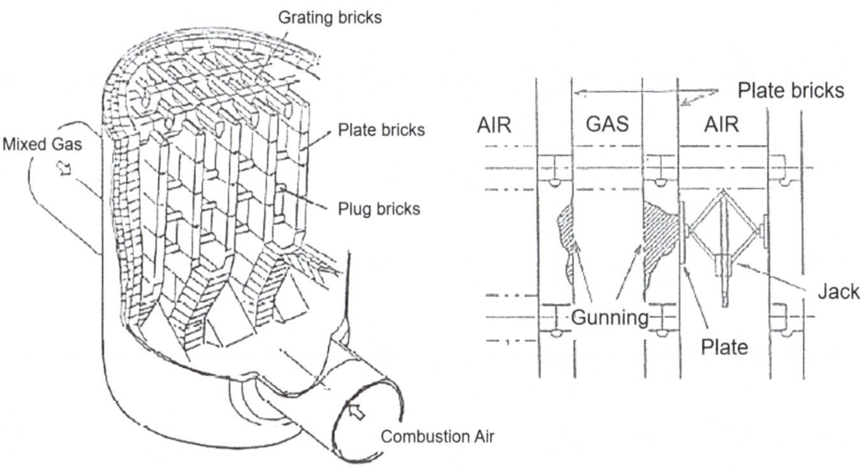

Fig. 4.20 Arrangements of grating bricks, plate bricks and plug bricks [13]

Methodology of Repair

With silica refractory in the stove, dome, top checkers and working lining of combustion chamber, in cold repair practice, the necessity of slow cooling, slow preheating and repair time can have a considerable effect on the productivity and stable operation of the furnace and therefore "Hot repair practice" must be considered. Atsushi et al. [13] had mentioned the hot repair practice to repair or complete replace of ceramic burner. In the repair, the stove contains silica refractory in the combustion chamber, checker chamber, dome and top few layers of checkers and this required that the silica refractory zone to be maintained above 750 °C, to provide a safe margin about the temperature at which silica phase changes occur. For short-time hot repair method, it is necessary to cool only the ceramic burner while insulating the silica refractory in the upper portion of the combustion chamber. The outline of the structure of ceramic burner is shown in Fig. 4.20. It was accordingly arranged to develop the heat insulating shield device that can be easily mounted inside the combustion chamber in a short time and to establish a short time hot repair practice.

Concept:

- Only the ceramic burner area is cooled while insulating the upper combustion chamber to protect silica bricks from rapid cooling. To maintain silica refractory above 750 °C.
- Man can enter in the ceramic burner zone and work comfortably to reline the ceramic burner.
- Installation of highly reliable heat protection device above hot blast connection pipe.
- Establish a short-time cooling method to cool the ceramic burner zone.
- Establish short-time brick laying method to repair gritting bricks of the burner.
- To complete the repair job within about 60 h.

References

1. Martin Gantenberg, Eric Schaub et al, "Comparison of different systems of Hot Blast Stoves", Proceedings, AISTech 2013, pp 509–518
2. Martin Gantenberg, Friedrich Eschmann, Eric Schaub, "Hot Stoves – More than 50 Years of Experience Design optimization and future concept", Proceedings, AISTech 2011, pp 95–103
3. Fu-ming ZHANG, Qimg – wu MAO, et al, "Dome combustion hot blast stove for hudge Blast Furnaces", Journal of Iron & Steel research, International, Vol 19, Issue 9, September 2012, pp 1–7
4. R. Gopalakrishnan, Sandeep Srivastava, I. Chatterjee, "Hot Blast Stove – Modernization and Advancement", International Conference on Advances in Refractories and Clean Steel making, India, June-26-28, 2013
5. Frank T Niemeijer, "Refractory properties of Hot Blast Stoves", AISE Steel Technology, May 2003, pp 53–58.
6. N.V. Pitak, V.L.Bulakh, "Differentiated use of Refractories in Blast Furnace Hot Blast Stoves", Refractories and Industrial Ceramics, Volume 27, issue 7–8, July 1986, pp 420–426
7. D. Frulli, "Production and properties of Refractory Raw Materials based on Andalusite and Mullite – Influence of impurities in Refractory behavior" Refractories worldforum, August 2016 [3]
8. Wolfgang Waltz, Pascal Hubert, "Basic principles for the usage of Andalusite cased shaped Refractories in Hot Blast stoves and Blast Furnaces" IREFCON 08, pp 66–72.
9. Tadahisa Arahori, Takao Suzuki, "Transformation of tridymite to cristobalite below 1470°C in silica refractories", Journal of Material science 22 (1987), pp 2248–2252.
10. Eric Schaub, Cristiano Castagnola, Johannes Munzer, "Hot blast stove developments" AISTEC 2017.
11. D J Radakovic, Y. Zhao, M Lucas, "Investigation of Bustle pipe Failure and repair", AISTech 2004 Proceedings, Vol I pp 111–118
12. Gordon Brown, " Hot repair of hot blast stoves", Report of VAI, UK
13. Atsushi Kondo, Kouji Doura, Hirofumi Nakamura, " Development of quick repairing technique for ceramic burner in hot stove of Blast Furnace", Iron making conference proceedings, 1997, pp 257–261.

Chapter 5
Refractory Practice in Electric Arc Furnace

5.1 Introduction

The electric arc furnace (EAF) was first commercially introduced in 1899. The history of EAF steel making is advanced by continuous improvements in efficiency and productivity, and it is able to make "tailor made" slag during the extended refining period by well-controlled production process. As a result, special and high alloy steel are exclusively produced by this process, and about 35% of world's steel is produced by EAF route.

The advantages of EAF route for making alloy steel include the following:

- A variety of grade of steel can be produced
- High precision in process control
- Relatively small size
- Cleaner environment.

Steel making through EAF is in a direct arc furnace, where current flows from the carbon electrode to the charge, and heat is transferred from the arc to the charge primarily by radiation, but a part of heat is generated in the charge. In modern EAF operation, only about 55% of the total input energy is used for melting and refining scrap or hot metal and the remaining 45% is lost to the slag, water cooling and with off-gas. Thus, though the EAF steel-making process has several advantages, due to high energy requirement, mainly in the form of electrical energy for arcing and energy losses, the process has not been adopted by many steel plants. The loss of energy is in the following forms:

1. Amount of energy loss in fume system, as unreacted CO and H_2 as well as sensible heat
2. Energy loss in cooling panels
3. Amount of oxygen in slag generation for purification
4. Carbon loss due to decarburisation

S. Biswas, D. Sarkar, *Introduction to Refractories for Iron- and Steelmaking*,
https://doi.org/10.1007/978-3-030-43807-4_5

Magnesite and Magnesia-C (MgO-C) bricks and monolithics are used in side wall and hearth in EAF [1]. Refractory wear rate in EAF has the largest extent compared to other types of steel-making equipment. Hence, it is necessary to think about the extension of EAF refractory lining life to maximal economic benefit. This aim is associated with the understanding the negative factors influencing low refractory life as explained below [2]:

- Very high temperature operation, 1600–1700 °C,
- Chemical interaction of lining refractory with slag,
- High thermal shock due to intermittent scrap charging, sometimes, scrap in wet condition,
- Direct impact of electric arc,
- Use of oxygen.

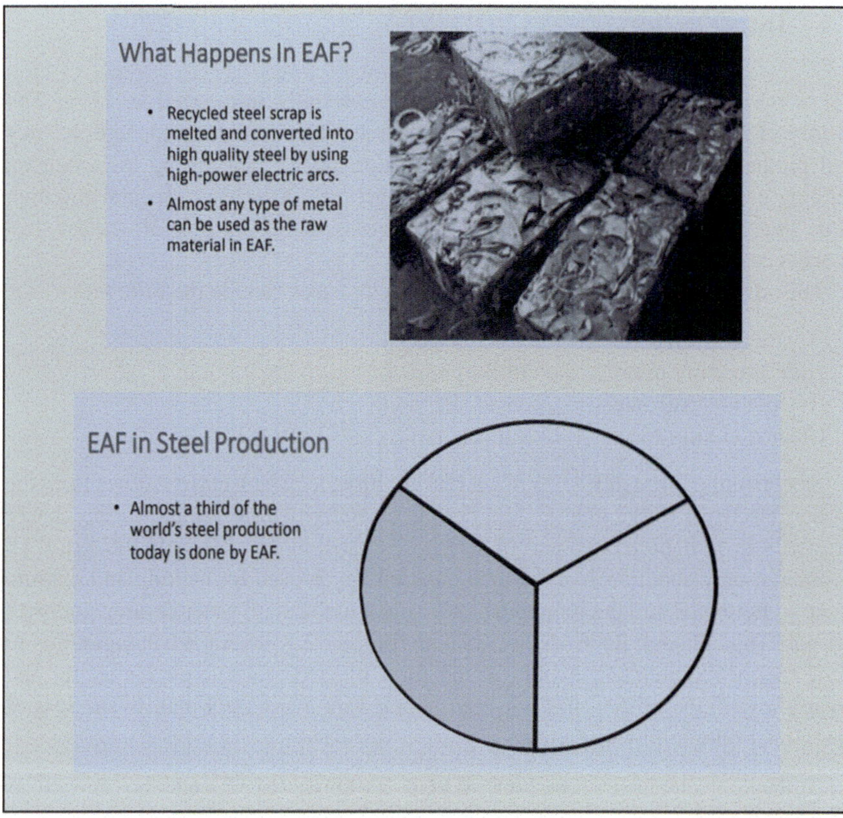

5.2 Features of an Electric Arc Furnace

The furnace is in cylindrical shape with a circular bottom, and the cylindrical side wall is sitting on the top of the bottom shell. Most recent design is the detachable bottom shell or split shell design. The cylindrical side wall of the furnace shell can be quickly decoupled and removed from bottom. The main advantage of this design is to change the side wall or bottom as per requirement and the downtime for relining is significantly reduced. After removal of sidewall steel shell, the bottom can be replaced or repaired independently [3, 4].

Water-cooled panel is provided outside of the shell in slag zone. The furnace roof is also consisting of water-cooled panels. The entire roof is made of refractory lining and the centre portion with the opening for electrodes is "called roof delta section". The furnace bottom consists of a steel shell with several layers of refractory. The different zones of an electric arc furnace are given below as shown in Fig. 5.1.

- Furnace roof,
- Cylindrical side wall,
- Bottom or hearth,
- Electrodes.

Challenges in refractory design	
Wear lining of walls	Usually MgO-C bricks in wall Hot spot with highest quality 100% fused MgO Dense pressed impregnated
Wear lining of bottom	Depending on steel grade For carbon steel production: MgO with high C ramming mass For alloy and stainless steel: MgO ramming mass and MgO-C bricks
Permanent lining	Usually burnt magnesite bricks For higher mechanical strength tongue and groove bricks Larger size MgO blocks
Roof/Delta lining	High alumina brick lining, electrode holes made of bricks, pre cast-pre fired delta blocks Use of low cement castable/ultra low cement castable Gunning of 80–100 mm over cooling panels

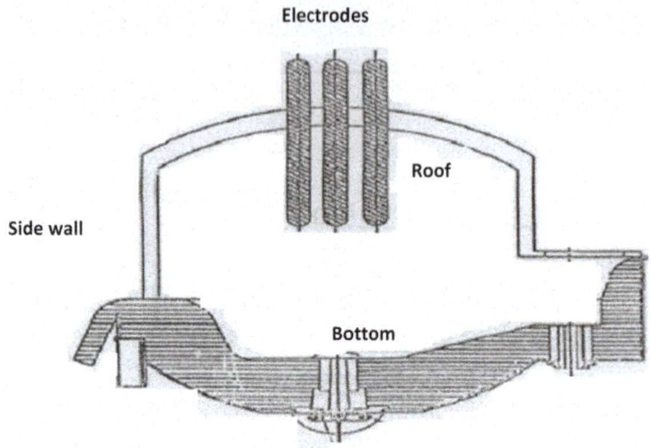

Fig. 5.1 Schematic diagram of electric arc furnace

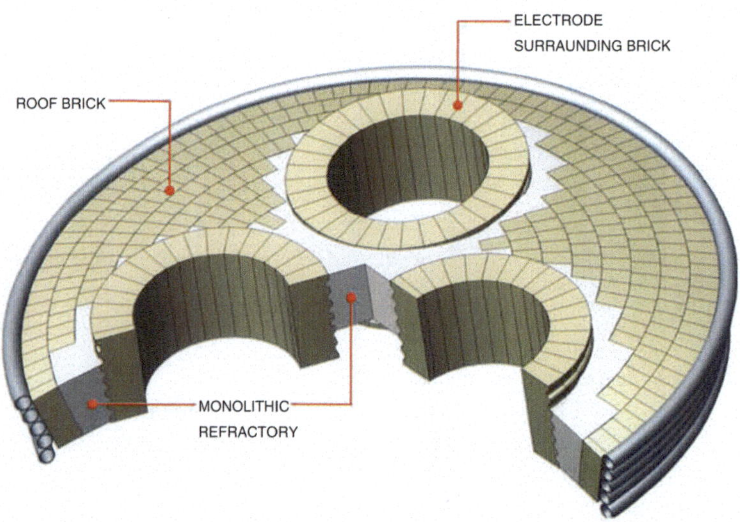

Fig. 5.2 Refractory lining in roof of EAF

5.2.1 Roof Construction and Refractory Lining

The roof of the arc furnace is exposed to severe stresses of thermal shock, chemical attack and mechanical impacts. Life of electric arc furnace depends on the refractory performance of furnace roof [5]. After the 1980s, the furnace capacity has significantly increased, and hence the working conditions of the furnace roof become more demanding and the quality of the refractory was also changed. The temperature of the roof often exceeds 1700 °C. It is exposed to corrosive slag and vapours, attack by FeO vapour to be considered in particular. Design of the roof is very critical in nature, which consists of three openings of electrodes and one additional opening for alloy addition or carbon addition. The design is shown in Fig. 5.2.

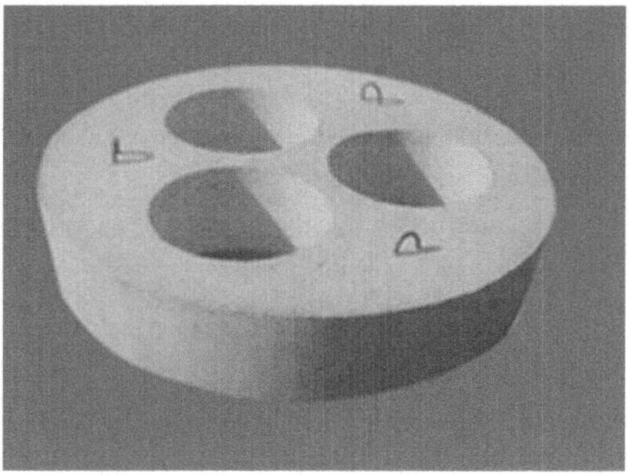

Fig. 5.3 PCPF block in delta section

In lining of EAF roof, silica bricks were used for long time. Later, due to increased operational demand, silica bricks were replaced by high alumina bricks with an Al_2O_3 content of 75–85%. High alumina bricks have become the main refractory material on arc furnace roof, it is reported that its life is 2–3 times more than silica bricks. Compared to silica bricks, the characteristics of high alumina bricks are as follows:

- High refractoriness, more than 1800 °C,
- Thermal shock resistance at all temperature range,
- Good slag resistance,
- Shorter preheating time.

With the development of large ultra-high-power electric arc furnace, the service life of high alumina bricks is reduced, magnesia and mag-chrome bricks are being used.

Initially, only shape bricks were used which were frequently changed for refractory failure due to spalling and falling of bricks in roof delta section, resulted in frequent stoppage of furnace and high repair cost. The lining in delta section was modified by using ramming and in later stage by using precast pre-fired (PCPF) blocks, as shown in Fig. 5.3. In roof and delta area, the main causes of wear are spalling, due to thermal fluctuation and corrosion for chemical reaction of slag with lining refractory. Absorption of slag and FeO fumes reacts with refractory to form low melting compound, and causes severe corrosion.

Overheating of steel shell was a major problem in EAF roof and use of water-cooled panel is a well-known practice to prolong campaign life [3, 6]. Castable or plastic mass is often placed around the perimeter of the precast deltas to lock the places with rest of the roof and with the shell, which creates a regular roof profile. Electrode ring bricks are installed around each mandrel and are held in place with steel bands. A castable or plastic is cast or rammed into place in the cavity between the outer rings and electrode ring brick.

In old design the cooling arrangement was provided along the roof frame, which supports the entire roof construction. In modern design, the cooling is extended to

Table 5.1 Properties of different refractories used in EAF roof

Type of refractory		High alumina bricks		Basic bricks	High alumina precast	
Area of application	Unit	Outer ring	Roof top	Entire roof	Delta section	
Chemical properties						
SiO_2	%	10.5–15.5	5.5–7.5	3.5–4.5	6.5–9.5	4.5–5.5
Al_2O_3	%	76–80	85–90	Nil	70–80	80–85
Fe_2O_3	%	1.5–2.5	1.2–1.5	1.5–2.5	1.2–1.8	1.2–1.8
MgO	%	Nil	Nil	62–65	2.5–3.5	Nil
Cr_2O_3	%	Nil	8.5–10.5	11.5–12.5	Nil	10
Physical properties						
App. porosity	%	18–19	13–16	12.5–14.5	4.5–6.5	4.5–6.4
Bulk density	g/cc	2.75	3.05	3.05	2.65	2.72
Cold crushing strength	kg/cm^2	440–520	700–900	850–960	>1200	>1200
Thermal expansion at 1500 °C	%	1.65	1.27	1.9	1.2	1.2
RUL	ta (°C)	>1550	>1700	>1600	>1600	>1550

the entire slag zone and almost the entire roof is constructed with water-cooling panels [4]. Currently, almost the complete roof is constructed with water cooling panels, and a thin protection layer of refractory is lined over the cooling panels by gunning castable refractory. The properties of different refractory materials used in lining furnace roof are listed in Table 5.1.

The damage of EAF roof refractory takes place due to the following reasons:

- Sudden change in temperature
- Effect of CO, CO_2 and SO_2
- Corrosion in presence of slag
- Radiation of arc
- Corrosion of dust
- Addition of hot metal.

At present, most of the steel plants are adding molten iron [5] in order to reduce cost. As the hot metal contains high Si, S and C, in the process of melting, they are transformed to SiO_2, SO_2 and CO/CO_2, respectively, which affect the lining refractory, and the lining life is reduced with the increase of those components in hot metal addition.

5.2.2 Side Wall

The side wall, the cylindrical part of the shell, extends vertically from the bottom of metal zone to the top of slag zone, including free board. Side wall refractory lining is experiencing severe thermal shock due to radiation of arc and corrosive action of hot metal and slag. It also experiences mechanical impact during charging. Depending on high operating temperature and corrosion on refractories, the side wall is divided into three different distinct zones, roof drip zone, hotspot zones and slag line. The most severe condition is in hotspot zone, which is near to arc, as well as it is corroded in the presence of slag. The main causes of refractory failure are as follows:

- High radiation from arc flame
- Arc position
- Thermal shock due to frequent temperature change
- Reflected heat from bath surface
- No slag protection coating layer

Nowadays magnesite, dolomite, chrome-magnesite and magnesia-carbon bricks are widely used in side wall. The side wall lining thickness is usually in the range of 450–600 mm, depending on the volume of the furnace.

In an EAF, slag line area and the hotspot in wall are the most critical zones. Special attention is taken in selection of refractories in hotspot areas as the life of furnace wall depends on the high erosion of hotspot areas. In the 1980s, MgO-C bricks are started to use in those zones to improve the refractory performances, and at present it is well accepted and widely used. A heat resistance thermal barrier anti-corrosive coating material is sprayed over the inner lining surface as a protective layer, resulted in reduction of specific consumption of refractories.

Above the slag line, the free board area experiences high-temperature arc flare, slag splashing, scrap impingement and flame impingement from burners or lances. In furnaces with water-cooled panels, magnesite carbon brick is the most common choice, with fused grain magnesite carbon brick in the hotspots.

In slag line, the type of refractory is selected on the basis of slag chemistry. To produce high carbon steel, lime rich FeO-SiO_2 slag is generated and highly basic high carbon content bricks are used, in stainless steel production, a more neutral fused grain magnesite chrome brick may be used. MgO-C refractory lining in side wall and monolithic lining in bottom is shown in Fig. 5.4. Properties of different quality bricks used in side wall are listed in Table 5.2.

Wear mechanism of EAF refractories

Thermal shock	Thermo-mechanical cracking During charging at around 350 °C to smelting at 1750 °C in 60–80 min
Oxidation	Attack of oxygen Oxidation in presence of FeO and MnO Steam from wet charge

(continued)

Fig. 5.4 Refractory lining in EAF side wall

Wear mechanism of EAF refractories	
Erosion	Hot abrasion in fast flowing molten steel and slag in tap hole and slag door Scrap charging causes wall damages Marangoni effect in turbulent slag line
Corrosion	Chemical attack Dissolution of MgO-C refractory in low basicity slag at beginning of smalting operation attack by gases, SO_2, SO_3, P_2O_5 etc.
Hydration	Disintegration of refractory in presence of water and steam from wet charge Water leakage from cooling elements during operation and maintenance

5.2.2.1 High Corrosion Resistance Material for Hotspot Area

The side wall is a highly corrosive area with high FeO fumes and basic slag tends to attack. Particularly at the hotspots of alternative current (hereinafter referred to as AC) EAFs, excessively high hot metal temperature promotes refractory corrosion because of slag properties as low saturation degree, high reaction rate coefficient and high mass transfer coefficient due to low viscosity. In addition, decrease in slag viscosity reduces the thickness of the slag-coat layer on the refractory surface, resulting in corrosion rate acceleration. Taking the configuration into account, application of durable refractory directly contributes to improved operation efficiency of EAFs through reduction in downtime for repair. High-grade MgO-C with enriched magnesia is used in that area.

Table 5.2 Properties of side wall refractory in EAF

		MgO-C unburnt bricks			MgO-Cr_2O_3 bricks (direct bonded)	
Properties		Type A	Type B	Type C	Type D	Type E
Application areas	Unit	Tapping hole	Hotspot	Slag zone	Slag line	Slag line tapping hole
Physical properties						
App. porosity	%	2.5	3.6	3.9	13–16	13–16
Bulk density	g/cc	2.98	3.04	2.83	3.25	3.27
Cold crushing strength	kg/ cm^2	450–500	550–650	400–450	860–900	940–980
Mod. of rupture at 1400 °C	kg/ cm^2	80–100	140–150	60–80	110–120	125–135
Chemical properties						
SiO_2	%	0.5–1.5	0.5–1.5	1.5–2.5	1.5	1.5–2.5
MgO	%	82–84	84–86	74–78	60–65	65–70
Cr_2O_3	%	NA	NA	NA	20–22	16–20
Fixed C	%	14–16	10.5–12.5	> 20	NA	NA

Conventionally, corrosion resistance of MgO-C bricks had been improved by using high-purity raw materials as well as optimizing metallic additives. The bricks have excellent corrosion resistance with improved thermal shock resistance. Different types of refractories used in side wall are listed in Table 5.2.

5.2.3 *Bottom of Hearth*

The bottom of the hearth is divided into two sections, sub-hearth and working hearth. Due to good resistance to slag steel and slag, magnesite or tar-impregnated magnesite brick are generally used in the sub-hearth, and this lining protects from breakout of the hearth, if the liquid penetrates the working lining. The sub-hearth is also constructed with few layers of high-temperature-fired magnesite bricks, along with high alumina bricks at the bottom of sub-hearth against the steel shell. Those additional lining prevents the shell from overheating due to its lower conductivity.

In working bottom, most electric furnaces are lined with a monolithic refractory lining of magnesite-based ramming mass or MgO-C bricks. In monolithic hearth lining, dry, vibratable high MgO ramming masses are used which develop strength during preheating of the bottom [6]. Fe_2O_3 is one of the impurities remain in ramming materials, which acts as a sintering agent and facilitates for hardening of the ramming mass. Large shape bricks and blocks are also used in hearth where the initial construction is totally made of carbonaceous bricks and during intermediate repair, ramming masses are used. In modern EAF operation, to homogenize the melt, and to attend uniform temperature distribution into the bath, gas stirring through bottom is a well-known practice. The inert gas is purged through porous refractory element attached in the bottom. Bottom refractory erosion was high due to turbulence created by purging gas.

5.2.3.1 Refractory Maintenance by Gunning

Gunning maintenance practices has improved the lining life as well as reduced the downtime for maintenance jobs. For repair of refractory lining, basic gunning material is in use. Gunning is used for repair in hotspots, tap holes and door breast areas. Refractory wear in AC furnaces is usually greater in the sidewall closest to the electrode, and by adopting the gunning maintenance the campaign is increased significantly. For manual gunning, human skill is a factor in the quality of the gunned patch, and gunning material consumption needs to be optimized to reduce the cost, which is based on expertise of the supervisors.

Conventionally, 40–95% MgO-based gunning material is used. Higher MgO content gunning material is used in higher operating temperature of high-power and high-capacity furnaces, while moderate temperature operations utilize lower MgO content gun mixes. In recent practice, the erosion of refractory wear profile is monitored by laser scanner, and the original refractory thickness is restored by quick gunning. The schematic diagram of wear profile is shown in Fig. 5.5.

Gunning repair becomes ineffective after several weeks of operation when in few critical areas the refractory remnant thickness becomes almost negligible. In most of such cases, the furnace is taken down for major repair and more than 30–60% refractory is replaced by new bricks and it requires long shutdown. During relining of the furnace or in major repair, the same refractory quality and thickness are reinstalled according to the initial refractory design. However, during minor repair alternate material with higher stick-ability material for thinner linings and lesser lining life is anticipated from the repaired lining, compared to life obtained from a new complete lining. One of the well-known technique for intermediate minor repair is gunning; for an example, applying MgO gunning mix effectively accommodates all the voids and cracks in the patched brickwork. The properties of gunning and patching refractories are shown in Table 5.3.

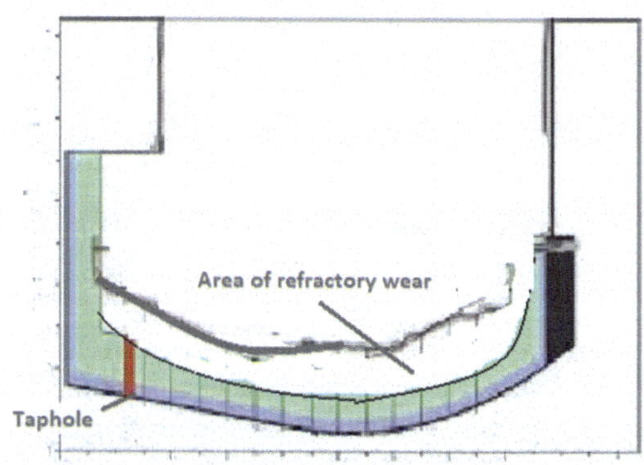

Fig. 5.5 Refractory wear profile by laser scanner

Table 5.3 Properties of gunning and patching refractory materials

Item		Gunning material (I)	Gunning material (II)	Thermal self-flow patching material
MgO min%		85	80	75
CaO max%		6	4	
C min%				10
Bulk density min g/cm^3	110 °C × 24 h	2.3	2.3	2.2
	1500 °C × 3 h	2.5	2.5	1000°C × 3 h
Cold crushing strength min MPa	110 °C × 24 h	8	8	12
	1500 °C × 3 h	25	25	1000°C × 3h
Modulus of rupture at RT min MPa	110 °C × 24 h	2.5	2.5	2
	1500 °C × 3 h	4	4	(1400 °C)

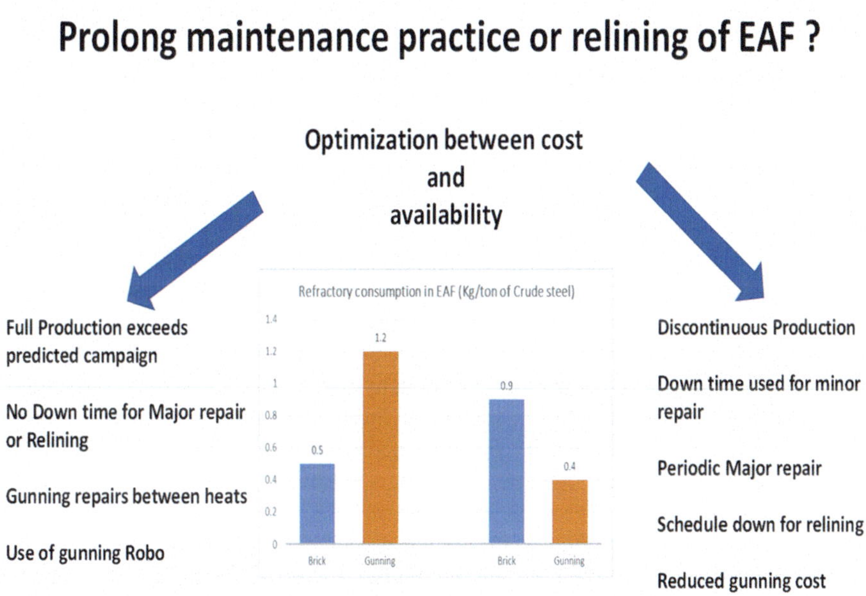

During major shutdown for brick replacement and patching jobs, the tap hole is completely replaced [7]. To extend the campaign and trouble-free operation, a proactive action plan is made for minor and major repair which is clubbed with major plant shutdown; Or, some operators have utilized the continuous patching concept where some brick are replaced every 2 weeks or even every week, and the complete lining is totally replaced only once or twice per year. These maintenance and patching decisions are usually dictated by the severity of the operating conditions as well as the company's economic status.

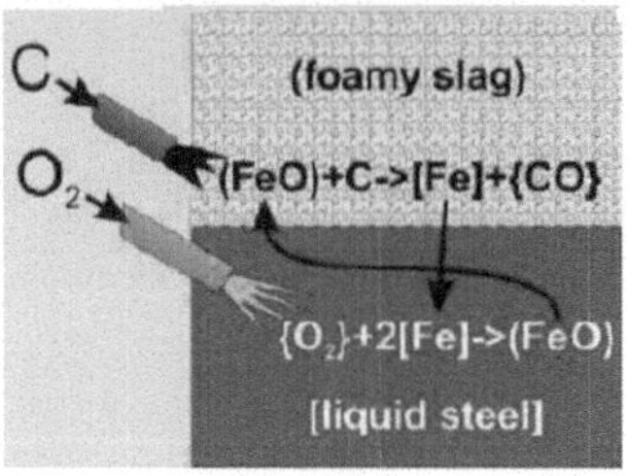

Foaming slag reduces refractory erosion - How it is formed ?
Slag foaming is induced by blowing Oxygen into the liquid steel and fine coarse coal injection into slag.
Foamy slag contains about 90% gas and 5 - 10 % solid Hence it reduces corrosion of refractory in contact of oxides.
To achieve good foaming, slag must have the right viscosity, which depends on slag composition and temperature.
The foamy slag is disposed off through furnace door. It is collected in slag pot below the furnace.

5.2.4 *Refractory Design in Bottom Tapping*

During past several years, the processes of EAF steel making had changed continuously, which increase the high demand of high refractory performances. Considerable changes have taken place after introducing gas stirring system through bottom as well as eccentric bottom tapping (EBT) for tapping the hot metal [8]. For effective homogenisation of the melt, faster refining, and to attend uniform liquid temperature in hearth, permeable blocks or porous refractory elements are introduced in the bottom to introduce inert gases for stirring the melt [9]. Although a tapping spout has been widely utilized, eccentric bottom tapping (hereinafter referred to as EBT) has been substituted for it in order to minimize slag contamination. Application of EBT is also advantageous for power consumption reduction, increase in alloy yield, decrease in flux consumption, and so on. Generally, MgO-C bricks through which a tapping hole is installed are vertically stacked together to form the EBT tapping tube. In tapping operation, downtime occasionally results from tapping tube opening failure caused by filler clogging. Clogging is induced by filler bridging, which is likely to occur at when filled sand is tightly packed in a worn joint of the stacked layer. Thus, a jointless monoblock structure is desirable to eliminate downtime resulting from clogging. If tapping is accompanied by splashing, smooth tapping operation becomes difficult. Since splashing is caused by adhesion and solidification of steel skull at the tip of the EBT tube, removal of skull adhesion, which requires downtime, is necessary. To cope with the issue, installation of Al_2O_3-C material that exhibits poor wettability to molten steel to the tip of the EBT tube was effective.

EAF with eccentric bottom tapping design, the tap hole is provided eccentrically. The tap hole is made of assembly of blocks as shown in Fig. 5.6. It is opened and closed by a lever during smelting and refining operation, and it is not required for tilting the furnace during tapping.

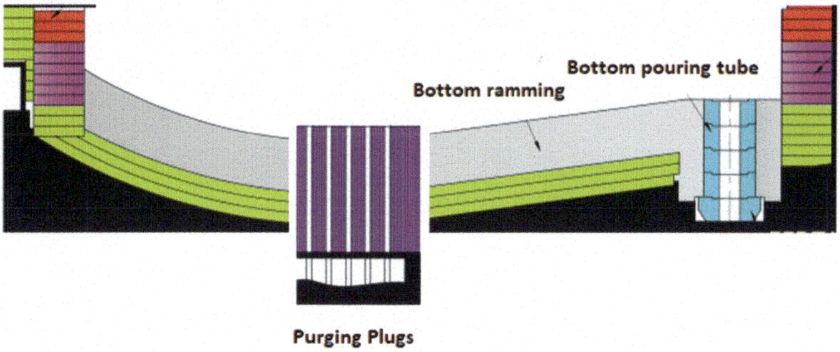

Fig. 5.6 Bottom assembly in EAF

5.3 Direct Current Furnace

Due to lower electrode consumption and stability of arc, electric arc furnace with direct current (DC) operation is becoming popular. In this type of furnaces, only one electrode is used, located at the centre of the roof, and it acts as a cathode. In this design, a conductive refractory is used in bottom, or a metallic element is embedded into the bottom refractory, which acts as anode. Different types of design for return electrode are as follows [9]:

- Use of metal pin electrodes with non-conductive refractories,
- Use of electrodes made of thick steel billet,
- Electrodes made of metal fins,
- Use of conductive bottom refractory.

In lining the bottom with conductive refractory, the refractory lining acts as the anode.

Multiple metal pins of 200–500 mm in diameter are used in pin-type electrode. These pins are installed vertically and penetrated into the refractory lining.

A six-inch-thick working lining consisting of carbon-bonded magnesia mixes containing 5–10 wt% carbon is used in a typical refractory lining design. These types of materials can be used either in hot or cold condition, depending on the types of bonds used. Three layers of magnesia-carbon brick are installed below ramming masses as intermediate layers. Herein, the residual carbon content of the bricks range from 10 to 14 wt%, and the billets are embedded in the bottom refractory. The rest of the hearth was lined with special magnesia-based ramming masses. The design of billet electrode embedded into refractory ramming mass exceeded 1500 heats.

In pin-type metal electrode design, multiple metal pins of 25–50 cm in diameter are used, and dry magnesia ramming mix is used for the entire hearth lining. This mix is rammed between the metallic pins. Alternatively, magnesia-carbon brick is used in the area around the anode. Adopting all those designs, a typical bottom life ranges from 2000 to 4000 heats is achieved; however, the campaign of furnace depends on quality of refractory used in slag zone, tap hole location, tap hole size, roof/delta construction, burner ports and expansion allowance provided in different locations.

5.4 Slag–Refractory Interaction

Oxidation of carbon in MgO-C refractory is one of the key factors for corrosion of refractory by hot metal and slag. Slag contains FeO, SiO_2 and TiO_2, which oxidizes carbon to form CO and the MgO-C refractory becomes unstable. The CO gas escapes out of the bricks forming channels, through which slag and metal enter into the refractory bricks and aggravate the corrosion reactions as shown in the following equations:

$$FeO + C = Fe + CO \tag{5.1}$$

$$SiO_2 + C = Si + CO \tag{5.2}$$

$$TiO_2 + C = Ti + CO \tag{5.3}$$

It is assumed that carbon is stable during EAF operation because of low partial pressure of oxygen in furnace atmosphere. It is also assumed that the brick hot face is usually covered by residual slag coating, resulting in suppression of oxidation of carbon. Oxidation phenomenon of bricks rarely found if the brick surface is coated by slag, and foamy slag plays a very important role for slag coating over the brick surface.

5.4.1 Corrosion of Roof Refractory Lining in Presence of FeO

The roof is lined with high alumina refractory with 80–85% Al_2O_3 and 4–8% SiO_2. The performance of high alumina refractory in the presence of FeO is not satisfactory due to low melting phases formation [10]. FeO reacts with SiO_2 to form fayalitic ($2FeO.SiO_2$), and it reacts with Al_2O_3 in the presence of excess SiO_2 to form iron–cordierite ($2FeO.2Al_2O_3.5SiO_2$), and the melting temperature is reduced to below 1300 °C as shown in Fig. 5.7. In order to extend the life, magnesia-chrome bricks with the combination of high alumina casted delta is in use.

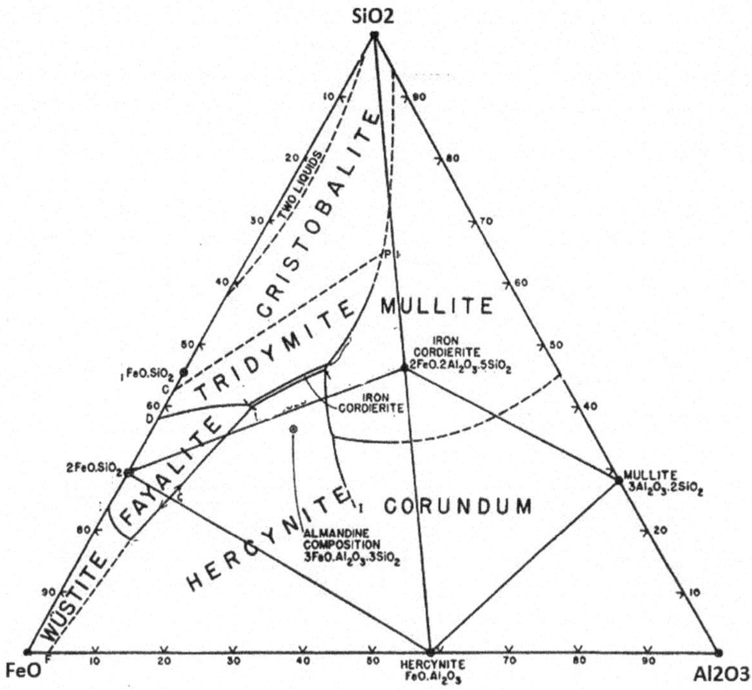

Fig. 5.7 FeO–Al_2O_3–SiO_2 phase diagram

5.4.2 Chemical Erosion in Presence of TiO_2

Analysis of after service MgO-C bricks, the presence of TiC and Ti-V-Fe alloys was observed in the reaction layer. During the smelting process, V_2O_5 and TiO_2 migrate in the brick and deposited around the particles of MgO. It is observed that the graphite elements have completely disappeared where the slag take up and just leaving rubble-like holes and a little round pore, however, areas where exit little slag have graphite residue and there are many metal particles with bright colour in the matrix of the bricks.

5.4.3 Reaction with EAF Slag

EAF slag compositions are described largely by its main components CaO–SiO2–FeO–MgO. Additional secondary components are typically 3–5% MnO, 2–5% Al_2O_3, 0.5–2.0% TiO_2 and 0.5–1.5% Cr_2O_3 [11, 12]. Those secondary components have very little effect on brick component of MgO concentration. It is observed that the change in basicity (CaO/SiO_2) has an impact on corrosion of MgO-containing brick. Wear of refractory is minimized when basicity is >2.0. At the beginning of the heat, aggressive acidic slag is formed in the EAF. The slag is consisting of olivine and wustite. This slag must be transformed as soon as possible (by an addition of dolomitic lime, MgO-briquettes, recycled refractories) to the slag of the required quality. Since lining is made of highly basic refractory, magnesite, Mag-Chrome or MgO-C, addition of lime (CaO) to neutralize the slag, it also helps in refining the hot metal. It is important to create a slag which has sufficient CaO saturation. The best case are the slags dually saturated with both CaO and MgO which satisfy not only the refractory wear decreasing and refining effectivity requirements, but also contribute to better slag foaming parameters. Slag saturation with magnesia plays the key role from the refractory-slag compatibility point of view. Solubility of MgO in the ternary CaO-FeO-SiO2 system is illustrated in Fig. 5.2. Some EAF shops operate with insufficiently MgO-saturated slags, so low refractory life (dissolution of lining to slag), high gunning mixes consumption and bad slag foaming quality are the consequences of this fact. Another common problem is a large content of iron oxide in the slag, which increases the total slag mass, decreases metal yield and at last destructs the lining (reaction between carbon of magnesia-carbon brick and iron oxide). Big importance should also be assigned to chemical interaction between MgO (of refractory lining) and aggressive acidic oxides—particularly silica or Al_2O_3, TiO_2, P_2O_5.

High SiO_2 in slag reacts with MgO of refractory to form low melting forsterite (2MgO. SiO2), and it also reacts with FeO at operating temperature to form the low melting fyalite ($2FeO.SiO_2$) as shown in Fig. 5.8 [13]. The melting point of those low melting compounds is 1480 and 1205 °C, respectively. Both the compounds form solid solution, resulted in further lowering the melting point.

The wear process uses to be the most intensive in the slag line area because of different slag and metal viscosity. Corrosion mechanism of magnesia-carbon bricks by slag (metal) in the EAF slag line location is based on the solubility and chemical interaction between slag–metal–refractory material (reaction with slag oxides, oxygen of metal) together with slag penetration into the pores and mechanical (erosive) destruction of lining surface (depending on the slag viscosity, wettability angle, interfacial tension between slag–metal–refractory).

5.5 State-of-the-Art Operating Practice and Refractory Performances

The slag foaming phenomenon has been one of the most discussed issues in the EAF studies during last few decades. May be the greatest benefit of foaming technology is a remarkable refractory wear decline, campaign duration prolongation and decrease of gunning mixes consumption. There was achieved more than triple increase of EAF sidewalls lifetime after the foaming slag practice implementation. The principle of mentioned excellent success originates in the fact that during slag foaming in the

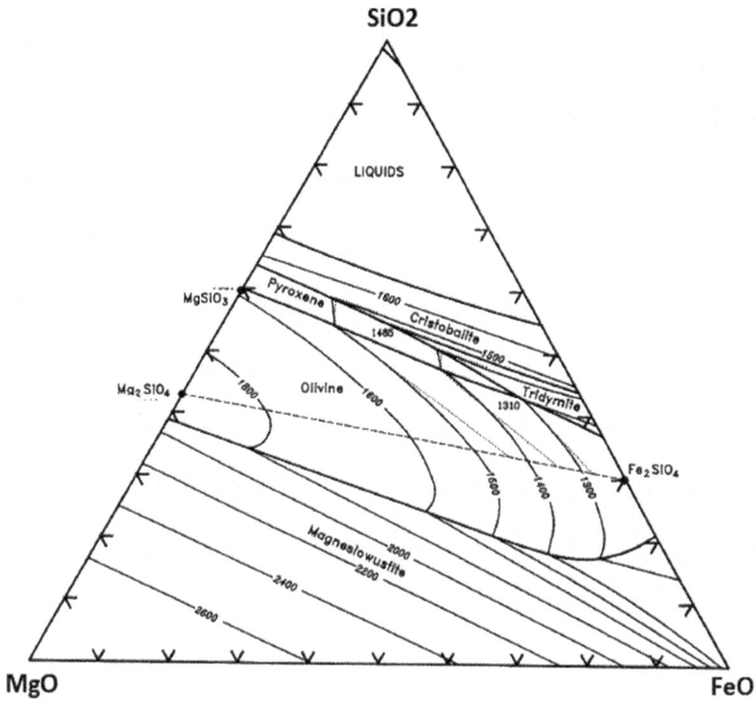

Fig. 5.8 Phase diagram of FeO–SiO2–MgO system

EAF a significant wear mechanism change takes place as follows: Refractory corrosion by foamy slag is inhibited owing to the presence of CO gas bubbles large quantity (lower reaction surface between slag and refractory), also arc is covered with slag bubbles, which protects the refractory from overheating [7].

The foaming slag operating practice has the following advantages:

- Reduction of tap to tap time, around 40% reduction, by raised melting capacity and energy reduction,
- Improved total economy of steel-making practice through EAF,
- Extended refractory campaign life.

The foaming system is based on the carbon injections directly applied into the bath [14]. The foamy slag is a result of massive CO gas generation within the slag, when injected either carbon into slag or oxygen in liquid steel bath.

$$2Fe + O_2 = 2FeO$$

$$FeO + C_s = CO_g + Fe$$

Table 5.4 Traditional to modern refractory practice for different zones of EAF [14]

Areas	Traditional concepts	Modern concepts
Hearth	Hearth refractory mixes—sintered MgO (80% MgO)	Sintered MgO with 86% MgO and microporous insulation
Side wall	MgO-C bricks—fused chemically bonded (MgO: 96.5% and C: 12%)	MgO-C bricks with 100% fused MgO, use of anti-oxidant, chemically bonded bricks (MgO: 98%, C: 12–16%), at hotspot, high-grade fused magnesia with microporous insulation
Free board	MgO-C bricks—fused chemically bonded (MgO: 96.5% and C: 12%)	MgO-C bricks with 100% fused MgO, use of antioxidant, chemically bonded bricks (MgO: 98%, C: 12–16%)
Roof	Water-cooled jacket with high alumina/magnesite brick lining	Water-cooled in complete roof with cast delta with tabular alumina-based no cement castable. Gunning in rest of the portion

Level of foaming depends on slag viscosity, surface tension and temperature. Those basic properties are governed by slag chemistry, basicity of slag and FeO content. Increase in FeO concentration improves slag foaming. The advantages of using foamy slag on refractory performances are as follows:

- Foamy slag is saturated with CO, which protects slag line refractory from wearing and oxidation.
- Direct carbon injection in the bath reduces carbon loss of MgO–C or AMC bricks.
- Foamy slag increases slag depth of arc, hence higher slag depth allows working with "Long arc", which is to be fully covered by slag foam, hence it protects direct corrosion of refractory in slag line and power transfer to the lining including the roof is minimized.

Continuous improvement in refractory quality has also brought a significant improvement in campaign life of EAF furnace as shown in Table 5.4.

References

1. Qin Jie, Qi Jian-ling, Analysis on the refractory Erosion of the Electric Arc Furnace and study on the measures of protecting the furnace lining, Advanced Materials Research Vols. 602-604 (2013) pp 2082-2086
2. Jorge Madias, Electric Furnace Steelmaking, Chapter December 2013.
3. Gregor Lammer, Akram Yaseen, Ronald Lanzenberger, Andreas Rom, Ashraf Hanna,Manuel Forrer, Markus Feuerstein, Franz Pernkopf, Nikolaus Mutsam, Advanced Data Mining for Process Optimizations and Use of AI to Predict Refractory Wear and to Analyze Refractory Behavior, AISTech 2017 Proceedings.
4. Stefano Miani, Alessandro Fornasaro, Luca Gemo, Mauro Milocco, Safe and Productive Electric Arc Furnaces, AISTech 2013 Proceedings,
5. Junichi Nakamura, Seikichi Higo, Yoshiyuki Udoh, Lifetime Extension of the Water-Cooled Duct, AISTech 2016 Proceedings.

6. Marshall Khan, Suresh Mistry, Vittorio Scipolo, Next-Generation EAF Optimization at ArcelorMittal Dofasco Inc., AISTech 2013 Proceedings.
7. Marcus Kirschen, Ashraf Hanna, Karl-Michael Zettl, Improvement of EAF Process and Refractory Consumption by Advanced Slag Modeling, AISTech 2015 Proceedings.
8. Borovský T., Kijac J., Domovec M., Analysis of the slag and metal influence on the life of electric arc furnace hearth refractory lining, Acta Metallurgica Slovaca, 15, 2009, 2 (77–85).
9. Terrence Koehler, Kristin Wallace, Frank Wheeler, Frank Atkinson, Yakov Gordon, David Rudge, and, Thomas Ma, Operational Benefits and Design Considerations for a Stationary Electric Furnace for Continuous Steelmaking, AISTech Proceedings, 2010,Vol 1.
10. F. López, A. Farrando, M. López, L. Picco, M. Loeffelholz, Slag Modeling for Optimizing the Use of Fluxes in a DRI-Based Steelmaking Operation, AISTech 2016 Proceedings.
11. Mohamme Md ERAIKI, Activity of Silica in the Slag of an Electric Arc Furnace Using Direct Reducedlron for Steelmaking, ISIJ International, Vol. 35 (1 995), No. 7, pp. 845–850.
12. Masakazu Iida, Atsushi Torigoe, Hisashi Tomiya, Hiroki Yoshioka, Recent Application Results of Refractories for EAF, AISTech 2017 Proceedings.
13. Dipl. Ing. Martin Novak, Dr. Michal Pribyl, Refractory wear in EAF viewed from The foamy liquid slag melting processes comparison Eurogress, Aachen, Germany, 2010.
14. Zushu LI, Kusuhiro MUKAI and Zainan TAO, Reactions Between MgO-C Refractory, Molten Slag and metal, ISIJ International, Vol. 40 (2000), Supplement, pp. sl ol–SI 05.

Chapter 6
Refractory for Hot Metal Transport and Desulfurization

6.1 Introduction

The transport of molten pig iron from the blast furnace to the steel-making shop is performed in open top transport ladle or torpedo ladle car. Capacity of open top ladle was 60–80 tons. The problems faced by the operators for transporting the hot metal through open top ladle, from iron-making to steel-making shop, are summarized below:

- Iron loss due to oxidation of Fe,
- Temperature loss of hot metal during transport,
- Poor logistics of transportation due to small ladle capacity,
- Low life of open top ladles,
- Safety.

To eliminate those disadvantages, and with increased productivity of hot metal, the open top ladles are gradually replaced by large size torpedo ladles of average carrying capacity of 200–460 tons.

In early days, when hot metal was transported by open top ladles, the hot metal was collected in a mixer of capacity 180–300 tons, and it acted as a hot metal reservoir to maintain the hot metal availability for steel-making processes. By 1970, the large capacity transport vessel, torpedo ladles are introduced, and at present it is used in most of the steel plants. In most of the steel plants worldwide, the torpedo ladle is not merely used as a transport vessel, but desulfurization also being carried out in it. The transport process is shown in Fig. 6.1.

About 70–90% S in blast furnace hot metal transferred from coke charge. In coke, S presents in three different forms, organic, sulphide and sulphates. In steel, sulphur lowers corrosion resistance toughness and electromechanical properties of steel. Deterioration in deep-drawing properties is also significant in the presence of high sulphur. Hence it is mandatory to reduce the sulphur level before further steel-making process and from technological point of view, much more favourable to

S. Biswas, D. Sarkar, *Introduction to Refractories for Iron- and Steelmaking*,
https://doi.org/10.1007/978-3-030-43807-4_6

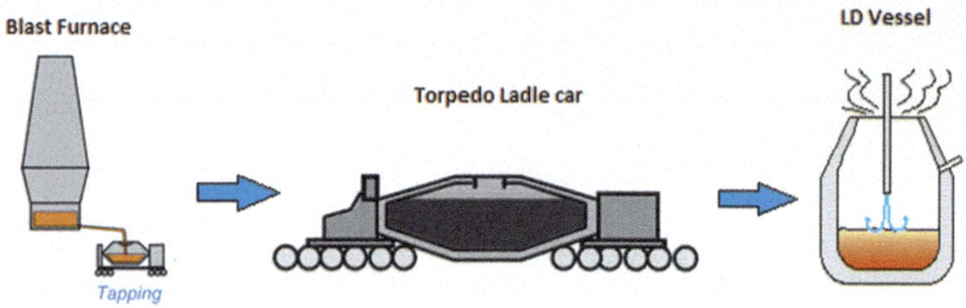

Fig. 6.1 Transport of hot metal

Fig. 6.2 Torpedo ladle car

carry out desulphurization in pig iron than in steel. Hot metal is transferred from torpedo ladle to transfer ladle for desulphurization before teeming hot metal in LD vessel. In most of the modern steel plants, desulphurization is being carried out in torpedo ladle also.

6.2 Torpedo Ladle Car

The ladles are called "Torpedoes" because they are shaped like a torpedo with conical ends and cylindrical barrel shaped at the central body, as shown in Fig. 6.2. These ladles have a central opening at the top through which the hot metal is tapped from blast furnace and is delivered into the transfer ladles after transporting to steel-melting shop as the entire ladle is slowly rotated using an electromechanical device. Torpedo ladles are placed on cars that are hauled by locomotives from the blast furnace to the steel-melting shops. Because of large thermal mass and a small mouth opening, the heat loss from such ladles during transit is significantly low. This is quite different from open top ladles, where excessive heat loss cannot be prevented, this is one of the prime reasons for phasing out of open top ladles.

6.2.1 Refractory Lining Practices

Refractory design and quality of refractory have undergone a continuous change in the years 1970–1990. The main reasons for the changes are severe working condition, increased hot metal-carrying capacity, high hot metal tapping temperature from blast furnaces and introduction of desulfurization in torpedo ladles. Different types of refractories are used in the following areas [1]:

- Working lining
- Intermediate/safety lining
- Insulation
- Monolithic lining in mouth.

The schematic diagram of refractory lining is shown in Fig. 6.3.

Challenges on Refractory Design of Torpedo Ladles [1]

1. The challenges are to transport hot metal with profound reliability from the safe hot metal transport point of view. Any leakage of hot metal or breakout during transit causes fatal accident, railway track jam by spillage of hot metal, severe equipment loss.
2. Temperature drop of hot metal during transport is a major issue. Heat loss by conduction through refractory lining to be minimized by proper selection of refractory quality.
3. Increase of shell temperature increases downtime for refractory maintenance and reduces availability. The shell temperature must be maintained within 300 °C at any point of time during transport of hot metal.
4. Online shell temperature monitoring is introduced in most of the steel plants worldwide to improve reliability.

(continued)

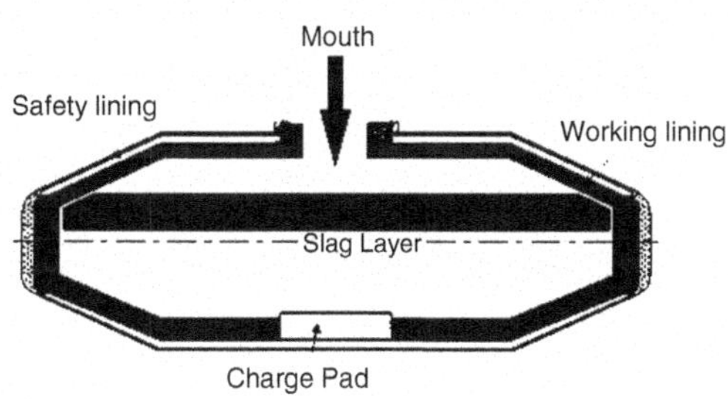

Fig. 6.3 Refractory lining in torpedo ladles [1]

5. Severe heat loss from hot metal was observed through mouth opening during hot metal transport; it is minimized by covering mouth by refractory-lined lids.

6.2.1.1 Old Working Lining Concept

Since 1970, alumina refractories were the main lining material in torpedo ladle cars. In past for many years, Bauxite-based refractory was in regular use. Under certain conditions of increased hot metal temperature and aggressive slag corrosion, phosphate-bonded and fired corundum-based bricks were used in 1970–1976.

During draining of hot metal, quantity of slag tapping to torpedo ladle is very less, but during transit the Fe and other elements dissolved in hot metal are oxidized and slag started generating [2]. The slag in contact with refractory surface infiltrates the porosity, up to the isotherm of metal solidification. The physico-chemical properties of the infiltrated layer change from the original refractory which results in change of thermal expansion characteristics and cause mechanical spalling of bricks. These impregnation-spalling phenomena appears in all the immersed parts of the torpedo ladles. This phenomenon is more prominent in aluminosilicate refractories. The penetration is higher when the lining is highly insulated.

If desulphurization operation is carried out in torpedo ladles, where lime and calcium carbide (CaC_2) are added in torpedo ladles, and the basicity of slag is increased to more than 2.0. Dolomite lining shows superior performance. Considering the effect of dolomite (MgO-CaO) refractories in contact with iron oxide, dolomite can dissolve approximately 20–22% FeO without a liquid-phase development at 1550 °C, and it is also a good resistant to high CaO-containing slag in the presence of SiO_2 and low Al_2O_3. Further, to improve corrosion resistance, carbon is often added in dolomite refractories. The carbon mainly present as graphite or pyrolysed carbon from resin or pitch. In MgO-CaO-C refractories, corrosion by iron oxide is reduced as carbon reduces some of the iron oxide. In addition, non-wettability of carbon limits the liquid slag penetration through pores as well as joints.

Why Refractory Deteriorates in Torpedo Ladles?

The cycles of filling and emptying of hot metal in the torpedo causes severe thermal shock to the refractory lining. Further, corrosion and erosion by hot metal and slag also contribute to the wear/damage to the refractory lining. The mechanism of wear of refractory lining can be explained as follows:

- Corrosion in working lining: Corrosion in working lining is caused by infiltration of FeO in slag reacting with refractory components viz. Al_2O_3

(continued)

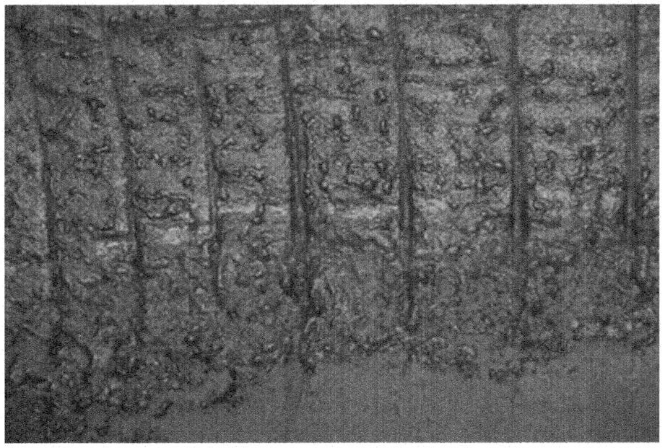

Fig. 6.4 A typical cobbling effect in torpedo lining

and SiO_2 to form low melting phases of hercynite ($FeO.Al_2O_3$, MP = 1550 °C) and fayalite ($2FeO.SiO_2$, MP = 1210 °C) and cordierite ($2FeO.2Al_2O_3.5SiO_2$, MP = 1320 °C).

- CaO in slag reacts with aluminosilicate refractory to form low melting compounds like anorthite ($CaO.Al_2O_3.2SiO_2$, MP = 1550 °C). Formation of these low melting phases.
- Spalling due to thermal shock: The working lining refractory is subjected to severe thermal shock during pouring and waiting period of empty torpedo for the next cycle causing spalling.
- Cobbling effect: It is observed that during operation for long campaign, the joints of the bricks opened up due to thermal stress, and the brick surfaces became curved which caused for change in dihedral angles and slag/hot metal penetrates within the bricks through joints. This phenomenon is known as "Cobbling effect", and it happens due to high compressive stress generated at mortar joints, as shown below (Fig. 6.4):

6.2.1.2 Present Refractory Lining Practice in Torpedo Ladle Cars

Use of zonal lining concept to the specific areas of the car is an important element for solving the severe refractory wear problem and has a direct impact on cost reduction of refractory lining. The stress on the refractory material depends on the zones within the ladle. In order to obtain even wear profile, which would reduce cost, improve service life and car availability; it is necessary to develop the most adequate refractory design. The different areas in the ladle such as impact zone, barrel area, venture area and neck area have different needs. Use of zonal refractory lining in a 300-t TLCs is shown in Fig. 6.5.

Charge Pad Area To meet the increased performance demand in modern torpedo ladles, in the early nineties, a new concept had developed, and on a global scale

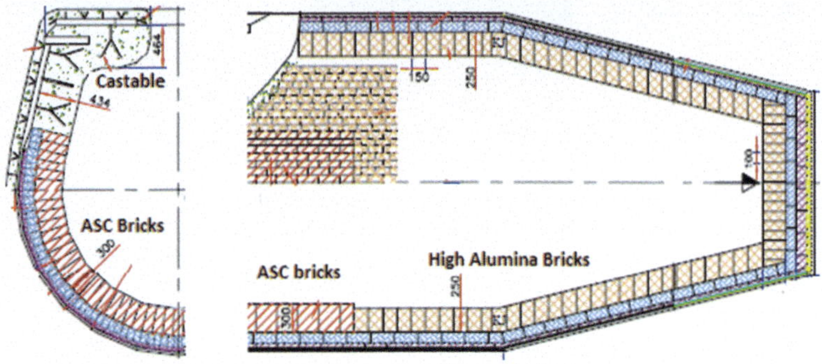

Fig. 6.5 Refractory lining pattern in 300-ton torpedo ladle car

unburned alumina-silicon carbide-carbon (ASC) bricks were introduced in most critical area of charge pad. This material has become the most durable refractory material for such application. The essential properties required for this application are as follows:

- Excellent impact resistance to resist impact of hot metal stream during filling.
- Corrosion-resistant towards hot metal and slag.
- Excellent thermal shock resistance.
- Resistance to penetration of hot metal in refractory bricks.

Addition of carbon in alumina refractory improves its thermo-mechanical properties with respect to corrosion resistance, non-wetting towards hot metal and slag and thermal shock resistance. The principal limitation of this type of refractory is related to oxidation of carbon. Oxidation resistance depends on temperature, oxygen partial pressure and quality of carbon. Usage of anti-oxidizing agents such as SiC, Al metal powder, Si–metal powder is extremely important as it protects carbon from oxidation and reduces porosity, resulting in reduction of hot metal penetration.

Barrel and Venture Area Aluminosilicate refractory is used in lining of barrel and venture area. Reaction of refractory with FeO in blast furnace slag leads to corrosion by formation of low melting phases. To reduce the effects of penetration of hot metal and corrosion, dense quality 85% Al_2O_3 refractory with mullite addition is used.

Use of carbon-containing refractory in barrel and venture area is also practised in many steel plants across the world. The lining concept proposes different quality bricks in slag and metal zones. In many integrated steel plants, the technology was adopted in the large size of 320–460 tons capacity torpedo ladle cars by lining them with alumina-carbon and alumina-magnesia-carbon bricks. However, in the present context, the techno-commercial benefits to be evaluated.

Back Up Lining The thermal conductivity of working lining, Al_2O_3-SiC-C charge pad in particular, is 2.5 times higher than the thermal conductivity of conventional fireclay refractory. To maintain shell temperatures below 300 °C, use of 42% Al_2O_3 bricks followed by high-strength insulation bricks and non-compressible ceramic

Table 6.1 Typical zone thickness and properties of different qualities of bricks

Quality of bricks	Al_2O_3-SiC-C	85% Al_2O_3	42% Al_2O_3	42% Al_2O_3	Insulating board
Area of application	Charge pad	Barrel & Venturi	Safety	Insulation	On shell
Lining thickness	300	250	114	37	12.5
Chemical properties					
Al_2O_3 (%)	72	85	42	45	50
SiO_2 (%)	13	12	53	50	35
SiC (%)	6.5	Nil	Nil	Nil	Nil
Fe_2O_3 (%)	0.8	1.5	1.2	0.8	–
C (%)	11.5	Nil	Nil	Nil	Nil
Physical properties					
AP (%)	7	20	17	N/A	N/A
BD (g/cc)	2.97	2.9	2.3	0.93	N/A
CCS (kg/cm^2)	400	500	300	15	N/A
PLC at 1000	N/A	±0.3	±0.5	−0.7	−0.1
PCE	N/A	38	32	N/A	N/A
RUL (ta)	N/A	1650	1450	N/A	N/A
Thermal conductivity	5.2 at 1000 °C	2.8 at 1000 °C	1.55 at 1000 °C	0.34 at 800 °C	0.15 at 800 °C

insulation board against the shell practised. The various properties of different qualities of refractory material are shown in the Table 6.1.

Operational Influence on Performance of Refractory

Performance of refractory depends on operational practice of torpedo ladle cars. Some important points related to the operational practice are as follows:

- Periodically (after transportation of every 20,000 THM) the mouth of the torpedo ladle needs to be cleaned, and the skull removed very carefully to avoid damage to refractory lining by impact of the cleaning equipment.
- After each heat, the remaining slag must be taken out to minimize corrosion of refractory.
- The liquid metal to be maintained above 1380 °C to avoid skull formation.
- The useful volume of the car to be kept within 80% of the available total volume to avoid spillage of hot metal and slag.
- Preheating of refractory lining is required if the refractory hot face temperature drops below 800 °C due to long outage of the torpedo ladle car.

Al_2O_3-SiC-C (ASC) Refractory for Torpedo Ladle

ASC bricks were developed for torpedo ladles to improve corrosion resistance against FeO, Na_2O and CaF_2 (for the process of desulphurization). ASC bricks are composed of alumina as base material, graphite, which provides resistance to spalling, corrosion resistance due to non-wetting and slag penetration, and SiC is added to protect graphite from oxidation, bonded with high carbon-containing pitch and resin, with some metallic powder as anti-oxidants to protect the carbon from oxidation. Main features of using ASC bricks are listed below:

1. ASC refractories have high application temperature as no low melting phase occurs between Al_2O_3 and C.
2. ASC refractory is a composite of Al_2O_3–C with SiC, and it has lower thermal expansion than alumina.
3. Graphite has very high thermal conductivity, which imparts high thermal conductivity in the Al_2O_3-SiC-C composites.
4. Thermal shock resistance of ASC is very high because the thermal expansion is low and the thermal conductivity is high.
5. ASC bricks prevent the penetration of slag and molten steel because of the non-wettability of carbon, similar to other graphite-based refractories.
6. Better ability to absorb stress, thus keeping down the amount of irregular wear due to cracks.

Currently ASC bricks are used in all areas of wear lining. But the composition of ASC bricks is different in different zones as shown in Table 6.2. This zonal lining is possible because of continuous strive for improvement in refractory quality and cost optimization, so there are products those have desired wear resistance characteristics for every part of the torpedo ladle [3].

Development in Torpedo Ladle Refractories

The wear rate of Al_2O_3-SiC-C refractory depends on the hot metal pretreatment methods, temperature of hot metal and time of treatment. Several developments in brick quality had been undertaken as follows [3]:

1. **Reinforcement of refractory matrix**—Alumina content of the matrix is increased so that reaction with penetrated slag would be minimum. To strengthen the matrix, aluminium-oxynitride (ALON) and beta-alumina are examples for alumina addition.
2. **Addition of metal powders**—It improves oxidation resistance in carbon-containing refractories. Al, Si, Al-Mg, etc., are used. In few brick compositions, more than one metal powder is added to improve properties in all temperature range, and the oxidation resistance has been improved more than if the metals are added individually. The addition of such metal powders also improves high-temperature strength, corrosion resistance

(continued)

Table 6.2 Tentative properties of ASC bricks

Properties		Charge pad metal zone	Charge pad slag zone	Rest of area	Rest of area
Quality of bricks	Unit	Al_2O_3-SiC-C	Al_2O_3-SiC-C + MgO	Al_2O_3-SiC-C	Al_2O_3-C
Al_2O_3	%	68–72	64–68	70–74	78–82
SiO_2	%	3.5–5.5	3.5–4.5	6.5–8.5	8.2–10.5
SiC	%	10.5–12.5	10.5–4.5	10.5–12.5	Nil
Fixed C	%	13–16	14–16	13–15	10.5–12.5
MgO	%	Nil	3.5–7.5	Nil	Nil
Porosity	%	3.5–5.5	5.5–70	3.5–4.8	4.5–5.5
Bulk density	g/cc	3.1	3.1	3.2	3.1
Crushing strength	Mpa	50–60	60–70	55–60	60–70
H. MOR at 1400 °C	Mpa	6.5–8.5	6.5–8.5	5.5–7.5	4.5–4.8
Thermal shock resistance at 1000 °C (air quenching)	Cycles	>20	>30	>20	>20
Thermal conductivity at 1000 °C	W.mK	5.2	6.5	5.2	4.8–5.0

and abrasion resistance. At high temperature, AlC and SiC are formed, and they improve the thermos-mechanical properties of ASC bricks.

3. **Formation of protective glass films**—Addition of optimum quantity glass or glass former makes a coating over carbon particle and over SiC grains, thus protects from oxidation. Borosilicate glass or phosphate glass are the examples for addition. This protective glass coating also prevents oxidation from back side (cold face) of the lining.
4. **Reduction of wear at brick joints**—To reduce the wear of mortar joints, an SiC added alumina-carbon mortar is developed with non-aqueous bond. MgO is also added in the mortar composition so that spinel ($MgO.Al_2O_3$) is formed with expansion at high temperature and hence shrinkage of mortar and joint opening at operating temperature was avoided.

Thermal Insulation in Torpedo Ladle Car

One of the major disadvantages of using high-conducting ASC bricks in torpedo ladles is high heat loss through the wall and results in high shell temperature during operation. To reduce heat loss and shell temperature while maintaining torpedo capacity, the intermediate lining thickness is reduced to accommodate the insulation layer. The properties of the insulating board in summarized in Table 6.2. The use of insulation board was expected to significantly improve the hot metal temperature during the residence time.

Effect of Insulation Lining in Torpedo Ladles

The effective lining thickness and lining of different layers to be optimized on the basis of shell temperature and thermal profile is shown in Fig. 6.5. The requisite of the brick lining design is to maintain the shell temperature below 300 °C to avoid overheating of the shell and its permanent deformation [4].

The thermal profile with change of insulation lining thickness is shown. With the insulation lining of 35-mm thick (Fig. 6.6a), the shell temperature is 242 °C and the interface temperature of ASC bricks and 42% Al_2O_3 bricks is 1100 °C and for insulation lining thickness of 25 mm, the shell temperature is higher by 14 °C, but

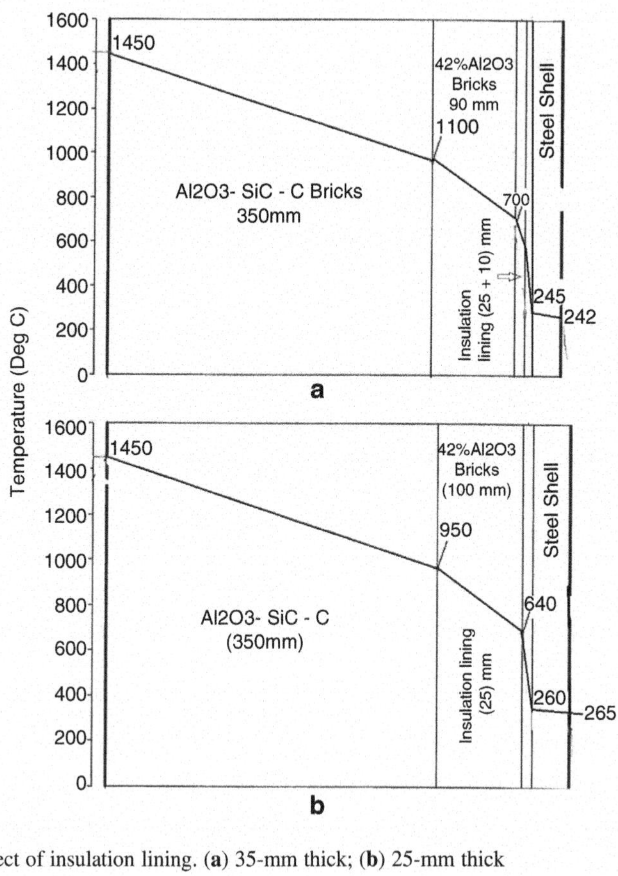

Fig. 6.6 Effect of insulation lining. (**a**) 35-mm thick; (**b**) 25-mm thick

the interface temperature is reduced by 150 °C (Fig. 6.6b). With higher insulation thickness, the penetration of hot metal and refractory corrosion would be much higher as the interface temperature is very close to the freezing point of hot metal (1150 °C). Hence, it is recommended to optimize the insulation lining thickness to avoid severe corrosion of bricks due to high temperature as well as to maintain the safe shell temperature.

6.2.2 Refractory Lining Design

Design of refractory lining is consisting of the working lining, intermediate lining or safety lining and insulation lining. The torpedo ladle is divided into the following zones:

1. Charge pad area
2. Venture or conical area
3. Barrel area
4. Mouth.

The working lining thickness of charge pad area is 50–80 mm higher than the rest of the areas to accommodate the additional load of hot metal pouring impact and oxidation resistance. The major part of that area is exposed to mouth opening and prone to severe oxidation when the ladle is empty. Few steel plants prefer to lay the charge pad bricks without using jointing mortar, and this type of bricks lining is known as "Dry-lining" [4, 5]. The key type (side arch) bricks lining in charge pad cannot lock in complete ring lining due to the mouth opening, and the bricks are locked in castable lining at the two sides as shown in Fig. 6.7.

To avoid dislodging of bricks, brick lining in "Bonded Design" is recommended in charge pad area. High level of masonry skill is required for laying bricks in bonded design. Ring lining may be followed in rest of the portion. Bonded lining and ring lining design are shown in Fig. 6.8.

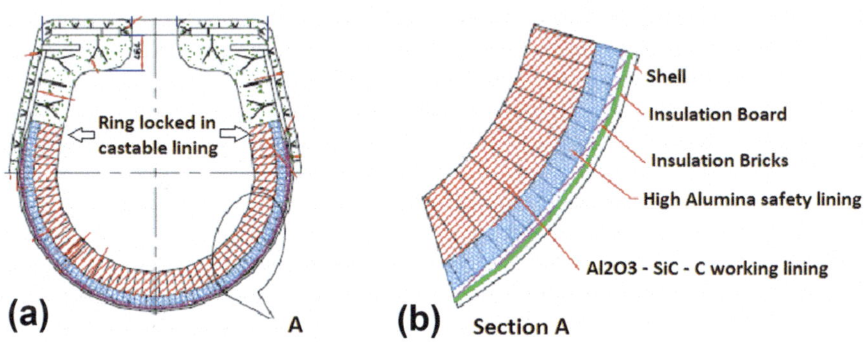

Fig. 6.7 (**a**) Refractory lining design in torpedo ladle. (**b**) Lining design in charge pad area

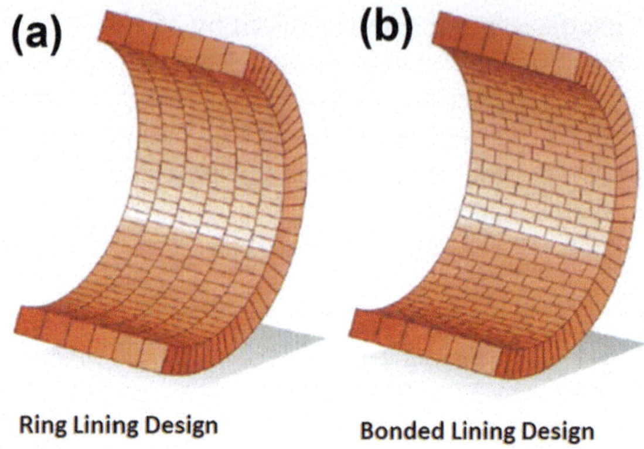

Fig. 6.8 (**a**) Ring lining and (**b**) bonded lining design for torpedo ladle

6.2.3 Refractory Maintenance Practices

Torpedo ladles are becoming very important for increased demand on maintaining availability for hot metal transport under increased productivity. An early unplanned outage of torpedo ladle results in a severe disturbance of the production processes. Hence the increase in campaign life is a paramount important, and the life of Torpedo ladles is enhanced by introducing critical maintenance practice. A schedule is followed for taking out torpedo ladles from the operation for inspection and refractory maintenance. The following activities are carried out for the maintenance:

- **Condition monitoring**: In last few decades, monitoring of shell temperature was the only available means to identify the refractory wear location. Regular shell temperature monitoring of loaded torpedo cars indicates the erosion of refractory. Programmed maintenance by hot gunning is done based on the thermography report on shell temperature. In recent practice, a laser monitoring system has been developed and in use worldwide. It is a mobile monitoring unit or fixed installed. Ferrotron, a division of Minteq International GmbH [2, 6], has introduced the laser monitoring unit, LaCam. Besides determination of the residual brick thickness, current units enable to detect the wear rate and wear speed of refractory. It is a very rapid measuring system, which is able to monitor all parameters for the 360° scan of the inner torpedo ladle within 40 s. A schematic picture of the laser measurement is shown in Fig. 6.9.
- **Gunning practice**: Hot gunning is done manually by inserting the gun pipe through the opening at mouth. The efficiency of gunning depends on the quality of gunning castable, performance of the machine used and water consistency of the gunning mix. The critically damaged refractory lining needs superior-quality gunning castable with excellent high-temperature adherence, while for regular application conventional high-lime-containing gunning castable is found satisfactory.
- **Shotcreting**: During shotcreting dense low cement castable is mixed with water and other liquid binders in a mixer machine. This mix is pumped and sprayed on

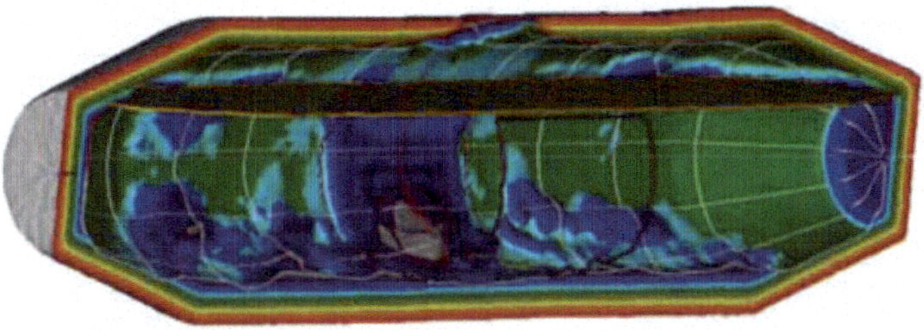

Fig. 6.9 Laser scanning of loaded torpedo ladle [2]

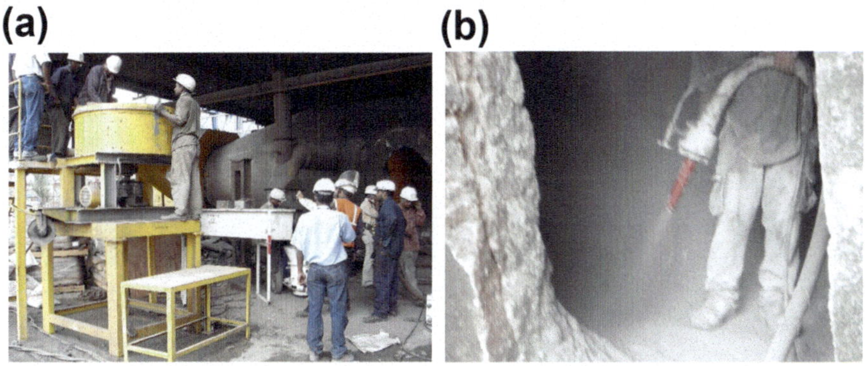

Fig. 6.10 (a) Arrangement for shotcreting. (b) Shotcrete in progress

the old refractory surface under high pressure to develop refractory thickness [7]. The process of shotcrete is shown in Fig. 6.10.

The shotcreting resulted in several advantages over conventional gunning processes:

- As in shotcreting process, the castable powder is premixed before pumping, rebound losses are only 2–3% compared to losses in gunning which is about 8% (in gunning, dry castable is pumped and water is added at the spray nozzle)
- A new refractory lining of 100–200 mm can be developed by the process of shotcreting, but in gunning only 50–100 mm can be lined. Hence, durability of shotcrete is much higher than the gunning process and results in significant reduction of downtime for maintenance can be obtained.
- The new refractory layer formed during shotcreting is denser than the new layer formed by gunning, which would improve penetration resistance and corrosion resistance.
- Downtime for gunning maintenance can be significantly reduced after shotcreting operation, resulting in reduction of gunning cost and improving availability.
- Significant increase in campaign life—comparative properties of gunning mix and shot-creting castable are shown in Table 6.3.

Table 6.3 Properties of gunning mix and shotcreting castable

Properties	Gunning mix	Shotcreting castable
Al_2O_3 (%)	84.2	60
Fe_2O_3 (%)	1.1	0.9
CaO (%)	–	1.4
Maximum service temperature (°C)	1700	1600
Bulk density (g/cc)		
At 110 °C/24 h	2.27	2.4
At 800 °C/3 h	2.51	2.35
At 1000 °C/3 h	2.57	2.35
Cold crushing strength (kg/cm^2)		
At 110 °C/24 h	350	500
At 800 °C/3 h	220	650
At 1000 °C/3 h	150	650
Permanent linear change (%)		
At 800 °C/3 h		(−) 0.4
At 1000 °C/3 h	(−) 0.3	(−) 0.5
Thermal conductivity(w/mK)		
At 800 °C	2.1	1.56
At 1000 °C	2.7	1.6
Hot MOR (kg/cm^2)		
At 1000 °C	65	80

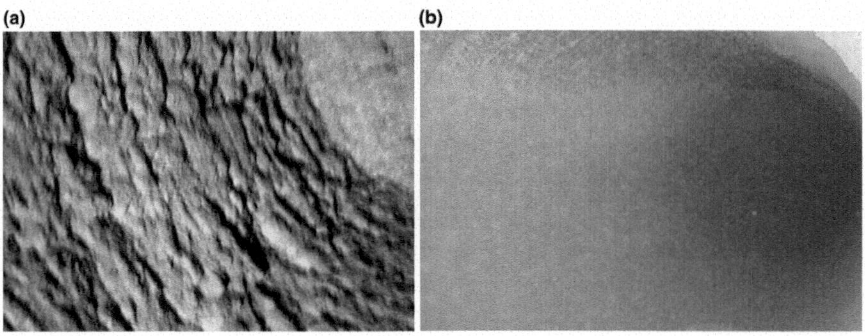

Fig. 6.11 (**a**) Before and (**b**) after shotcreting in torpedo ladles

After shotcreting, additional life of 850–1000 trips (about 264,000–300,000 THM) can be achieved. Figure 6.11 shows the condition of torpedo ladle in before and after shotcreting process [7].

Future Challenges

Refractory lining in TLCs face severe wear and higher shell temperatures, leading to lower availability and reduced campaign. With the present increased productivity, generation of hot metal is all set to go up, hence maintaining availability of TLCs and logistics to optimize fleet size would be critical. To meet the requirement and yet contain the fleet size, it is a major challenge to increase ladle life and reduce down time for refractory maintenance.

The factors for attaining higher life of torpedoes are as follows:

- Upgradation of refractory brick quality to superior wear resistance and thermal shock resistance,
- Installation of online thermal scanning system

Effective gunning maintenance by use of equipment like shooters. However, the above would depend on technical feasibility and also would require additional investment for installation of equipment. Further, the feasibility study of introduction of shot-creting maintenance practices and intermediate repair of brick works to extend the campaign beyond 2000 trips is needed.

6.3 Desulphurization in Hot Metal Ladle

Over the past few decades the demand of steel around the world had been increasing, coupled with an improvement of clean steel. One major process to improve the quality of steel is to reduce its sulphur content. Sulphur is one of the most detrimental elements in the steel-making processes, affecting both internal and surface quality of finished steel. High sulphur content in steel affects bendability, ductility, toughness, formability, weldability and corrosion resistance of the steel. Thermodynamically it is beneficial to remove sulphur from hot metal than from steel in convertor operation. The sulphur is removed from hot metal before charging it in convertor. For the reduction of sulphur content in steel, two fundamental ways are possible: production of pig iron with low sulphur or desulphurization in hot metal ladle [8–10].

Lime has always played a vital role in desulphurization. Its low cost and availability make it an attractive reagent. CaO reacts with S to form CaS, and in presence of Si in hot metal, CaO oxidizes Si to SiO_2, and on further reaction with CaO, it forms $CaSiO_4$ and S is removed from hot metal.

$$2CaO + 2S = 2CaS + O_2$$

$$Si + O_2 = SiO_2$$

$$2CaO + SiO_2 = 2CaOSiO_2$$

However, there are some critical disadvantages, during desulphurization process; lime particles are being covered by calcium sulphide and di-calcium silicate and

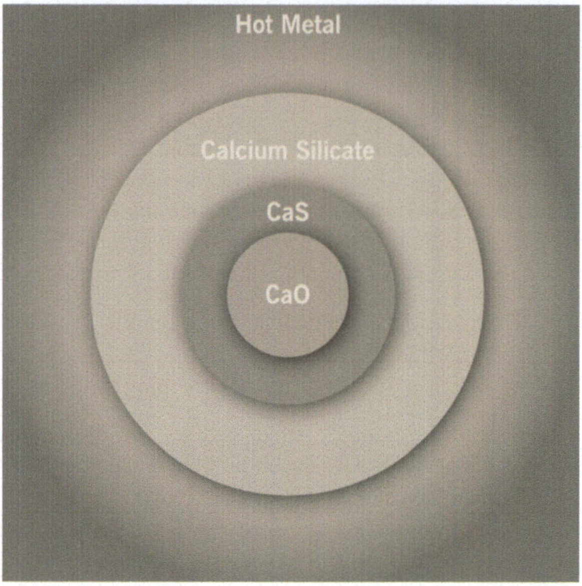

Fig. 6.12 Lime reaction and products [8]

forming a thick barrier at lime and hot metal interface, as shown in Fig. 6.12; in order to reduce this growth, the grain size of lime is restricted to 45-micron maximum [11].

The other reagents used are CaC_2 and combination with magnesium. This is carried out by blowing inert gases as a carrier of the powders. Through monolithic lance into the hot metal bath in hot metal ladle. Co-injection of lime and magnesia is one of the most commonly used for desulphurization processes with increased productivity and amount of low sulphur grade steel required. Due to lime addition and carry over slag from blast furnace the slag volume increases and due to falling of temperature skimming of slag became difficult [12]. Total Fe content in total slag is as high as 75% in the form of iron droplets dispersed in the slag, which is one of the highest yield losses in the steel-making process. In order to generate more fluid slag and to increase the Fe yield, slag-modifying agents are often added directly in the hot metal ladles. The slag-modifying agents are fluorspar, dolomitic lime, sodium carbonate, potassium chloride, cryolite, etc. All those fluxing additives have an adverse effect on corrosion of refractory lining. Hence, SiC-added refractory is in use to protect such corrosion. Hot metal ladle is a steel vessel of a truncated cone shape, which is lined with refractory materials.

6.3.1 Refractory Lining Practice

Several different concepts are undertaken for hot metal ladles. Depending on the operating condition, economy and plant conditions, the best lining concept is developed and it differs plant to plant. Typical lining refractory are unburnt

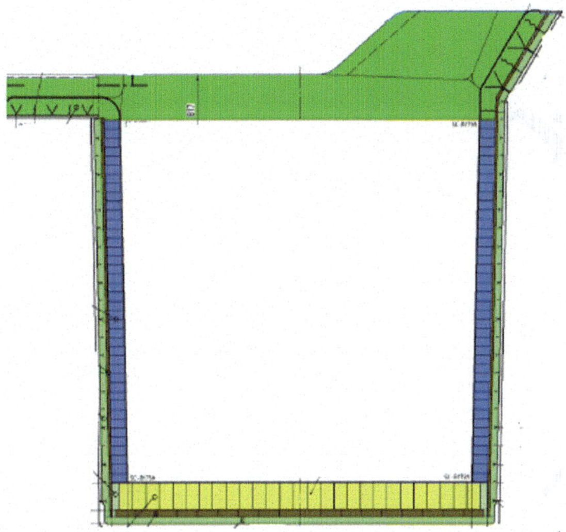

Fig. 6.13 Hot metal ladle

Al_2O_3-SiC-C or MgO-C bricks in working lining; in the case of desulphurization process, high alumina refractory is used in intermediate or backup lining and insulation castable or bricks against the steel shell. A schematic diagram of hot metal ladle is shown in Fig. 6.13 [13].

6.3.2 *Wear Mechanism*

As mentioned in Sect. 6.2, the typical pretreatment fluxes are lime based materials; CaO and CaC_2 and soda-based materials; soda ash (Na_2CO_3) represents the soda-based fluxes; lime and fluoride and calcium carbide represent the lime-based fluxes [14, 15]. Presence of CaO and CaF_2 have a little corrosion effect on carbon and SiC-containing high-alumina refractories compared to soda-based fluxes. The base material of Al_2O_3-SiC-C and MgO-C bricks is in the form of corundum and periclase, respectively. Corundum reacts with Na_2O as a dissociation component from soda ash (Na_2CO_3) and transforms from α Al_2O_3 to β Al_2O_3. Hence, Na_2CO_3 is penetrating an Al_2O_3-SiC-C brick other than oxidation of graphite. MgO is not attacked by lime-based fluxes, so in the case of MgO-C bricks, the soda-based fluxes react with graphite, and periclase remains unattacked. Hence, oxidation of carbon in MgO-C bricks limits its use.

Na_2CO_3 is the strongest oxidant for graphite in corundum-based or periclase-based bricks. It oxidizes graphite at 1300–1400 °C. Even compared to CaF2, Na_2CO_3 is more detrimental to the carbon-containing bricks. Vapour pressure of the fluxes at 1400 °C are as follows:

Na_2O: 10^{-1} atmosphere

CaF_2: 10^{-5} atmosphere
CaO: 10^{-9} atmosphere

Therefore, Na_2O is much more volatile than either CaO or CaF_2. Under commercial condition, soda ash is more volatile than lime or CaC or CaF_2, penetrate the lining bricks deeper to oxidize graphite. The reactions of fluxes with graphite at 1400 °C are as $Na_2O + C = CO + 2Na$ [10, 15].

Evaluation of Lining Practice for Higher Campaign

- It is presumed that oxidation of graphite will not be carried out by CaO and CaF_2. Hence when lime-based flux is used, it is not oxidizing graphite, but soda-based fluxes attack graphite severely.
- Lime-based fluxes attack lining bricks only on the surface of hot face, but soda-based fluxes easily penetrate into the bricks by oxidation of graphite.
- In soda-based fluxes, Na_2CO_3 reacts with corundum of Al_2O_3 and graphite simultaneously, causing the transformation from α Al_2O_3 to β Al_2O_3 and absorbs Na_2O. This converted β Al_2O_3 deposits on the hot face surface, and penetration of Na_2CO_3 is restricted. Hence further oxidation of graphite into bricks is arrested. In MgO-C bricks, Na_2O is not reacting with MgO (Periclase), and it easily penetrates deep into the bricks, further oxidides graphite.
- When lime-based flux is used as a major part of the fluxes, MgO-C bricks are suitable to use as the working lining, as CaO and CaF_2 are not taking part in graphite oxidation.

(continued)

Table 6.4 Different properties of hot metal ladle refractories

Properties	Unit	Al_2O_3-SiC-C (slag zone)	Al_2O_3-SiC-C (metal zone)	80% Al_2O_3 bricks (safety lining)	45% Al_2O_3 bricks (safety lining)	Castable (backup)
Al_2O_3	%	72–82	81–86	78–84	45–50	50–60
Fe_2O_3	%	0.8–1.2	1.1–1.5	1.5–2.0	1.5–2.0	1.5–2.0
SiO_2	%	5.5–7.5	7.5–8.5	8.5–10.5	20–25	20–25
SiC	%s	8.5–11.5	5.5–7.5	Nil	Nil	Nil
Fixed C	%	6.5–8.5	6.5–7.5	Nil	Nil	Nil
Bonding	Resin/ceramic	Resin	Resin	Ceramic	Ceramic	Hydraulic/ceramic
Bulk density	g/cc	2.9–3.1	3.05–3.15	2.7–2.9	2.1–2.3	1.9–2.2
Porosity	%	5.5–7.5	5.5–7.5	14–18	16–22	9.5–10.5
Cold crushing strength	N/mm^2	80–85	90–100	70–80	40–55	45–65

- In the case of soda-based flux, Al_2O_3-SiC-C-based refractory is suitable to use. Addition of SiC in corundum and graphite-based refractory improves corrosion resistance and oxidation resistance. At 1400 °C, SiC oxidizes in the presence of soda ash as $SiC + 3Na_2O = SiO_2 + CO + 6Na$. The newly generated SiO_2 forms an impervious coating over hot face refractory and resists further penetration of Na_2CO_3 into the bricks [13]. The properties of different quality bricks used in hot metal ladles are shown in the following Table 6.4.

References

1. Joao R.C Filho, Manoel Poubel Bestos, Gerson Correa Filho, Performance of Al2O3-SiC-C for Torpedo car lining, UNITECR 93, pp 1632–1640.
2. Rolf Lamm, Laser Measurement System for the Refractory Lining of Hot Torpedo Ladles, AIST.org March 2013.
3. Tetsuo Hirota, Masayuki Sakaguchi and Yukio Oguchi, Deformation behaviour under load of Al2O3–SiC-C Bricks for Torpedo Car, Taikabutsu Overseas Vol 15, No 2, pp 42–47.
4. D Gruber, T Auer, H Harmuth, Thermal and thermos-mechanical modelling of a 300t torpedo ladle, 9th Biennial Worldwide Congress on Refractories, pp 896–899.
5. N Sahoo, S K Choudhry, Development of improved quality Alumina silicon carbide carbon bricks for hot metal transfer and Torpedo Ladles, UNITECR 07, pp 381–383.
6. Dr. Ing. Manfred Koltermann, Torpedo ladle Refractories in West Germany, Taikabutsu Overseas, Vol 5 No 2, pp 35–40.
7. Satoru ITO, Takayuki Inuzuka, Technical development of Refractories for steel making processes, Nippon Steel Technical Report, No 98 July 2008.
8. Koichi Takahashi, Keita Utagawa, Hiroyuki SH|日ATA, Shin-ya KiTAMURA, Naoki KIKUCHI and Yasushi KISHiMOTO, Influence of Solid CaO and Liquid Slag on Hot Metal Desulfurization, ISIJ International, Vol.52 (2012), No. 1, pp. 10–17.
9. Brock Gadsdon and Xingguang Han, Hot metal desulphurisation: benefits of magnesium lime co-injection, Raw Materials and Ironmaking, Millennium Steel, 2010 pp 31.
10. Barron, M.A., Hilerio, I. and Medina, D.Y. (2015) Modeling and Simulation of Hot Metal Desulfurization by Powder Injection. *Open Journal of Applied Sciences*, **5**, 295-303.
11. Zushu Li, Mick Bugdol and Wim Crama, Optimisation of hot metal desulphurisation slag in the CaO/Mg co-injection process to improve slag skimming performance. Technical Report, Tata Steel RD&T, Swinden Technology Centre, Rotherham, United Kingdom.
12. Masahito Suitoh, Kanji Aizawa, Masahiro Ariyosi, Ryoji Nagai, Hiroshi Nishikawa, Shigeru Omiya, Total Hot Metal Pretreatment System at Kawasaki Steel, KAWASAKI STEEL TECHNICAL REPORT, No. 24 (April 1991).
13. Hiroshi Kyoden, Kenji Ichikawa, Teiichi Fujiwara and Yuji Yoshimura, Wear mechanism of Refractories by Hot metal pretreatment flux, Taikabutsu Overseas, Vol 7, No 2.
14. JAN LASOTA, Refractory Linings of Pig Iron Transfer Ladles, MATERIA£Y CERAMICZNE/CERAMIC MATERIALS/, 63, 3, (2011), 692–695.
15. J. KIJAC, M. BORGON, desulphurization of steel and pig iron, METALURGIJA 47 (2008) 4, 347–350.

Chapter 7
BOF Refractory

7.1 Introduction

The operating conditions prevail the prime importance when refractory lining life of an LD vessel is considered. Lining life can be defined as numbers of heats processed in a vessel during a campaign. Optimizing the lining life is of a prime importance in steel shop, and it is not only depending on number of heats produced but also includes other factors such as vessel availability, hot metal availability, grade mix, cycle time, output steel chemistry, maintenance practices and overall cost. Lining life is also influenced by type of bricks used and hot metal chemical composition. Zoning lining of vessel based on the study of corrosion characteristics has become an important factor to consider improving lining life and cost-effectiveness. Zoning is used by using higher quality refractories in those areas that require them to prevent severe corrosion and erosion. Trunnions are typically the highest wear areas in a vessel due to inability of slag coat during normal operation and maintenance practices.

In the beginning till 1976, the main refractory used in BOF was dolomite-based bricks, such as tar-bonded stabilized dolomite or high-fired magnesia-enriched dolomite bricks. Magnesia-dolomite bricks are produced by enriching dolomite with MgO to increase the resistance to hydration and slag attack, and to raise its hot strength, thus improved the refractory lining life of vessels.

Present Refractory Practices
During 1980s, for increased productivity, high demand of stringent steel quality and emphasis on reduction of gas emission, it was felt to improve refractory quality. MgO-C bricks were introduced for its better corrosion resistance and thermal shock resistance compared to dolomite bricks

(continued)

S. Biswas, D. Sarkar, *Introduction to Refractories for Iron- and Steelmaking*,
https://doi.org/10.1007/978-3-030-43807-4_7

[1]. The range of use of MgO-C bricks was rapidly expanded to include the whole lining of the vessel and that practice is continued to the present.

MgO-C bricks have excellent corrosion and spalling resistance because they are a composite material containing magnesia, which is high corrosion-resistant for high basicity slag, and graphite, which improves spalling resistance. Use of graphite has the following advantages:

1. Non-wetting towards liquid steel and slag,
2. High thermal shock resistance,
3. High spalling resistance,
4. Stable volume at high temperature.

The properties of magnesia aggregates have been studied frequently as the wear of the aggregate correlates closely with the corrosion of MgO-C bricks. The chemical composition and crystal size are considered to be key properties of MgO aggregates.

Bricks composed of low B2O3 and high CaO/SiO2 ratio with **high-purity MgO show** superior corrosion resistance.

The nature and quantity of fluxing oxides (SiO_2, Al_2O_3, Fe_2O_3) as well as MgO/CaO ratio had a decisive effect on the physical and chemical refractory properties in service. Such refractories thus contain five significant components; they can usually be represented in the solid state as CaO-MgO-C_3S-C_4AF since their molar ratio of Al_2O_3/Fe_2O_3 is approximated to 1.0 and the molar ratio of CaO/SiO_2 is >3.0. The eutectic temperature of the quaternary system in between 1300 and 1550 °C. In the dolomite lining the damage due to thermal and structural spalling with high metal penetration, low hot strength and hydration was observed as big problems to achieve higher life.

7.2 Operating Conditions and Refractory Lining

All basic oxygen steel-making processes, top, bottom and combined blowing, are the oxidizing process in which oxygen is purged and elements, dissolved in hot metal, such as C, Si, P and some S are oxidized to form CO, CO_2, SiO_2 and P_2O_5. Manganese and iron are oxidized to MnO and FeO. The CO and CO_2 gases are escaped out and other oxides form acidic slag. All those slag-forming oxides react with basic refractory and responsible for severe corrosion.

Lime is added to the molten bath and it neutralizes the acidic effect of slag, and at the later stage of blowing operation, basic slag is formed. The added lime is fluxed and dissolved by the siliceous slag and the resultant basic slag acts as a refiner to absorb sulphur and phosphorus from the molten steel. The resultant basic slag is less

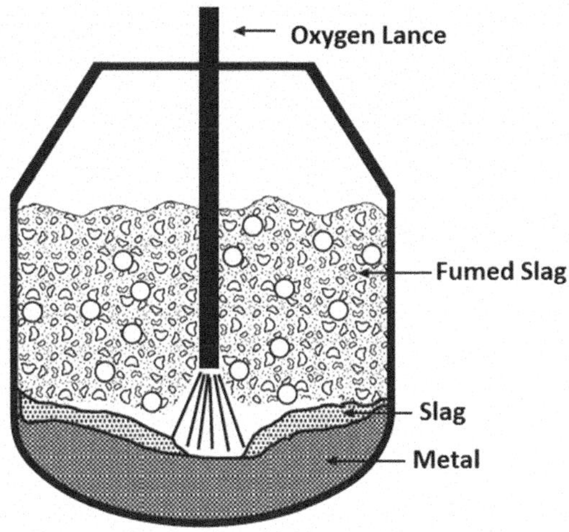

Fig. 7.1 Oxygen blowing and slag foaming

corrosive to the basic lining refractories. The refractory lining life in BOF depends on the following operating conditions:

- Oxygen blowing
- Chemistry of hot metal
- Quantity and time of lime addition
- Lance position
- Other additions, such as MgO, dolomitic lime, CaF2, etc.
- Slag volume and its viscosity
- Operating temperature.

After scrap and hot metal are charged, the furnace is set upright and the oxygen is supplied through a water-cooled lance. The oxygen blow times typically varies from 13 to 25 min under different shops' practices. The oxygen is added in three stages, depending on the position or height of lance tip from liquid bath and varying oxygen blow rate. In the first stage, the tip of the lance is positioned at about 3.5–3.8 m from the bath height. The initial slag is formed rich in SiO_2 and FeO. During initial blow, basic refractory lining, MgO-C brick, comes in contact direct to oxygen and also with acidic siliceous slag and suffers from severe oxidation and corrosion. To protect the lining refractory, slag coating over MgO-C bricks is a regular practice in all steel plants. Burnt lime and dolomitic lime are also charged to absorb SiO2 and FeO and hence corrosion of bricks is minimized.

In the second stage of blowing, lance is lowered by 0.75–1.0 m, and slag starts to foam due to reduction of FeO in the slag and CO formation. As the blow progressed, CaO dissolved in slag and the slag volume increased, as shown in Fig. 7.1.

Finally, in the third stage of blowing the lance is placed at 1.8 m above the bath and FeO content in the slag increases because the rate of de-carburisation decreases. The presence of oxides in slag at different stages of blowing is shown in Fig. 7.1. Increase in % MgO is contributed from refractory or addition of MgO or dolomitic

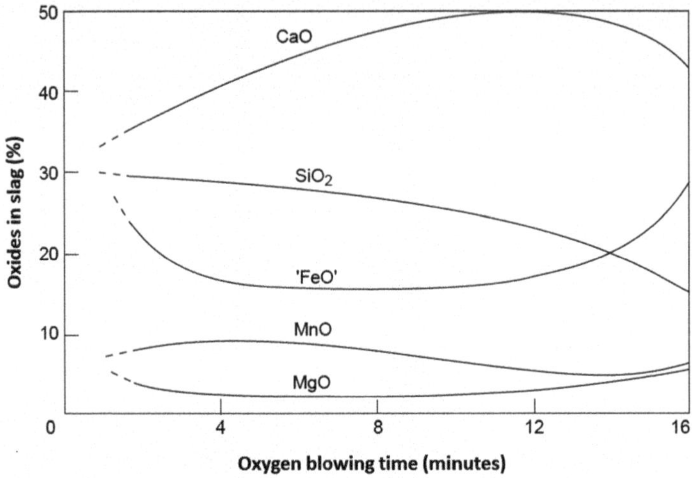

Fig. 7.2 Change in slag composition with time of oxygen blowing

lime. During the blow, the temperature of the melt increases from 1350 to 1700 °C and the slag temperature would be about 50 °C higher, which has a direct impact on refractory lining life.

Generally, lime is added within few minutes after the start of oxygen blowing, so that lime mixes and dissolves into oxide elements that are being generated in the presence of oxygen. However, lime does not dissolve immediately, and it takes a considerable period of blow. During that time, slag basicity remains low in the presence of SiO_2 in slag and is detrimental to the lining refractories. This delay is caused by formation of dicalcium silicate ($2CaO.SiO_2$) shells around CaO particles.

Despite the high basicity of the steel-making slag in BOF, a chemical gradient exists between the slag and the MgO in refractory lining. At steel-making temperature, the condition allows for 6–8% MgO from refractory to dissolve in final turned down slag, as shown in Fig. 7.2. If such a slag is saturated from external source, then MgO would not dissolve in slag from refractory lining. This is the basis to add MgO or dolomitic lime in vessel slag during refining processes.

Operational Practice and Refractory Wear

Unless basicity of the slag is raised quickly, the reaction of acidic slag with basic refractory results in severe corrosion of refractory. This mechanism is shown in Fig. 7.3. The figure shows that in a slag containing CaO, SiO_2 and FeO at 1600 °C, Conversely, MgO dissolution decreases with solubility of CaO in slag. Hence every effort to be made to develop steel making practice of adding optimum quantity of slag within stipulated time after blowing of oxygen starts.

(continued)

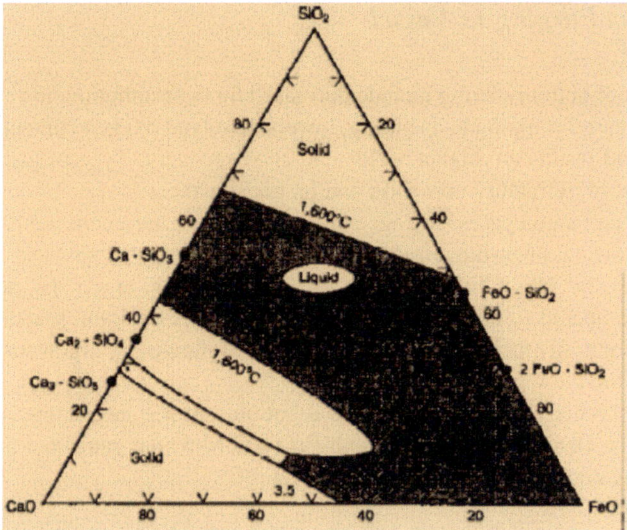

Fig. 7.3 Phase diagram of CaO-FeO-SiO_2 ternary system

Use of high porosity soft burnt lime has been identified as an important element to add for achieving early dissolution of lime in slag. Use of fluorspar flux (CaF_2) is common as CaF_2 dissolves dicalcium silicate ($2CaO.SiO_2$) shells and thus accelerates lime dissolution. The optimum operational practices are summarized as follows:

1. Timely addition of lime with optimum quantity,
2. Fluxing the melt by adding fluorspar to improve lime solubility,
3. Addition of soft burnt dolomite with manganese ore,
4. Addition of Al_2O_3 containing flux to dissolve lime,
5. Practice of staggered addition of lime and delayed scrap addition and rapid FeO generation by soft blowing. In the staged lime addition, lime is added incrementally or continuously with slow addition rate so that lime would be dissolved continuously.
6. Fluxing CaO with FeO slag generated in soft blowing practice (high lance height or reduced oxygen blow rate) is common in most of the steel plants. High FeO in slag also helps to reduce slag viscosity and helps to dissolve lime.
7. MgO can dissolve in basic slag of about 6–8%. To reduce MgO dissolution from refractory lining, external addition of MgO or dolomitic lime in slag would help to reduce corrosion of MgO-C bricks [1, 2].

7.2.1 Gas Purging in Vessel

The benefits of bottom stirring include decreased flux consumption, lower FeO level in slag, decreased slopping and spitting, lowers dissolved oxygen content in finished steel. The end results are higher yields and alloy recovery. Type of gas used has a direct impact on refractory corrosion and its performances.

Nitrogen and argon gases: Nitrogen and argon are the most commonly used gases for bottom stirring. From the refractory erosion viewpoint, both gases are chemically inert towards the vessel lining. Based on the plant data, argon has no effect on refractory compared to nitrogen, due to its inertness. A common practice is to use nitrogen gas initially for 60–80% of oxygen blow, followed by argon stirring for the last 20% of the blow period.

CO_2 gas: When CO_2 gas is used as a bottom stirring gas, molten iron at the vicinity of the element is oxidized. Both FeO and MnO are generated according to the following equations:

$$\mathrm{Fe} + \mathrm{CO_2} = \mathrm{FeO} + \mathrm{CO} \quad (1)$$

$$\mathrm{Mn} + \mathrm{CO_2} = \mathrm{MnO} + \mathrm{CO} \quad (2)$$

The FeO and MnO thus produced is corroding the MgO-C refractory and causes extensive deterioration by oxidizing the graphite in brick.

CO gas: In few steel works, CO gas is used as a stirring gas. It is a reducing gas, does not react with iron as well as it has no corrosion effect of MgO-C lining refractory. In terms of bath agitation, carbon monoxide has a similar stirring effect as nitrogen and argon. But it is combustible, toxic and hazards for the safety point of view, which limits the use of CO gas.

7.2.2 Refractory Design in Vessel

MgO-C brick lining has become standard practice in BOF all over the world. Design of refractory bricks is also a critical factor of special consideration to achieve higher campaign life [3, 4]. The general sub-division of areas within a BOF vessel is shown in Fig. 7.4. The main areas are as follows:

- Top cone
- Barrel
- Tapping Zone
- Trunnion
- Lower cone

→ **Zone 1**

- Bottom → **Zone 2**

7.2.2.1 Zone 1

Bricks at the cone of BOFs easily drop off during service and during jam removal. Repairing is necessary in each time and service life often deteriorates due to bricks dislodging. Construction of brick work plays very critical role in this conical area. The bricks fall in the lower part of the cone lining due to stress generated during operation. The stress is increased due to high thermal expansion of the bricks at elevated temperature. Since last three decades, lot of modification had been done in brick lining design. Three types of models had been developed in top cone area, without change of lining pattern in barrel and rest of areas. The conventional lining is

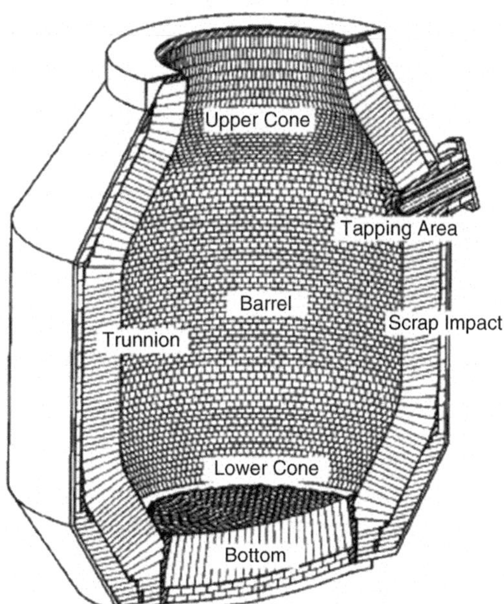

Fig. 7.4 Different zones within BOF vessel

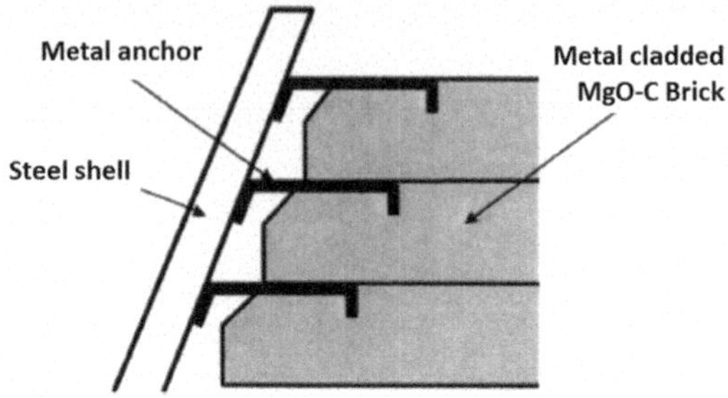

Fig. 7.5 Brick fixing with metal anchors

"Horizontal brick lining", and improved lining are "forward inclined" and "Backward inclined". The slop angle of forward inclined and backward inclined is kept within 5–10°.

Feature of the horizontal lining is that the stress at lower part of the cone near the barrel was higher. The stress increased in each operating cycle and attend a constant level after about 10–20 cycles. The stress attends the peak during each cycle just before tapping when the temperature in the vessel is highest. In other words, when the thermal expansion of the bricks was highest, the peak stress was produced. Once the bricks in lower part of the cone damaged, bricks above those regions fall off. Hence failure of top cone lining bricks takes place at early stage of the campaign.

Features of the backward and forward inclined lining are that neither lining had high stress at any point of time in cyclic operation. In both the lining design, it was found possible to relax bending in cone section. The brick movement in backward and forward design was uniform, and stress is not generating high, which results in stable lining construction and hence possibility of damaging the bricks was minimized. It was also reported [5] that backward inclined lining was performing better than forward inclined lining.

At the Kimitsu and Muroran works, Japan [6], a steel anchored brick design was implemented to arrest falling of bricks in top cone; as shown in Fig. 7.5, each brick was fixed to the steel shell by a hooked metal anchor. The bricks were wrapped with metal case so that the case fused and adhere to each other and prevent falling. Additional measures were taken to use non-slip coating applied on the brick surface, and small amount of powdered pitch was added to the brick material to improve resistance to spalling.

7.2.2.2 Zone 2

Bottom construction may be of two types, removable bottom and fixed bottom. In the case of removable bottom, special consideration to be taken in joining the removable bottom with rest of the lower cone lining attached with barrel, as shown in Fig. 7.6. The joint between the removable bottom stationary portion must be replaced each time that a new bottom is installed.

Fig. 7.6 A typical schematic of removable bottom lining in BOF

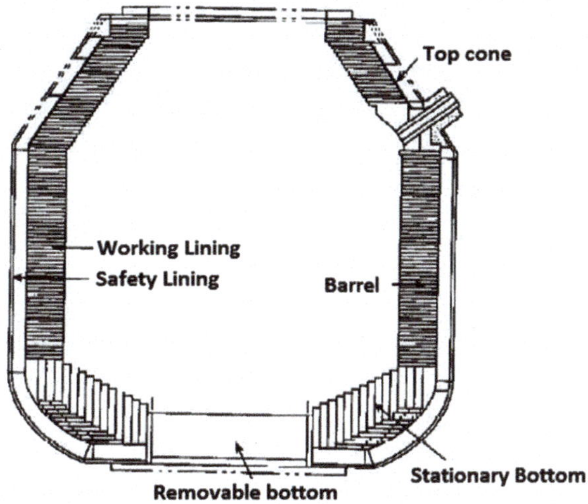

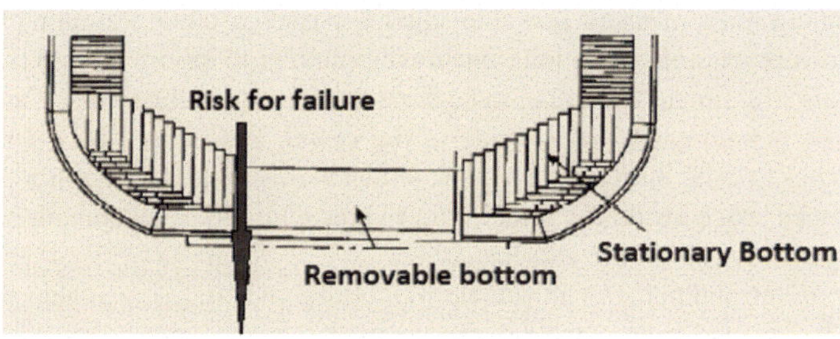

Fig. 7.7 Position of straight joints

On the initial or cold lining, this filling can best be done from inside of the vessel, by ramming or pouring a self-flow castable refractory with associated vibration. On replacement bottom, the need for a rapid bottom change does not permit proper joint installation from inside of the vessel. In this case the joint material is installed by either hot patching or gunning practices. The gunning is preferred as the gunning material has developed strength after drying in full depth, but hot-patched material develops full strength only after considerable operation.

Bottom Joint to Ensure Workmanship

1. The gap between the removable bottom and stationary bottom with integral part of barrel lining is 80–150 mm and the full depth is 1.5–2.0 m which is almost a straight joint, as shown in Fig. 7.7. Very critical to fill the joint without any air pocket left. Proper working skill is required for filling the gap, otherwise this may lead to breakout.

(continued)

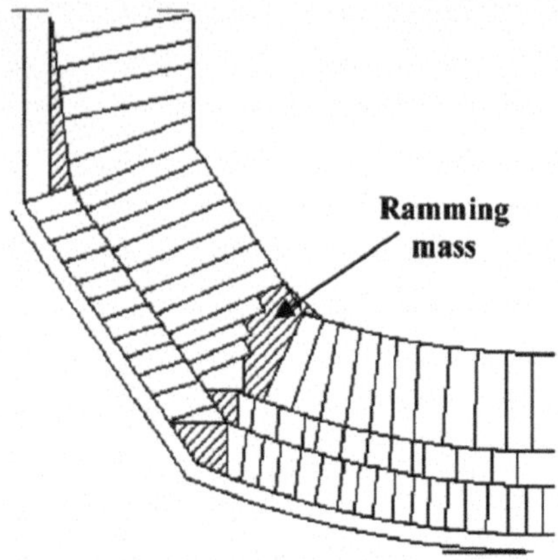

Fig. 7.8 Bottom fixing by ramming mass

2. Gap may be filled up by ramming the MgO-based carbon-containing ramming mass or using flowable MgO-C-based castable. Ramming to be done very carefully with long pneumatic rammer to ensure that no air gap left after ramming. Ramming to be done from inside of the vessel. Castable can be poured easily from inside of the vessel. But in both the operation man has to enter inside of the vessel, which involves safety hazards as it is a confined space inside the vessel. Proper precaution to be taken with proper oxygen level inside the vessel.
3. At present pumpable or injectable monolithic material is available which can be injected from outside of the vessel.
4. Self-flow MgO-based castable is preferred over ramming masses as the early strength is developed at the time of drying in using castable or injection material, but in the case of ramming masses, full strength is developed during operation.

Fixed bottom design is practised in most of the steel plants worldwide. The schematic diagram of fixed bottom lining is shown in Fig. 7.8.

The bottom used to be changed with the complete relining of the vessel. The joint is filled by ramming with MgO-based carbon-containing non-aqueous ramming masses. The lining in bottom consists of three layers of bricks. Hot face lining is made of corrosion-resistant fused magnesia-based MgO-C bricks containing 10–20% carbon, intermediate lining by sintered magnesite bricks and safety lining, by magnesia-chrome bricks. In lower cone and side wall, MgO-C with 14% C is used.

Development of radial bottom design has eliminated the joint and turns into a smooth radial lining with bricks without any joints [7]. This design is the best precondition for reaching an increased lining life of vessel. The radial round corner

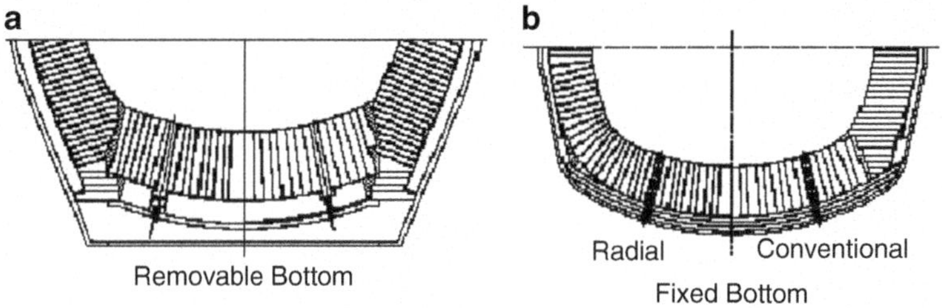

Fig. 7.9 (**a**) Removable and (**b**) fixed Bottom

design can avoid the stress concentration by changing the direction of stress to the linear direction of the barrel. The conventional fixed bottom design and the radial fixed bottom design are shown in Fig. 7.9. The maximum wear of the bricks takes place at the corner of the bottom due to stress concentration at the point in change of lining direction and change in brick shapes. To make the wear uniform, the numbers of bricks are divided uniformly, and the surface of the hot face and cold face of the bricks was designed to conform to an idealized sphere shape as shown in Fig. 7.9b.

7.2.3 Refractory Design in Tap Hole Sleeve

Tap hole sleeves based on resin-bonded magnesia carbon refractory containing fused MgO around 15% C and metallic additives are state-of-the-art refractory design. The main causes for wear in tap hole sleeves are as follows:

- Severe erosion at operating temperature due to steel flow,
- Oxidation by dissolved oxygen in liquid steel,
- Oxidation between the melting sequence,
- Thermo-mechanical spalling when steel started flow through tap hole at the beginning of tapping.

MgO-C refractory with an excellent hot erosion resistance, oxidation resistance and excellent thermal shock resistant is used. Properties of tap hole sleeve refractories are shown in Table 7.1. To achieve the superior corrosion and hot abrasion resistance, refractory has a carbon-enriched matrix, and the enrichment has been done by post-impregnation by carbonaceous materials or special polymers.

The design of tap hole sleeve has a vital role in achieving longer life. During last three decades, continuous modifications had been done in design as well as material and installation processes development. Earlier long sleeves had been manufactured and those sleeves had been replaced by small-size sleeves, assembled inside the mother block. The advantages of the improved quality and designed tap hole sleeve sets were analysed as follows:

Table 7.1 Properties of tap hole sleeve refractories

Characteristics	Unit	Type A	Type B
Type of magnesia		Fused	Fused + sintered
Bond		Resin	Carbon
Carbon content	%	14–16	15–17
Antioxidant		Yes	Yes
Properties			
Coked porosity	%	8.5–9.5	7.5–8.5
Cold crushing strength (after coking)	N/mm^2	35–38	42–46
Hot Mod. of rupture at 1400 °C	N/mm^2	15–17	20–23
Thermal shock resistance at 1000 °C (air quenching)	Nos. of cycles	>100	>100

Use of a gunning machine to fill up the annular space with a superior quality gunning mass in contrast to the conventional slurry-pouring technique. A sleeve design based on the following criteria:

- Greater material density
  - The pressing machine is able to press smaller lengths with more force/unit area.
  - The result is denser material compared to the long-pressed single stretch of sleeve.
- Less material segregation
  - Small sleeve lengths have less material segregation which leads to less variation in density, chemical composition and strength across the tap hole length.
- Resistance against thermal spalling
  - The fast heating and cooling create thermal stress between particles, and this stress initiates dislocations.
  - In the longer tap hole sleeve, the dislocations expand and create external and internal cracks. In assembled small-size sleeves, the probability of cracking is eliminated as shown in Fig. 7.10.
  - In the case of the smaller sleeve lengths, the dislocations are accommodated in joint points between the sleeves which reduce the potential of thermal stress.
  - The inner surface of the channel is as smooth as for a single sleeve tap hole due to the special construction of joints. Pre-assembly ensures minimized effort for installation and maximum performance. The special form-fitting joint design prevents the steel from penetrating possible voids behind the sleeves (Fig. 7.10). Key to the performance in terms of the virtually joint-free surface and reduced infiltration is the superiority of the Isojet conical fitting, as this sleeve assembly is popularly called.

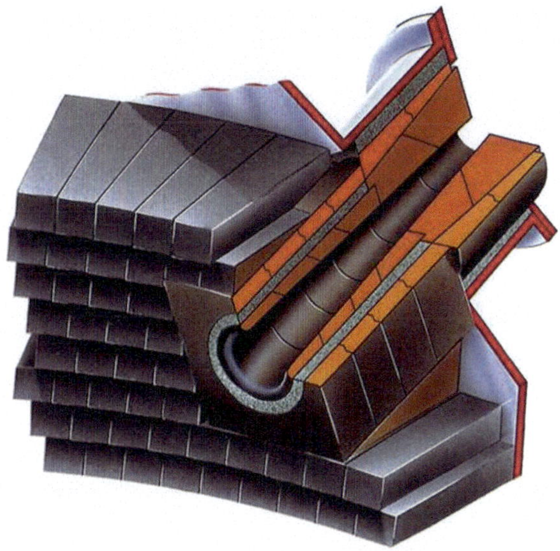

Fig. 7.10 Arrangements of tap hole sleeve

7.2.4 Wear Mechanism

The local corrosion of MgO-C refractory is considered as one of the important factors for shortening the life of the BOF. Reactions between the MgO-C bricks, molten slag and steel have a close relation to the local corrosion of the lining refractory [3]. At high temperature, two critically important failure mechanisms to be responsible for the wear of refractory lining in BOF vessels are chemical and mechanical wear. Various areas of the lining are being affected differently by the wear phenomena listed below:

1. Chemical
   (a) Slag corrosion, dissolution of MgO in slag,
   (b) Reduction and vapourization of magnesia in contact with carbon,
   (c) Oxidation of carbon.
2. Mechanical
   (a) Thermal shock during start and end of each cycle operation results in generation of stresses,
   (b) Mechanical impact due to scrap charging in vessel,
   (c) Mechanical erosion and abrasion caused by turbulence created by slag and liquid steel movement, gas formation and slag foaming.

3. Corrosion—It may cause by mechanism like such as the following:
   (a) Dissolution of refractory in contact with a liquid,
   (b) Reaction with a solid, liquid or a vapour,
   (c) Penetration of liquid or vapour.

   In most of the cases, those mechanisms occur in combination with one another.

7.2.4.1 Reaction of Refractory with Slag

Dissolution of refractory at slag–refractory interface in controlled by chemical reaction and diffusion. When a refractory (oxides, carbides) dissolves in slag, the reacted products are expected to be distributed into the melt uniformly. This is possible if the diffusivity of the reacted product is faster than generation of products. The dissolution rate of a refractory brick is controlled by a chemical reaction by the following equation:

$$r = K\frac{sa}{so}Cm \quad (3)$$

where, r = rate of corrosion, K = rate constant, Sa = actual surface area of refractory (cm^2), So = apparent area (cm^2) and Cm = concentration of refractory oxides in the melt.

Shape of Vessel and Refractory Wear

Comparing the two vessel designs as shown in Fig. 7.11, it is noticeable that some relationships between lance blowing parameters may defer significantly [3]. Like L/L0, which may determine the nature of the blow, affecting yield by

(continued)

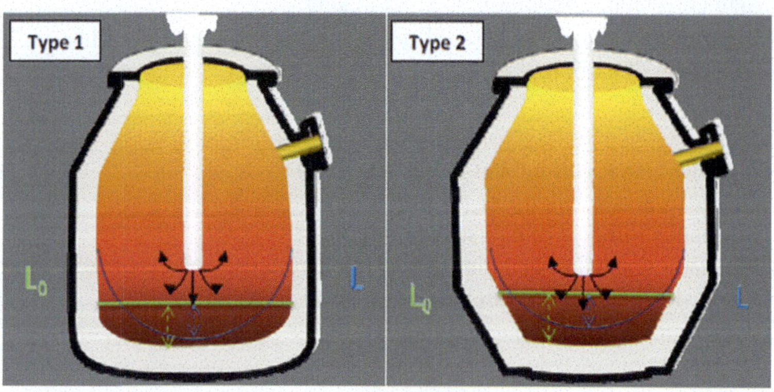

Fig. 7.11 Vessel design and dimension. (**a**) Type 1. (**b**) Type 2

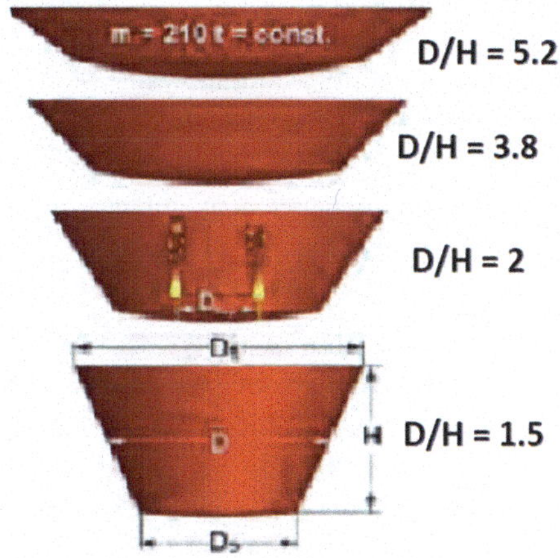

Fig. 7.12 Optimization of bath diameter to bath height ratio

oxidation of bath or even dead zones when the lining wears unevenly. Convertor shell designs have evolved from type 1 to type 2 throughout the years as the knuckle-shaped vessel would offer better L/L0 ratio even when the vessel lining gets thinner and the bath weight changes with the same charge weight. In the case of type 2, the refractory erosion is much less, which ensures long campaign life.

However, as the lining becomes older, uneven wear combining the slag buildup significantly changes the lining profile. As shown in Fig. 7.12, studies have confirmed that optimum ratio of bath diameter to bath height (D/H) between 2 and 3 should be good both in the terms of mixing efficiency and minimum refractory wear rate in the knuckle and barrel.

Porosity in the bricks, open pores in particular, would increase the actual surface area, hence presence of porosity increases the ratio of Sa/So and results in increase in the rate of corrosion. After initial reaction, a situation appears when rate of removal of reacted product from interface is slower than the rate of chemical reaction. A saturated boundary layer of reacted product is formed on the hot face of the refractory bricks and then the dissolution rate depends on rate of diffusion and the situation is expressed as follows:

$$R = D\frac{Cs - Cm}{d} \tag{4}$$

where, R = rate of refractory wear, D = diffusion coefficient (cm^2/s), Cs = saturation concentration of refractory oxide in melt (g/cm^3), Cm = actual concentration of refractory oxide in melt (g/cm^3) and d = effective boundary layer thickness (cm).

Both the above mechanisms depend on type of slag and chemical composition of slag and refractory. The lining life of MgO-C bricks can be dramatically improved if the dissolved MgO in BOF slag reaches or exceeds the saturation point of MgO in slag at operating temperature. Severe corrosion observed when the MgO in slag falls below 6.0% and the saturation of MgO in slag is 8.0%. Presently, the slag composition of BOF operation can be optimized as high CaO/SiO_2 ratio (35–40% CaO and 10–15% SiO_2), low FeO (18–20%) and high MgO (8–12%). Fused MgO grains resist slag attack better than sintered MgO grains.

7.2.4.2 Reduction of MgO

The graphite used in MgO-C bricks has contributed to the improvement of corrosion resistance by preventing slag penetration. However, graphite is easily oxidized in steel-making condition, and due to oxidation of graphite, the characteristics of the brick are greatly deteriorated. The causes of oxidation are (1) gaseous phase oxidation due to O_2 and CO_2, (2) liquid phase oxidation due to FeO and slag and (3) reaction between MgO aggregates and graphite.

MgO can coexist with graphite up to 1800 °C under normal atmospheric condition, as shown in Fig. 7.13, but under vessel operating condition, the reaction of MgO with C starts at 1650 °C as per the following equation:

$$MgO + C = Mg(g) + CO \quad (5)$$

However, it is considered that Mg (g) produced in the inner part of the brick would condense at the hot face of the brick, due to oxidation and prevent further reaction between MgO aggregate and graphite at operating condition of steel making [4].

It is assumed that the vapour Mg is oxidized and concentrated at the brick hot surface to form a dense magnesia layer, suppressing the MgO reduction reaction by carbon. It is also assumed that the dense magnesia layer plays a protective role in preventing the corrosion of brick caused by slag and oxidation of carbon, contributing to refractory lining life improvement.

Fig. 7.13 Equilibrium relation with MgO and C

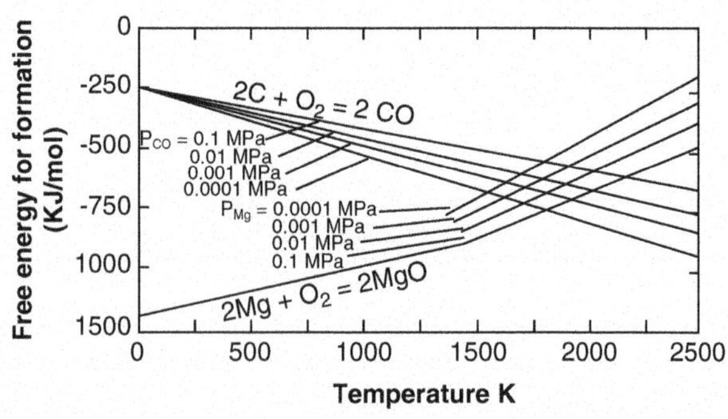

7.2.4.3 Penetration

Penetration without dissolution is not considered as corrosion. Usually, dissolution of refractory takes place after penetration of liquid or vapour into refractories. Further, dissolution without penetration is superficial. Penetration of any fluid can be considered as penetration of mercury in a capillary tube. Rate of penetration can be expressed as follows:

For a laminar flow of liquid in a tube, as per Poiseuille's law, rate of volumetric flow is derived as follows:

$$\frac{dV}{dt} = \frac{\pi \cdot \Delta P \cdot r^4}{8\mu l} \tag{6}$$

If the depth of penetration is l, and then

$$\Delta P = \frac{2 \cdot \gamma \cos\theta}{r} \tag{7}$$

Combining the above two equations,

$$\frac{dl}{dt} = \frac{\pi \cdot 2\gamma \cos\theta \cdot r^4}{8\mu l\pi, r^2 \cdot r} \tag{8}$$

$$\frac{dl}{dt} = \frac{r \cdot \gamma \cos\theta}{4\mu l} \tag{9}$$

where, dl/dt = rate of penetration, r = radius of capillary, γ = surface tension of fluid, θ = contact angle between solid and fluid, l = length of penetration and μ = viscosity of fluid.

Both physical penetration and chemical reaction are favoured by effective slag–solid wetting characteristics and low viscosity of slag. Therefore, control of slag chemistry is of paramount importance to control corrosion.

Surface energy of BOF slag is generally of the order of 400–500 ergs/cm^2, surface energy of magnesia is 1000–1400 ergs/cm^2, so the molten slag penetrate easily into MgO-based refractory. However, addition of carbon in the form of graphite, carbon black, tar and resin prevents wetting by molten slag. Liquid steel has a high surface energy around 1800 ergs/cm^2 which is higher than magnesia bricks, hence, penetration into bricks is not likely. However, in actual practice, it is observed that in metal zone also penetration takes place as the penetration of liquid depends on grain size, grain size distribution and pore size also. When the diameter of open pores exceeds the critical value above which severe penetration takes place. In the case of blast furnace hot metal, the critical pore size diameter is 1 μm, and >95% of the pore size in carbon blocks used in hearth lining is kept less than 1 μm to prevent penetration. The critical diameter of steel can be determined from the graph recommended by Mr. Gerald Routschka [8], shown in Fig. 7.14.

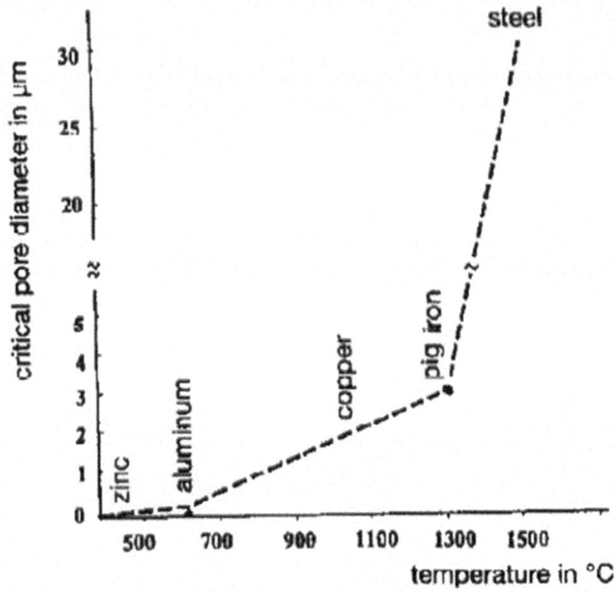

Fig. 7.14 Critical pore diameter in refractory at the melting temperature for the infiltration of melts

7.2.4.4 Thermal Stress and Spalling of Refractories

Cracking of refractory is the result of excessive thermal expansion at the hot face caused by the preheating procedure after relining of the vessel. The crack in bricks also occurs because of thermal cycling between heats. When the bricks in the lining are exposed to the fast temperature increase, the hot face would expand more than the remaining part of the bricks, and the thermal stresses would cause separation or cracking of hot face area from the remaining part of the bricks. The cracking takes place parallel to the brick hot face.

Corrosion Resistance Phenomena

The coexistence of MgO and carbon in MgO-C bricks causes the reduction of MgO by carbon and destroys the bonding system of the bricks. Under operating condition of BOF, the reaction of MgO with carbon starts above 1400 °C as shown in the following equation:

$$MgO + C = Mg_{(g)} + CO_{(g)} \tag{10}$$

The reaction products are escaping out of the system. Along with the progress of the MgO-C reaction, the brick structure deteriorates and wear of

(continued)

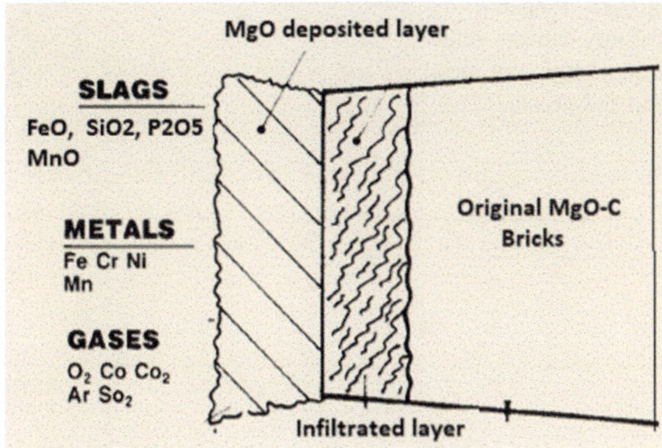

Fig. 7.15 Corrosion mechanism of MgO-C bricks

bricks occurs rapidly. However, there are assumptions and different school of thoughts as mentioned below:

1. The magnesium vapor is oxidized in the presence of FeO and other oxidizing elements and concentrated on the brick hot face to form a dense impervious layer, as shown in Fig. 7.15. This newly formed dense MgO layer suppresses the further reaction and prevent corrosion.
2. Initial reaction with MgO and C generates CO and Mg gas, which deteriorates the brick structure and porosity of the bricks increases, which allows more slag to penetrate, before formation of dense MgO layer and hence the brick structure destroys within few initial heats.
3. The dense MgO layer would be unstable due to turbulence of slag and liquid steel and severe turbulence during bottom purging. The dense layer may also be chipped off due to mechanical spalling caused by difference in density between brick and dense layer and fresh surface exposed to slag and corrosion continues.

7.2.4.5 Impact Caused by Charging the Scrap and Hot Metal

The refractory in charge pad area damages due to impact of the heavy metal scrap and the tapping stream of liquid hot metal during charging of the vessel. The damages take place when the impact stresses exceed the compressive strength of the bricks. Figure 7.16 demonstrates the plausible charge pad zone where the refractory may damage due to excessive impact exerted by charging materials.

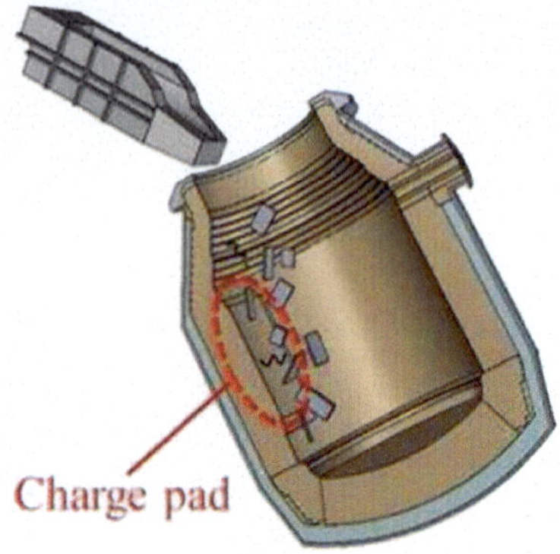

Fig. 7.16 Probable refractory damage zone during charging of metal scrap and stream

7.2.4.6 Mechanical Erosion and Abrasion

Certain areas of the BOF lining are exposed to severe mechanical abrasion. In the cone area, mechanical wear is caused by the high velocity gases carrying the particles of slag and metal. Severe wear can also occur during de-scaling of the slag build-up in the mouth area, when the skulls are removed mechanically or by melting the metal jam by lancing. The wear in cone due to those factors can be minimized with the introduction of higher percentage of graphite in refractories. Due to non-wetting characteristics of graphite, slag would not adhere to the lining, and it is easily removed without damaging the lining. The charge pad, in addition to the damage caused by impact of solid scrap charging and tapping hot metal into vessel, can also be worn out by scrap sliding down its surface. In combined blowing BOF, severe mechanical erosion takes place at the bottom of the vessel due to turbulent movement of liquid metal and slag around the tuyeres, caused severe erosion at the stadium area of the bottom.

Thermo-Mechanical Stresses and Expansion Joints

The thermal and permanent expansion of the refractory lining which is confined and restricted from expanding by the steel shell of a BOF can cause very high thermos-mechanical stresses when proper expansion allowance is not provided in the lining. The stresses formed in BOF lining can be expresses as follows:

$$\sigma = \alpha \cdot E \cdot \Delta T \tag{7.11}$$

where, σ = Thermo-mechanical stress, α = Linear thermal expansion coefficient, E = MOE, T = Temperature

If a typical data is used for BOF refractories in the above equation, stresses of several thousand kPa would be developed in a BOF lining.

(continued)

For example,

$\sigma = 11 \times 10^6$ °C, $E = 48.2$ GPa, and $\Delta T = 1400$ °C, the stresses of the restricted lining can reach 740 MPa

If proper expansion allowance is provided in the lining so that the magnesia refractory can expand freely and the stresses can be lowered to the safe range and thus cracking and spalling of bricks can be avoided. Hence, providing expansion joints in brick lining in BOF is a very critical task to extend lining life.

MgO-C Refractory for BOF

The MgO-C bricks are widely used as a lining refractory for BOF because of the advantages of excellent corrosion resistance to slag-containing high CaO/SiO_2 ratio, resistant to slag penetration for the non-wetting character of graphite addition and spalling resistance. However, it has a major disadvantage over oxidation resistance. Presence of carbon and carbonaceous binder makes it weak in oxidation resistance in an oxidizing atmosphere and in the presence of FeO and MnO, which poses a major problem in using those bricks in lining BOF. The different methods for improvement of oxidation resistance in Mgo-C bricks are as follows [9–11]:

Densification by Optimizing Grain Size and Their Distribution

Since the largest component of Mag-C brick is the magnesia grain, the composition and properties of the grain play an important role in the characteristics of the brick. There are many types of magnesia grain available today, with widely differing properties and prices. The best choice for any particular purpose cannot possibly be made commercially; the operator and the refractory expert must be in perfect communicative harmony if success is to be achieved.

In terms of magnesia grain, the higher quality grain for withstanding basic slags, erosion, abrasion, temperature, etc., is the most important factors to decide. The grain density, size and chemistry are vital. In terms of chemistry, the lime/silica ratio of the grain is important in excess of 2:1 $CaO:SiO_2$ to ensure the formation of dicalcium silicate, a high melting point phase. Some MgO grains have a ratio as high as 6:1, but these then become more susceptible to hydration. Low basicity will result in low melting point phases and the loss of hot strength can be catastrophic. The amount of secondary minerals formed in the grain is also important, so the overall SiO_2 should be as low as possible (<0.3%). High Boron content is also very critical for example and will destroy the grain's hot strength.

Grain density can vary from 3.2 to over 3.5 g/cc. Low grain density means high porosity making the grain susceptible to slag penetration. Large grain crystallite will normally outperform low crystal size due to a reduction in interstitial porosity thereby reducing the chance of slag penetration into the grain boundaries and by lowering the susceptibility of the MgO to reduction by the C present in the brick during high-temperature service. Large crystallite size is generally considered to be over 140 μm in size. Fused MgO grain can exceed 1000 micron.

Addition of Antioxidants

As an antioxidant agent, metal Al, Si, Al–Mg and glass-forming agent like B_4C are used as depicted in Table 7.2. Addition of metal powder and other agents prevents oxidation and improves toughness in MgO-C refractories. In the case of MgO-C bricks, the following influences can also be predicted:

(a) Wear resistance decreases because of forsterite ($2MgO.SiO_2$) formation as a result of the reaction between MgO and SiO_2 in matrix.
(b) Wear resistance decreases because of $3MgO.B_2O_3$ formation as a result of reaction between MgO and B_4C glass.

Effect of Al Metal Addition as Antioxidant
The effect of aluminium antioxidant is demonstrated as [9] aluminium vapor diffuses through the MgO-C refractory, reacting with carbon and N_2 to form $Al_4C_{3(s)}$ and $AlN_{(s)}$ whiskers, respectively. Aluminium compounds have higher specific volume than the base refractory materials, and formation of those compounds are thermodynamically stable at wide temperature range. Al_4C_3 is practically formed above 750 °C. This compound results in MgO/graphite binding and reduction of brick permeability. Formation of aluminium compounds by reaction with carbon, N_2 and oxygen involves large specific volume and thus the micro-pores inside the bricks are partially filled up by those compounds, resulted in diffusion rate of gases. At high temperature up to 750 °C, Al in liquid form reacts with oxygen directly to form Al_2O_3, and above 900 °C, Al_2O_3 reacts with MgO to form $MgO.Al_2O_3$ spinel. The reactions are shown as follows:

$$2Al(l) + O_2(g) = Al_2O(s) \tag{12}$$

$$Al_2O(s) + O_2 = Al_2O_3(\text{presence of excess air}) \tag{13}$$

$$MgO + Al_2O_3 = MgO.Al_2O_3 \tag{14}$$

The rate of oxidation of aluminium in the presence of carbon and N_2 decreases above 750 °C. At higher temperature, Al(g) reacts with N_2 to form AlN (s) whiskers

Table 7.2 Important antioxidant, their phase formation and influence on the refractories

Additive	Phase formation	Remarks
Al metal	Al_4C_3, MgO. Al_2O_3 spinel	High temperature strength (H.MOR), wear resistance and oxidation resistance increase
Si metal	SiC, $2MgO.SiO_2$	Wear resistance decreases due to formation of forsterite
B_4C	$3MgO.B_2O_3$	Wear resistance decreases due to formation of flux

and with carbon to form plate like Al_4C_3 (s) as well as spinel as per the following equations:

$$2Al(g) + MgO(s) + CO(g) = MgO.Al_2O_3 + 3C \quad (15)$$

$$2Al + N_2 = 2AlN \quad (16)$$

$$2AlN + MgO + 3CO(g) = MgO.Al_2O_3 + N_2 + 3C\left(\text{above } 900^{\circ}C\right) \quad (17)$$

$$4Al + 3C = Al_4C_3 \quad (18)$$

$$Al_4C_3(s) + 2MgO(s) + 6CO(g) = 2MgO.Al_2O_3(s) + 9C(s)\left(\text{above } 900^{\circ}C\right) \quad (19)$$

The spinel formation results in volume expansion, porosity reduction and pore filling, which results in improved oxidation resistance.

7.3 Zonal Lining Concept

Initial top-blown BOF employed relatively simple zoning during the 1960s and 1970s since there are only three types of refractories available for BOF applications. A tar- or pitch-bonded magnesia bricks was manufactured utilizing two different qualities of magnesia grains. The lower quality was composed of a natural magnesite containing 90–92% MgO and the higher quality was at that time consists of 94–96% MgO with sintered pariclase grains. Both the varieties were bonded with coal tar or petroleum pitch and carbon additives of 4–6%. The third quality was the sintered magnesite composed of 94–96% MgO and then vacuum impregnated with a low quinoline insoluble coal tar or petroleum pitch to reduce or seal the surface porosity of the bricks.

With the three available brands, the lower quality tar bonded bricks were used in less critical areas with respect to corrosion, such as cone and bottom. The higher quality pitch bonded bricks are used in barrel including trunnion area.

In modern design, when the raw material technology was combined with the addition of different additives, antioxidants, the total range of refractory properties could be tailored to produce the required mechanical and chemical properties at application temperature. During last decade, from only three varieties of bricks, more than 30 compositions had been developed and one may choose the best quality bricks out of them. Normally, 5–10 qualities have been selected to line BOF in different zones to cover most current operating practice and associated with corrosion mechanism. The zoning of MgO-C refractory in working lining at three critical zones are explained below:

7.3.1 Bottom

Bottom bricks including tuyere blocks are exposed to significant temperature changes because of cooling action by bottom blowing gas and high temperature of liquid steel. Accordingly, it is assumed that those bricks are subjected to high-temperature abrasion due to turbulent flow of liquid steel. Therefore, bricks containing 20–25% more carbon than brick of other parts and development of refractories which have both spalling resistance and dense structure with proper grain size distribution to avoid liquid penetration.

7.3.2 Charging Side Wall

The charging side wall is directly impacted by mechanical and thermal shock during charging of scrap and hot metal. The stresses in the impact area of a BOF require the use of bricks with high fracture toughness or damage tolerance as well as optimal thermal shock resistance. Use of antioxidant and other additives help in developing the required properties. It has been reported [10] that MgO-C bricks of which the brick structure is stiffened by significant improvement in hot modulus of rupture (H MOR) by utilizing the reaction between Al-alloy powder and CaB_6, resulted in very good durability. With the increase of H MOR the hot abrasion resistance increases as shown in Fig. 7.17.

7.3.3 Tapping Zone

While tapping the tap pad in under very high stress depending on the steel grade being produced, very high temperature of up to 1780 °C and low viscus steel and slag rich in FeO or oxygen can be in contact with the refractory lining for a longer

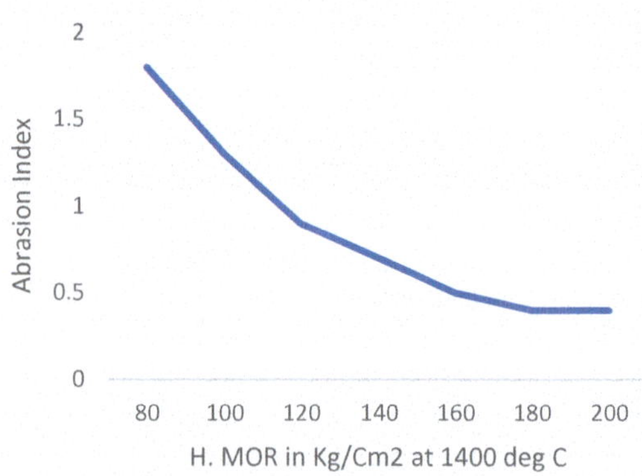

Fig. 7.17 Abrasion resistance increases with increase in H MOR

period of time. As a consequence, de-carburization occurs and low melting FeO-containing compounds are formed. Hence the refractory used in this area is effectively protected by using corrosion-resistant fused magnesia containing >98% MgO.

7.3.4 *Trunnion*

The trunnion area is not coming in contact of liquid steel and slag because of low filling height of the vessel. In consequence, slag coating is not possible in this zone. However, it is exposed to high temperature and gases and, because of that oxidation is increased compared to top cone. By raising the carbon content up to 15% with fused magnesia improves oxidation resistance. High resistance to oxidation of the flake graphite addition ensures a delayed loss due to burning.

Zonal Refractory Lining in a BOF
The different zones of a BOF is shown in the following Fig. 7.18. Suggested BOF lining concept detailed in Table 7.3.

7.3.5 *Mouth and Cone*

It is considered that peeling off the refractory due to stress fracture caused by expansion of bricks in the barrel lining and the loss of refractory due to mechanical stress by the removal of adhering metal are the main damage mechanisms for the mouth and cone. This area also suffers from oxidation due to presence of CO and

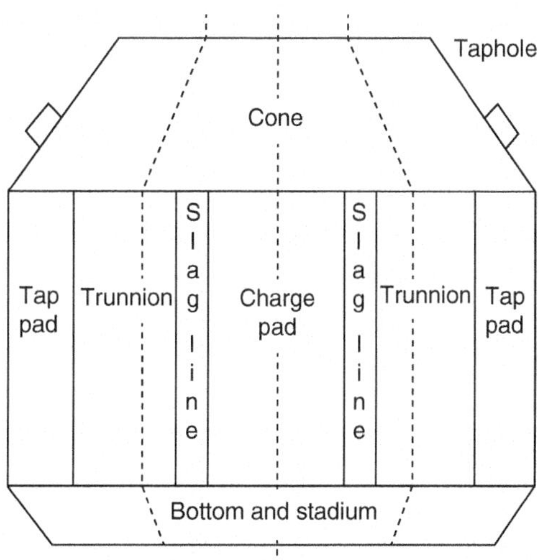

Fig. 7.18 Zonal refractory lining practice in BOF

Table 7.3 Different classes of refractories with respect to area of BOF

Area	Operational parameters	Type of refractory
Cone	Oxidation in the presence of CO_2, O_2 and FeO dust	97% fused and sintered magnesia with 10% carbon
Trunnion	Oxidation in the presence of CO_2, O_2 and FeO dust and mechanical stress. No slag coating possible	97–98% fused magnesia, 14% carbon and metal powder addition
Tapping side	High erosion due to liquid flow, corrosion due to molten steel and slag	98% fused magnesia, metallic additives and 10% carbon
Charging side	Severe thermal shock and mechanical stress induced by the contact and impact with scrap and hot metal	98% fused magnesia, metallic additives and 8% carbon
Side wall slag zone	Corrosion in contact of slag	97% fused magnesia, 10% carbon
Bottom	Thermal shock and erosion due to heavy turbulence of liquid metal due to gas purging	98% fused magnesia, metallic additives and 14% carbon
Tap hole sleeves	Hot abrasion, severe corrosion and thermal shock	98% magnesia, metallic additives and 10–14% carbon

Table 7.4 Typical properties of MgO-C bricks at different zones of a BOF

Properties	Unit	Top cone	Lower cone	Charge pad	Barrel	Bottom
Chemical						
MgO	%	95.5	96.2	92	85.5	97
Al_2O_3	%	5.5	0.4	6.2	7.5	0.3
Fe_2O_3	%	0.6	0.8	0.6	0.8	0.6
CaO	%	1.2	1.4	1.1	1.4	1.2
SiO_2	%	0.8	1.1	0.8	4.7	0.8
Cr_2O_3	%	Na	Na	Na	Na	Na
C	%	10	10	14	14	10
Physical						
Bulk density	g/cc	3.02	3.04	3.01	2.96	3.05
Porosity	%	3	3	3	3	3
Cold crushing strength	Mpa	45	40	35	35	40
CCS after coking	MPa	40	30	30	30	30
Thermal expansion at						
1000 °C	%	1.1	1.1	1.1	1.1	1.1
1400 °C	%	1.6	1.8	1.6	1.6	1.7
Thermal conductivity at						
1000 °C	W/mK	8.5	7.5	9.5	9.5	7.5
1400 °C	W/mK	6.5	6.2	8.2	8.2	6.2

oxygen. Refractory bricks made of dead burnt magnesia (DBM) containing >96% MgO and 5–10% carbon have shown superior durability and the properties of MgO-C bricks used in different zones of LD vessel are shown in Table 7.4.

7.4 Vessel Relining

Relining of a vessel is a very critical job with respect to lining the proper quality bricks in proper zones, maintaining homogeneity in lining profile and monolithic lining by ramming and casting. Workmanship plays a vital role to confirm perfect brick works. At the beginning of the reline, marking the furnace wall with the position of different zones that are being installed, such as trunnion, tap pad, charge pad, dome blocks and lower cone, is desirable. The foreman must ensure that the correct number of courses are being laid within the zone area.

7.4.1 Safety

After tearing out and after fixing the vessel in vertical position, the vessel tilt drive must be locked and the locks must be left till the relining jobs completed. If there is a possibility of gas flow, a positive isolation to be made for all incoming gas lines, prior to entering in the vessel. A continuous gas monitor to be installed inside the vessel to ensure gas-free condition. After fixing reline tower and brick lifting machine, a safety deck to be installed over the vessel to prevent from falling any material inside, while men are working. If access to the vessel is made by a ladder on the tower or brick-lifting machine, safety guard or cages must be installed along the entire length of the ladder to prevent falling off the ladder during ingress or egress. Precaution must be taken in handling of the lining MgO-C bricks; due to the presence of graphite, the bricks are very slippery and slip fall causes serious injury. The reline machine requires less manual handling of the bricks. Proper handling of bricks should be highly emphasized during relining.

7.4.2 Tear Out and Profiling the Old Lining

After the last heat is tapped, any skull present is removed from the top ring, and the vessel is then cooled down. Before the tear out begins, a profile should be taken of the location and size of the ruminant bricks. During dismantling of old refractory, a remnant refractory thickness profile to be recorded. Thus the profile taken is used for future study and modification of lining design. The profile includes the bottom, knuckle, barrel and cone area of vessel as shown in Fig. 7.19.

The profiling is important to understand the wear pattern at different zones at the end of full campaign. It is also important for design, material upgrading and brick sizing that will be used in future linings to increase lining life as well as lining cost control efforts.

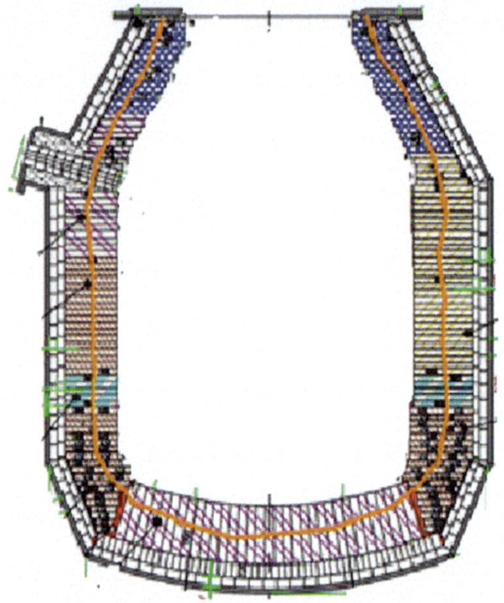

Fig. 7.19 Wear out profile of refractory

7.4.3 Bottom Lining

The bottom safety or tank lining is consisting of dome bricks made of hydration-resistant sintered or burned magnesite bricks laid in a concentric ring design. A heat set periclase mortar is used in lining the safety bottom. In a typical vessel, using multi-layer safety linings, there can be a total of 61 courses in the safety bottom, 30 courses in the lower layer and 31 courses in the upper layer. Bottom lining is started by keeping the space for fixing the bottom purging elements. All the bottom bricks are laid with 3 mm mortar joints.

7.4.4 Bottom Wear Lining with Herringbone Design

The Herringbone construction of the bottom wear lining is started by first determining the centre line of the vessel between the charge pad side and the tap pad side. A key wedge brick is the preferred brick shape to use in the wear lining. For the purpose of illustration, a brick having 76 mm thickness and 152 mm width across the inside face is being used. Two courses are laid with 152 mm width facing from tap to charge pad. Then the herringbone construction can be started. The grade, quality and length or height of the bricks are determined by the wear pattern of the bottom of a vessel. Use of knuckle bricks of various sizes is necessary to match with bottom. All the voids are rammed by MgO-based ramming masses. In the case of using MgO-C

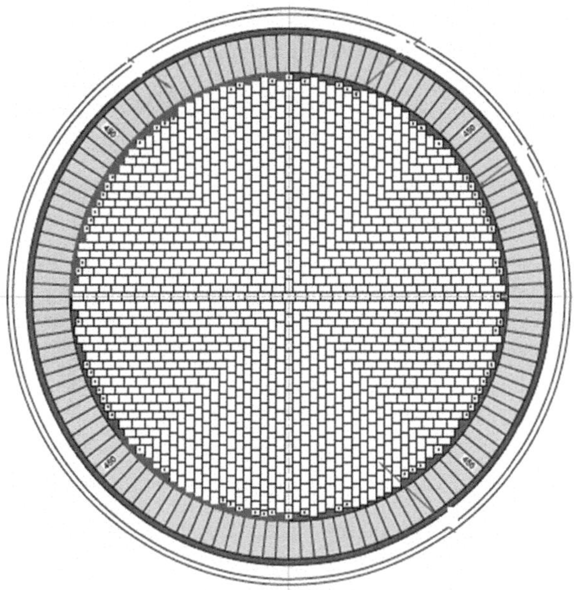

Fig. 7.20 Herringbone refractory lining in vessel bottom

bricks, providing optimum expansion joints is necessary. The herringbone bottom design is shown in Fig. 7.20.

7.4.5 Bottom Wear Lining with Concentric Ring Design

Concentric ring design is the most preferable for bottom blown convertors. The lining starts by setting the positions of purging plugs in the exact locations. Measurement to be taken at various points to ensure that the rings are running uniformly around the bottom. Concentric ring construction can be installed with full mortar joints or dry setting depending on shop practice. In this design, knuckle skew shape are normally used. The knuckle brick can be lined on the top of the knuckle skew.

7.4.6 Barrel

Before installing the barrel, the different zones in the vessel to be marked properly as per drawings. Also, notes can be included to show the size and quality of bricks used in each of those areas. As brick rings or courses are installed, different qualities of refractory are required to start the particular zones. The barrel is installed in circular course by course, through each zone to completion. Expansion allowance if designed must be adhered to. Tight fittings of key bricks require skill and are dependent upon the design of the lining and installation practices. Starting the course of lining and tightening the key area are predetermined as per design, for example starting the lining in tap pad zone and fixing key in charge pad zone.

7.4.7 Cones

The transition from barrel to cone is very critical. It should be 76 mm steps from the last course of the barrel to the first course of the cone, and continued in every course in cone area, as shown in Fig. 7.21. The 76-mm step is important as pinch spalling during burn-in or operation of the vessel can result if the steps are too large. The transition must be keyed properly. Expansion in this area is, again, a matter of design and may be dependent on whether or not expansion is causing a spalling or cracking problem in the cone or if the cone is water-cooled.

Standard Operating Procedure for Refractory Lining in BOF

1. While the vessel is being de-bricked, the inside condition of the worn-out refractory lining is inspected visually and the specific observations are made and noted. The full old safety lining and steel jams are to be removed. Before the de-bricking starts, a final wear profile must be taken.

(continued)

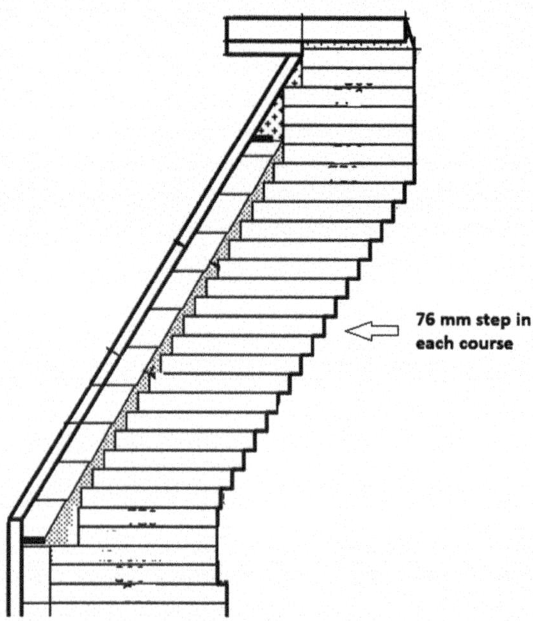

Fig. 7.21 Refractory lining in cone area

2. The vessel is put in vertical position and the pre-assembled tap hole block consisting of a square mother set and an exchangeable circular inner sleeve is inserted in its position. The necessary bolts are fixed.
3. Oxy meter, CO gas detector, rescue system, Oxy Pac, fire extinguisher, emergency light must be at site. The platform is to be tightly held with chain block to vessel hood to avoid any movement and derailment during use.
4. People should be careful during handling of the bricks specially the heavier bricks like 1050 mm and 900 mm bricks. People should be careful during the use of pneumatic and other tools. Only trained masons and craftsmen should be allowed to use the tools and tackles.
5. Put one layer of safety lining in chrome-magnesite quality in the bottom. Put one layer of dome blocks of magnesite quality keeping places vacant for fixing bottom purging elements.
6. Lay top bottom with Mag-C bricks and bottom purging elements as per the drawing. Make a pad with MgO-based ramming mass upto the level of dome blocks and lay first four courses of side wall in Mag-C quality as per drawing mutually agreed by the supplier and the department.
7. Ram MgO-based ramming mass between top bottom and side wall. During vessel relining the following parameters are to be maintained:

    (a) Profile of the first bottom is smooth.
    (b) Joints between two adjacent bricks is not more than 2 mm.
    (c) Side ramming is to be tight enough so that if pressed with thumb there should not be any impression left on it.

8. Start sidewall lining with MgO-C bricks following the relining pattern and complete it as shown in the drawing mutually agreed by supplier and the department and fill the gap, if any, between the face lining brick and safety lining with basic filling mass as per the need.
9. During vessel relining, please follow the following:

    (a) All the basic bricks are to be laid dry or with mortar as per shop practice.
    (b) Gap between two bricks should not be more than 2 mm.
    (c) Adequate moisture to be given in preparation of ramming mass for top tightening.
    (d) Ramming mass may be prepared in pan mixer also in place of manual mixing.

10. The lining of courses to be continued up to the top ring and the top layer of brick in cone lining to be tightened with the top ring plate.

7.5 Refractory Maintenance Practice

Refractory maintenance practice is of paramount importance to maintain availability, extend campaign life and optimization of cost. Materials for maintenance are very important as the volume of the maintenance materials consumed over a campaign may be up to two times the volume of original brick. Maintenance helps to balance the wear of refractory lining due to excessive wear in an area of repair whether the damage done mechanically or chemically.

A typical cost curve is shown in Fig. 7.22 [12]. The brick cost shows the cost of lining if the gunning maintenance is not considered. Gunning cost is the cost for refractory maintenance by gunning, and total cost is increasing from brick cost when the vessel is becoming old. There is a possible end of campaign in the absence of gunning maintenance at a life of 4000 heats. The life is extended to 12,000 heats by maintenance by gunning, patching or by slag splashing. The example shown in Fig. 7.22 is not unique and for various reasons, gunning rates may increase example of considerably higher specific cost, stages of the campaign.

7.5.1 Gunning

In the tapping side, charging side, slag lines and trunnions, the material of choice for repair of a BOF is high MgO-containing mix. High MgO is normally more than 85% MgO, medium purity is 65–85% MgO and low purity is less than 65% MgO. If the tapping side, the charging side or trunnion areas of the vessel, is merely eroded or slightly dished, a typical high MgO containing gunning castable with low rebound loss is used. Density, strength and rebound loss are the main criteria for the selection

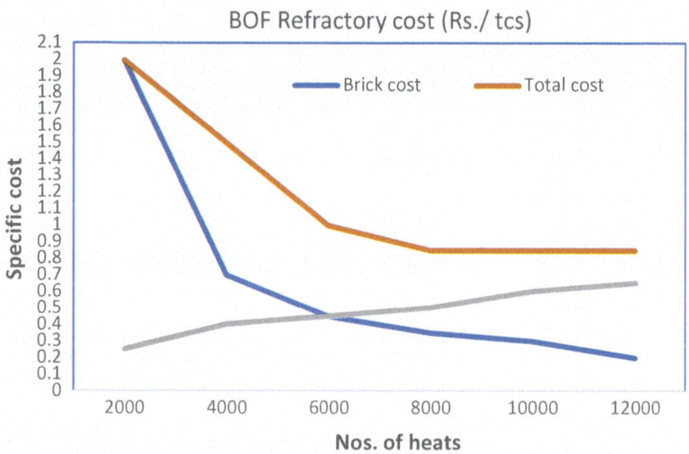

Fig. 7.22 BOF cost estimation for original lining and repair

of material for gunning repair in a vertical side wall. Stickability of the gunning mix with old worn-out refractory surface must be very good for higher durability. The requirement for BOF gunning maintenance is as follows:

1. Maintenance of pre-wear areas (balanced lining)
2. Increase of safety
3. Extension of campaign lifetime
4. Reaching of planned lining life
5. Production planning

There are nearly as many variations of vessel gunning programmes as there are shops employing the programmes, but, generally two types of programmes are employed:

1. Application of gunning materials to patch holes and balancing the lining profile as needed.
2. Programmed application of gunning material according to a predetermined schedule.

The later type is commonly referred as a preventive maintenance programme especially, as applied early in the BOF campaign. The availability of laser remnant thickness measurement system indicates when the maintenance to be initiated to maintain a predetermined optimum lining thickness up to the end of the campaign. A typical example of specific consumption (cumulative) of gunning material in a 160-ton capacity BOF is shown below: up to 1000 heats: 0.30 kg/ton of crude steel, 1000–2000 heats: 1.10 kg/ton of crude steel 2000–4000 heats: 2.25 kg/ton of crude steel and 4000–6000 heats: 2.75 kg/ton of crude steel.

7.5.2 *Patching*

Severe wear of the tap pad, charge pad or bottom refractories cannot be repaired by slagging or gunning. It requires to pour the compatible patching castable. Such patches may require 250–600 Kgs of material, even some times more than 1.0 tons. High MgO-based materials are used for pouring and patching. The patching material must have the following properties:

1. High mix density and very good stickability with old refractory lining,
2. Flowability of more than 80% to spread over the affected area,
3. No shrinkage at operating temperature,
4. Excellent thermal shock resistance,
5. Excellent corrosion resistance.

The refractory material to be applied is mixed with water in a large mixer to a flowable consistency. Mixing of excess water to be avoided because complete drying of the patch is required before taking the vessel in operation. Any moisture retains into the mix may cause serious explosion and dislodge of patching material.

Fig. 7.23 A typical vessel image (**a**) after gunning and (**b**) after patching of refractory

The mixer castable is transferred from the mixer to a pan and poured into the vessel. The vessel is rocked to place the material in the desired area. The material must be thoroughly dried prior to charging the vessel again. A gas flame may be required for drying. Durability of the patch material is 30–150 heats, depending on the quantity of material used and severe corrosion in affected area. Panels (a) and (b) of Fig. 7.23 show the condition of vessel refractory lining after gunning and patching, respectively.

7.5.3 Slag Splashing

The process of slag splashing was developed to extend the refractory life of BOF and hence, cost reduction. Today all major steel plants worldwide employed this maintenance practice in order to increase convertor life. The slag splashing technique involves the usage of high-pressure nitrogen blowing through the oxygen blowing lance to splash slag on to the vessel refractory. The slag forms a coat of after splashing, over the old refractory lining, cools, solidifies and creates a solid layer of slag that creates a protecting layer over refractory lining. By introducing the splashing technique, the normal converter life has increased to 4000 heats. In few shops it is reported as high as 30,000 heats. The efficiency of slag splashing depends on physical and chemical properties of slag at the end of the blow, and temperature plays a vital role. The melting behaviour and development of the coated layer over old refractory are a function of slag composition and splashing parameters. The effectiveness of the splashing process depends on input variables, blowing practice, addition pattern, vessel geometry, nozzle design and bottom plug configuration, etc. It experiences that the physical, chemical and thermal variables are responsible for the formation of good slag coating. To obtain a uniform coating thickness and eliminate bottom build-up, it is important to maintain the slag composition in a well-defined range with MgO saturation, along with temperature control and lance characteristics.

For formation of effective slag coating, slag characteristics are very important. During splashing, the slag must enter into cavities, depressions, created in the refractory lining and form a solid well sintered layer. This solidified layer prevents the direct contact of refractory bricks with liquid steel during blowing, thereby preventing the erosion and oxidation of carbon in the MgO-C brick. To form a uniform sintered layer slag is melt in splashing condition and it also adheres to the refractory surface. The following conditions need to be satisfied for a good slag splashing:

1. Optimisation of the parameters required to obtain a slag lining over the worn-out lining; the parameters are gas flow, lance height and slag composition.
2. The formation of slag coating provides adequate protection to the refractory lining from oxidation and corrosion, and the effectiveness of coating formation depends upon melting and mineralogical composition of slag.

7.5.3.1 Characterisation of Slag for Slag Splashing

A most suitable slag for good coating formation and splashing contains high melting and low melting components. The two important phases are defined as follows:

1. A low melting phase, which helps to spread the slag over the lining and fill up the cracks and pores of eroded old lining.
2. A high melting phase, which provides the necessary protection for the refractories.

An effective slag suitable for slag splashing contains 10–15% FeO, in the form of calcium ferrite ($CaO.Fe_2O_3$) and enhanced MgO content to maintain a slag basicity of around 2.5.

7.5.3.2 Conditioning of Slag with MgO

The MgO (the major component of basic refractories) content in slag plays a significant role for corrosion resistance and improves refractoriness, which has a direct relation to the durability of new slag coating. Increasing temperature and iron oxide content of the liquid increase the diffusion rate of lime. Lime is fully incorporated in the slag near the end of the oxygen blow. MgO can be enriched also by adding dolomitic lime. When CaO and MgO from fluxes could be incorporated near to the saturation limits into the liquid slag, refractory dissolution by slag would be minimum. Addition of calcined dolomite through the bulk material charging system, at the very beginning of the blow, aided complete MgO incorporation into solution. Magnesite chips and MgO-based gunning mass are also used along with scrap as per the requirement. The above pattern ensured good dissolution of all additives. These processes of addition fluxes have resulted in consistently developing a suitable slag for effective splashing.

Slag Splashing and its Benefits

1. The benefits of slag splashing to the steel-making processes are as follows:
2. Much longer refractory life with effective reduction of refractory cost can be obtained. The extended life as reported is 30,000–60,000 heats [13].
3. The use of slag composition for splashing operation, very close to MgO saturation limit, reduces the corrosion of refractory in.
4. Yield improvement arises from an increase in volume with a decrease in "slopping".
5. During operation after slag splashing, which contain high MgO content and also CaO content (in case of dolomitic lime addition), less CaO is needed in BOF addition during blowing. Hence cost of CaO is reduced.
6. Cost of gunning material consumption has been reduced when slag splashing is introduced.
7. Faster phosphorus removal is possible as the melting of the low melting phases of the slag causes rapid dissolution of CaO (from slag coating) by SiO_2 in BOF slag.
8. There is no significant difference in S, P, Mn and nitrogen contents of the steel when slag splashing is applied.
9. Frequency of vessel refractory relining is significantly reduced by adopting slag splashing and hence yearly outage time of vessel is also reduced. A typical reduction in outage time for a 160-ton capacity vessel is reduced from 60 days/year to 20 days/year.

7.6 Modern Refractory Practice to Prolong Life

Pure MgO is extremely refractory, having a melting point of 2800 °C. However, the amount and type of impurities present in raw materials form low melting phases and reduce the dissociation and softening temperature of MgO. For this reason, emphasis is given on total amount of impurities present and the proportion of CaO and SiO_2 in magnesite raw materials. Over the past four decades, the level of impurities has been reduced from 6–8% to less than 2%.

Of the impurities present in refractory magnesia, CaO and SiO_2 are the most important and usually they are most abundant. In magnesia grains, the CaO and MgO combine with SiO_2 to form silicate compounds. The types of silicate phases depend on CaO/SiO_2 ratio. For a number of reasons, dicalcium ($2CaO.SiO_2$ or C_2S) and tricalcium silicates ($3CaO.SiO_2$ or C_3S) are the most desirable phases in magnesite grains, used in manufacturing modern BOF refractories. The CaO/SiO_2 molar ratio is 2:1 in dicalcium silicate and 3:1 in tricalcium silicates. Dicalcium silicate (C_2S) has a high melting point, 2130 °C, and it is a stable phase and does not form any low melting compound in the presence of MgO. However, in presence of high SiO_2, the C/S ratio drops below 2:1, and the silicate phases change to dicalcium silicate to merwinite ($3CaO.MgO.2SiO_2$ or C_3MS_2) and then to merwinite alone,

Table 7.5 Different classes of magnesia used for BOF refractories

Properties	Unit	Type 1	Type 2
MgO	%	98.5	97.5
CaO	%	0.95	1.8
Al_2O_3	%	0.11	0.2
Fe_2O_3	%	0.16	0.5
B_2O_3	%	0.003	0.01
SiO_2	%	0.3	0.5
CaO/SiO_2	Ratio	3.2:1	3.6:1
Bulk density	g/cc	3.53	3.5
Crystal size	Micrometer	>780	>800

then to merwinite to monticellite ($CaO.MgO.SiO_2$ or CMS) and then at a ratio of 1:1, only phase present is monticellite. In the presence of merwinite and monticellite, the melting point is reduced to 1480–1650 °C, depending on the proportion of low melting phases present, which adversely affects the refractoriness of magnesite raw materials used.

Of the other impurities, Boron (B_2O_3) is very important to control. With high C/S weight ratio of more than 1.87:1, very high hot strength, hot modulus of rupture (H MOR) can be obtained, but in the presence of B_2O_3 the strength is reduced significantly. In general, the B_2O_3 content in the brick had to be less than 0.05% to maintain high hot strength. In modern refractory practice, the B_2O_3 content in magnesite raw material is maintained as low as <0.01%.

The two remaining impurities are Al_2O_3 and Fe_2O_3. If the C/S ratio is less than 2:1, these two impurities usually combines with MgO to form spinels, $MgO.Al_2O_3$ and magnesia-ferrite, $MgO.Fe_2O_3$. These two spinels are forming continuous solid solution. However, with the increase of Al_2O_3 or Fe_2O_3, part of the spinels tends to dissolve in the low melting silicates, merwinite or monticellite, which causes volume expansion and reduces refractoriness.

If the $CaO/SiO_2 > 2:1$, the excess CaO combines with Al_2O_3 and Fe_2O_3 to form calcium aluminates and calcium ferrite, which have low melting points. Thus, the amount of these two impurities needs to be maintained very low. The tentative properties of fused magnesia used in modern upgraded MgO-C bricks for BOF application are shown in Table 7.5.

7.6.1 Source of Carbon

Graphite is the main source of carbon, and other sources are carbon black, pitch and carbon-containing binders. Graphite is classified as flake graphite, amorphous graphite and highly crystalline graphite. Amorphous graphite is cryptocrystalline graphite. Sometimes it is a result of graphitization from glassy carbon, under high temperature and pressure. There are different properties among amorphous graphite. The properties that are important are carbon content, true density and chemical composition. The flake graphite structure is very similar to fish scales, flake graphite occurs in large flakes, small flakes and thin flakes. The general properties, which are important for refractory applications, are ash content and its composition, carbon content, size of flakes true density, surface area and degree of graphitization. The rate of oxidation of MgO-C bricks largely depends on surface area and true density.

7.6.2 High Crystalline Graphite

In modern refractory practice, high crystalline graphite is largely used in MgO-C bricks. This type of graphite exhibits one very important property, which is not available in other types of graphite. This type of graphite is highly flexible under extremely high pressure, the graphite becomes almost plastic and begins to flow into any void and pores present in the brick matrix. High crystalline graphite is the only variety of graphite that can be compressed to approach the true density of graphite.

The high crystalline graphite is primarily restricted to use in commercial manufacturing of bricks due to the limited availability and very high cost.

7.6.2.1 Use of Nano-Carbon

Carbon is used in refractory for its excellent thermal shock resistance and corrosion resistance towards BOF slag and steel. But it has an inherent drawback of oxidation. In BOF operating condition, the oxidation is very severe due to oxygen lancing and presence of high FeO in slag. Other drawbacks to use in BOF refractory are high thermal conductivity, which results in heat loss and increase in shell temperature. One of the modern refractory practice is to attempt to reduce total carbon content. Development of nano-carbon containing MgO-C bricks with reduced total carbon content has improved oxidation resistance and reduced thermal conductivity. Addition of nano-sized carbon would bring down the total graphite content of conventional magnesia carbon bricks having around 14–9% C. Nano-carbon has a very high specific surface area of 120 $m^2.g^{-1}$ compared to flake graphite (5 $m^2.g^{-1}$). Due to the very high surface area, nano-carbon has higher reactivity with anti-oxidant to form oxidation-resistant products faster [14], thereby arresting brick failure. The nano-carbons fill up the pores in the graphite matrix leading to matrix consolidation and reducing gas permeability into the brick structure. Nanomaterial has a lubricating effect on the matrix in which they are dispersed and thus nanoparticle absorbs stresses arising from thermal expansion of the bricks during service, thus preventing thermal spalling.

References

1. M A Serry, and M.S. Attia, Thermal Equilibrium and Properties Of some MgO-Dolomite Refractories, Br. Ceram. Trans. J., 84, 1985, pp 142–145.
2. Mohamed A. Serry and A Barbulescu, Thermal equilibrium of Magnesia–Dolomite Refractories within the system CaO-MgO-C2S-C4AF, Transaction of British Ceramic Soc, Vol 80, No 6, 1981, pp 196-201.
3. Kenji Anan, Wear of Refractories in Basic Oxygen Furnaces (BOF), J of technical Association of Refractories, Japan. Vol. 21, no 4, pp 241–246.
4. H. Jansen, MgO-C Bricks for BOF lining, Institute of Materials, Minerals and Mining 2007.
5. Takashi Miki, Mitsuo Satoh, Shigeki Uchida, Minoru Satoh, Backward inclined lining, a new design for BOF cone linings, Shinagawa Technical report, Vol 50, 2007, pp 21–30.
6. Saturo ITO, Takayuki INUZUKA, Technical Development of Refractories for Steel making processes, Nippon Steel Technical Report, No 98, July 2008.
7. Dr. Hans Ulrich Marschall, Christoph Jandl, Design Evaluation of BOF-linings with Aid of Thermomechanical Simulation, AISTech 2011 Proceedings–Vol 1, pp 1223–1230.
8. Gerald Routschka (Editor), Book on "Pocket Manual Refractory materials", 1997.
9. S.K. Sadrnezhaad, Z. A. Nemati, S. Mahshid, S. Hosseini, B. Hashemi, Effect of Al antioxidant on the rate of oxidation of carbon in MgO-C refractories, J. Am Ceram. Soc, 90, [2] 509–515 (2007).
10. Shigeyuki Takanaga, Wear of magnesia-carbon bricks in BOF, Taikabutsu Overseas, Vol 13, No 4, pp 8–14
11. Ritsu Ebizawa, Wear of refractories in combined blowing convertor, Taikabutsu Overseas, Vol 13. No 4, pp 15–19
12. Pitágoras Gomes de Lanna, Dave Ehrhart, Mateus Vargas Garzon, Thiago Avelar, Mark Loeffelholz, BOF Refractory Linings: Balancing Brick Life and Gunning/Hot Repairs to Maximize Performance, AISTech 2014 Proceedings, pp 1301–1309
13. Kenneth C. MILLS, Yuchu SU, Alistair B. FOX, et al. A review of slag splashing, ISIJ International, Vol 45 (2005), No 5, pp 619–633.
14. Kalyani Ravi, Subir Biswas, Sanat Hazra, Use of low carbon content Nano-carbon added Magnesia carbon in steel ladle metal zone at Tata Steel, Jamshedpur, TAIKABUTSU, VOL 71, NO 4, 2019, PP 158–164.

Chapter 8
Refractory for Secondary Refining of Steel

8.1 Introduction

The selection of most suitable refractory for lining steel ladle depends on a series of factors, namely steel-making processes, steel quality, ladle availability, plant operation logistics and cost. In the course of processing, the ladle for secondary metallurgy lined with refractory material is filled with liquid steel and different types of slags. Various additives and treatment time cause further chemical and thermal stresses on the working lining of treatment ladle. The design of refractory lining consists of working lining, permanent lining and backup insulation. The design also consists of two distinct zones, lining in slag zone and lining in metal zone. The slag line is the region that comes into contact with various types of slags, which can produce aggressive behaviour in terms of corrosion and erosion at the refractory interfaces. There is also a potential for oxidation depending on slag chemistry and quality of refractory used. The side walls undergo thermos-mechanical movement during operation, which can result in high erosion in the working lining. Refractory lining in different zones in steel ladle is shown in Fig. 8.1.

Refractory materials of secondary refining ladles are exposed to two main wear factors [1]:

1. Erosion at high temperature due to gas purging effect
2. Corrosion caused by slag refractory interaction at the time of processing the heats.

Slag composition depends on the different alloying additives and additives for killing processes used, silicon-killed steel, aluminium-killed steel, Al/Si—bi-metal addition and calcium-treated steel are the examples for different additions, which forms slags of different chemistry and they have a direct impact on refractory wear, in slag zone as well as in metal zone.

S. Biswas, D. Sarkar, *Introduction to Refractories for Iron- and Steelmaking*,
https://doi.org/10.1007/978-3-030-43807-4_8

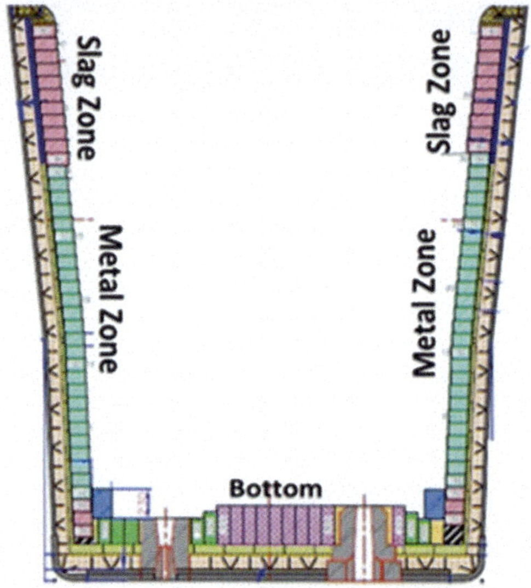

Fig. 8.1 Different zones in steel ladle

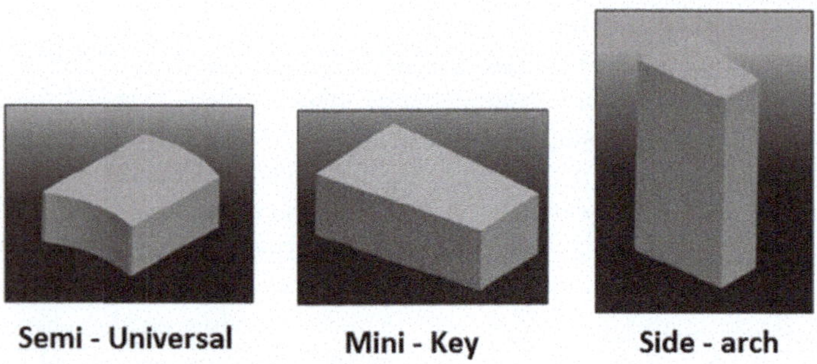

Fig. 8.2 Different shapes of ladle lining bricks

8.2 Refractory Design in Steel Ladles

There are several ladle shapes in existence today, each with some advantages and also disadvantages, but their use would be decided on plant operating practice and logistics. There are three main types of brick shapes available, mini-keys, semi-universal ladle bricks and side arch, as shown in Fig. 8.2. Semi-universal bricks are generally available in 100 mm course height, with curved ends, making them suitable for spiral lining construction. In this type of lining, it is common to build a starter ramp with the sets of universal bricks with varying course thickness, or the ramp can be made with castable lining. But the castable lining has a chance for high erosion; hence, the ramp built with varying thickness in semi-universal bricks is preferred. This shape is popular due to the quick installation time, the easy and flexibility of the built. There are very few vertical joints, but it is not a tight built.

Hence this design is suitable for smaller capacity ladle. The lining can become unstable, with a loss of key in thin or worn-out lining, and is susceptible to cracking in the middle of the brick due to hoop stress induced in service.

Mini-key bricks is described as a brick in side arch, single tapered, with a 150 mm tentative width dimension in the middle of the brick. The construction comprises with the combination of bricks of different taper angles. Thus, the combination of different taper bricks enables to turn the ring. This is an easy and flexible build, having good lining stability, which makes it suitable for lining larger size, 300 tons ladles and oval shells. There is consistent course height, with less potential for metal penetration and joint erosion. In the ladle lining, carbon-containing bricks such as MgO-C, Al_2O_3-MgO-C and CaO-MgO-C bricks are used in working lining. However, the applications of those bricks accompanied an issue of high heat loss from ladle shell, increase of shell temperature. In order to reduce the heat loss and durability of lining, an intermediate lining with high-alumina bricks or high-alumina castable lining is used. In many shops worldwide, an insulating lining is also introduced above the steel.

Is the Use of Insulation in Ladle a Feasible Solution for Energy Saving?
The backup insulation lining generally results in the increase of the wear rate and severe slag infiltration into bricks of working lining, because temperature of the working lining increases. As a result of the temperature rise in the working lining, the amount of heat storage in the ladle lining increases, which increase chemical corrosion of the working lining that may reduce the campaign life. Hence use of insulation lining in ladle backup may not be a feasible solution for energy saving in many steel plants; however, it depends on the shop respective operating practice.

In order to solve the issue, in modern refractory practices, magnesia–carbon bricks with low thermal conductivity is developed and in use. Use of nano-size carbon is able to reduce the total carbon content in MgO-C ladle bricks and hence reduces the thermal conductivity. Conventional MgO-C ladle bricks contain 10–12% carbon and by adding nano-carbon the total carbon content is reduced to 6–8%. The change of thermal conductivity by using nano-carbon is shown in Fig. 8.3 [2].

8.2.1 *Volume Stability/Expansion/Shrinkage*

Length and volume expansion value at different temperature is one of the important parameters to decide the expansion allowance in refractory lining. They can be influenced deliberately by using a certain type of bonding and additives. Thus, volume of the bricks can be changed, expands or contracts with change of temperature. However, it can be considered also that the actual expansion or shrinkage is altered by infiltration of slag or liquid steel. Once the molten charge has been tapped

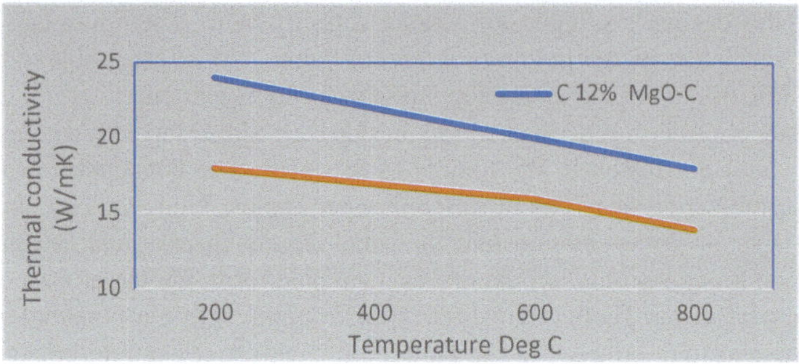

Fig. 8.3 Change in thermal conductivity with respect to graphite and nano-carbon

Fig. 8.4 Change of thermal stress in steel ladle working layer with time

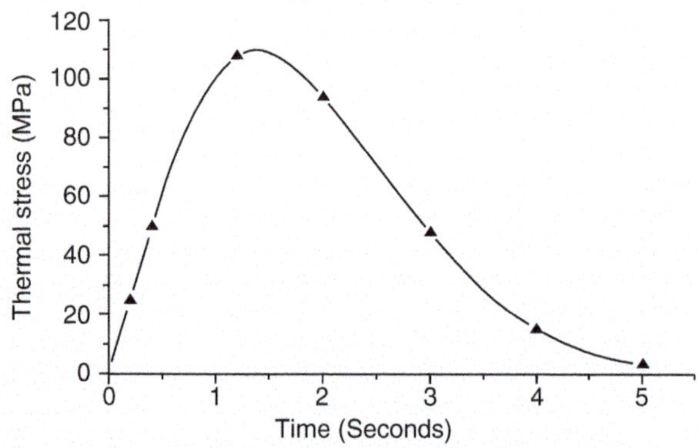

off, the ladle would be empty and the lining will shrink as soon as the refractory materials cool down. During heating up the ladle before taking it to operation, the key bricks are pushed backward for expansion and to adjust the expansion, a loose mass is provided at the back of the working lining.

Thermal Stress in Steel Ladle Refractory Lining

During tapping of liquid steel into ladle, the refractory in face lining expands and it is under compression and the cold face lining remains under tension. During emptying of the ladle, the stress in bricks under compression in face lining is released. Due to this repeated heating and cooling, vertical permanent crack is generated in face lining. During heating the ladle (during tapping) the thermal stress at the working layer receives its maximum value approximately within 1.5 s after metal pouring and then goes down as shown in Fig. 8.4.

Gradually the thermal stress decreases as the temperature difference between metal and lining surface decreases, which results in relaxation of

(continued)

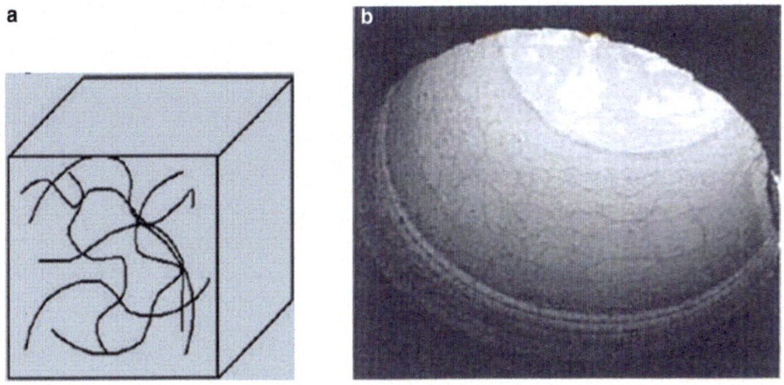

Fig. 8.5 (**a**) Crack formation due to stress, (**b**) cracks in steel ladle

thermal stress; it also reduces the velocity of lining heating. Typical compressive strength of refractory is 30–50 MPa; hence, as the thermal stress generated within 1.5 s of liquid steel pouring exceeds the compressive strength, microcracks develop at hot face of the bricks.

Results of compressive stresses

Numerous surface cracks grow; crack sizes are about 5 mm deep. These cracks appear around 1–3 s from the moment of metal pouring, as shown in Fig. 8.5.

A dry mix is backfilled between face lining and intermediate lining, consisting of a lightweight refractory mix. It is recommended to fill in the mix in layers and compress it. In metal zone, face lining with dolomite bricks, dolomite-based backfill mass is used, for MgO-C face lining, magnesia-based mass is used. In slag zone special quality backfill mass is used. As the backfill layer of 70–80 mm thick is provided to absorb the stresses of face lining bricks due to expansion, the sintering characteristics of the backfill mix is very important; it should not fully have sintered and it must have a property of very slow sintering and be flexible enough to adjust the expansion and contraction of the face lining bricks.

8.2.2 Stress During Cooling

During rapid cooling (in between fills) after working layer temperature falls below than deeper layer temperature, tensile stress begins to grow in the working layer. The thermal stress becomes more than material strength within 7–10 s from the cooling starts, as shown in Fig. 8.6, and the bricks crack. This tensile stress grows permanently for a long time.

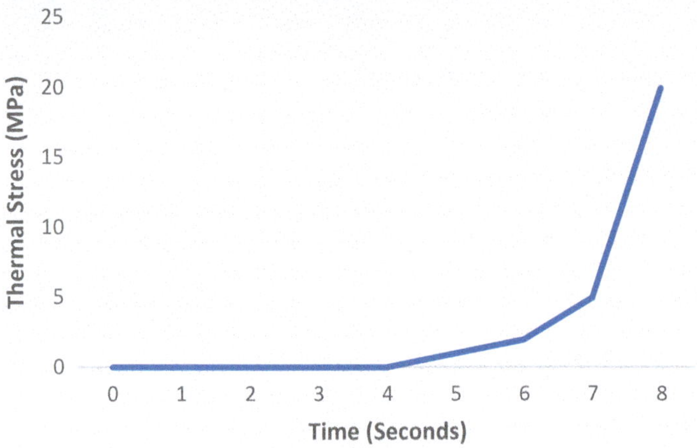

Fig. 8.6 Thermal stress at working layer

Fig. 8.7 Vertical crack formation in ladle side wall

8.2.3 Results of Tensile Stress

1. The tensile stress can reach the value of more than 300 MPa after around 15 min, but it is not possible as cracking in breaks appears.
2. The crack growth will absorb the excess stress (compression of cracked bricks during cooling) for creating new crack surface.
3. Deep cracks appear commonly at places where stress concentration exists, such as brick corners, for example. In some cases, the crack depth can become more than the thickness of the bricks and results in complete breaking of bricks and dislodging, as shown in Fig. 8.7.

8.3 Ladle Refractory Lining for Silicon-Killed Steel

The slag generated from silicon-killed steel is siliceous, which has V (CaO/SiO_2) ratio is 2.0–2.5, and therefore dolomite-based refractory is the best suitable material for lining in metal zone. Tar- or resin-bonded Dolo-C bricks perform well provided there is good ladle utilization and thermal management. It has been known that in certain favourable conditions, dolomite-C refractory can extend ladle performance to more than 100 heats. If dolomite is used in an intermittent operating plant, this can then result in the lining cracks upon cooling, creating the potential for metal penetration and excessive damage to the safety lining, thus compromising the ladle security. Magnesia-carbon is compatible to use in slag zone and bottom, but it may not perform in metal zone as it will not pick up a protective coating, which is a notable characteristics of dolomite bricks.

8.3.1 Slag–Refractory Interaction

8.3.1.1 Slag Interaction with Dolomite Refractory

The SiO_2-rich slag often forms a white coating of di-calcium silicate on the dolomite refractory. During casting of steel, as the steel level descends, the small dolomite crystals of 15 μm size readily form high-melting compound like di-calcium silicate (C_2S) due to their high reactiveness towards the encountered slag. The SEM microstructures of after service dolomite are shown in Fig. 8.8. The dolomite region very close to the hot face, as shown in Fig. 8.8, has large areas of C_2S appearing white and segregated dark periclase crystals [3]. No presence of free lime could be seen in the microstructure. A substantial reduction in periclase crystal size could be due to corrosion by the slag. Figure 8.8 shows the microstructure of after service dolomite refractory away from the hot face. The microstructure reveals modification of the bonding phase from lime matrix to C_2S phase for periclase crystals. Penetration of slag into the refractory is believed to have resulted in the change. The growth of periclase crystals even after long exposure to high temperature was found to be minimal, which might be due to bridging by the C_2S phase.

8.3.1.2 Slag Interaction with MgO-C Refractory

Figure 8.9 represents a micrograph of the cross-section of MgO-C brick after use. It is seen that a slag infiltrated layer is covered by an outer layer containing many MgO "islands" of various sizes. At the interface between slag and brick, MgO grains were found to be detached from brick and dispersed in slag (Fig. 8.9a). Further, a decarburized refractory layer between the magnesite refractory and the slag infiltrated layer could be seen from the micrograph. Oxidation of graphite takes place

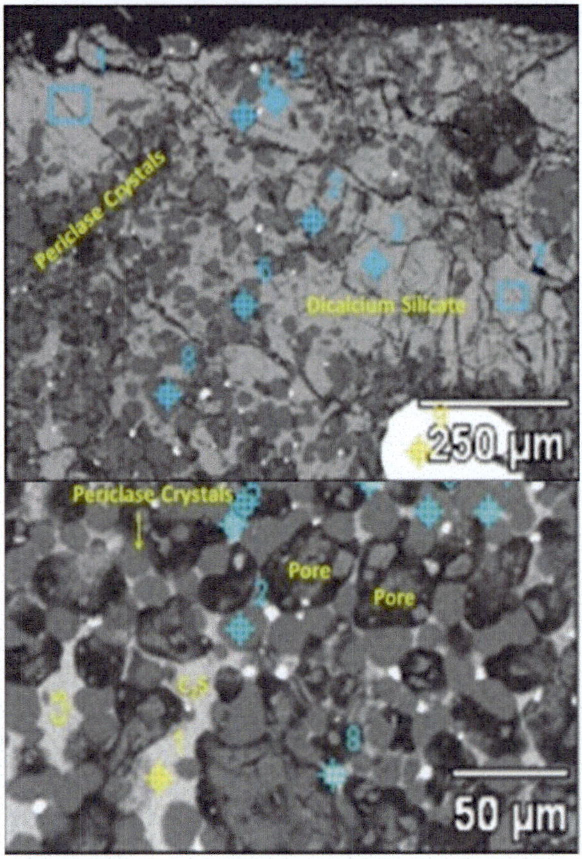

Fig. 8.8 Microstructure of dolomite refractory

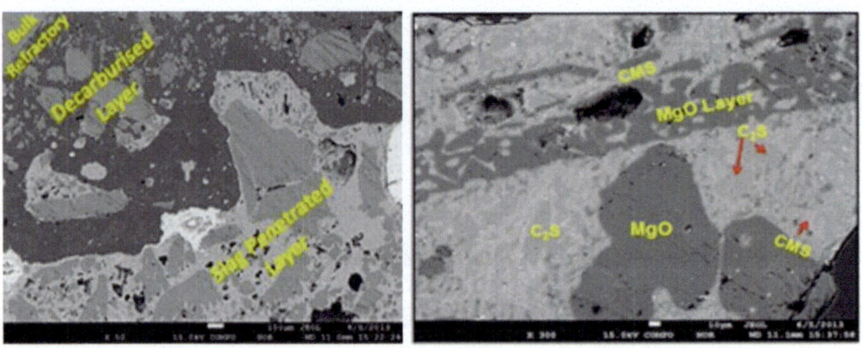

Fig. 8.9 Microstructure of MgO-C bricks

Table 8.1 Phases identified in micro analysis (EDS)

Phases	Type of sample	CaO	Al_2O3	MgO	SiO_2
$CaO·MgO·SiO_2$ (CMS)	MgO-C Brick (After service)	50–52	0	22–25	20–22
MgO		0	0	100	0
$2CaO·SiO_2$ (C_2S)		65–70	0	0	30–35

during ladle preheating which increases bricks porosity and allows slag penetration into the refractory. This in turn leads to dissolution of MgO grains in the calcium-silicate slag and their dispersion in the slag. A bridge of MgO phase could be seen near the hot face which can be attributed to precipitation of MgO phase due to faster cooling of slag compared to infiltrated slag (Fig. 8.9b).

It is interesting to note that the slag consists of two phases, a dark grey phase and a light grey phase. The two phases are uniformly distributed, and composition analysis revealed that the composition of the infiltrating slag was the same as that of the slag in the outer layer consisting of CaO-MgO-SiO_2 (CMS)-based monticellite phase with MgO content as high as 22–25 wt% (Table 8.1). Since the slag is usually unsaturated with MgO, the dissolution of MgO into the adhering layer would result in the formation of CMS phase leading to corrosion of the refractory at high temperatures. It is expected that the time for MgO dissolution is very short and the dissolved MgO would precipitate when the temperature decreases. This would explain formation of a bridge of MgO at the hot face. The dark grey phase was identified as crystals of dicalcium silicate (C_2S). Presence of iron layer at the outermost portion of the sample was seen which had come from the steel during casting of the last heat. Oxidation of this iron layer has resulted in the formation of its oxide with time or during sample preparation. Thus, when MgO-C refractory comes in contact with siliceous slag, no silicate coating is formed, and further low-melting phases of monticellite are formed, which increases the dissolution of MgO-C bricks, resulting in low campaign life. Dolo-C bricks are best suitable refractory to use in producing silicon-killed steel.

8.3.2 *Prospective Dolomite Refractory*

During past years, tar- or pitch-bonded dolomite is used in large extent in steel ladle metal zone. Due to ever-increasing environmental problems, it became more and more difficult to manufacture and use pitch-bonded products. Pitch belongs to the group of components which can cause cancer, if the content of benzo[a]pyrene exceeds 50 ppm. Pitch binder usually used in brick making contains very high level of benzo (A) pyren, and hence lot of precautions to be taken at the work place including additional health control of the workers. In the past years, many efforts had been made to develop alternate binder to manufacture Dolo-C bricks.

Today, a lot of resin-bonded dolomite bricks are in service. The base of the binder is a thermos-setting phenolic resin.

Dolomite, containing less than 1.0% silica, is more thermodynamically stable than high alumina or even magnesia chrome refractories. Therefore, a dolomite lining provides lower level of oxygen activity in the melt, which can lead to an increased alloy yield in silicon-killed steel, particularly for those alloys with high oxygen affinity.

The recent development is MgO enriched dolo-carbon bricks used in steel ladles. The enrichment of dolomite with MgO improves its performance in lining ladle furnace. The MgO/CaO ratio and the amount of fluxing oxides, SiO_2, Al_2O_3 and Fe_2O_3, are the main factors affecting the performance of MgO enriched dolomite-carbon refractories. The matrix of the dolomite is made of CaO and MgO is spreaded over the calcia (CaO) rich matrix, which is not saturated; hence, addition of MgO in dolomite bricks tends to improve corrosion resistance of dolomite bricks. The addition of MgO also improves hydration resistance of dolomite bricks.

Dolomite Refractory for Ladle Metal Zone

Dolomite refractories can generally be considered in two types depending on the bond system: tar or pitch bonded and resin bonded bricks. There is no doubt that in recent years there has been a trend towards the use of magnesite or magnesite–carbon bricks, due to the increase in the percentage of con-cast steels and need to improve ladle availability. Nevertheless, a dolomitic lining proved the best suitable refractory for treating silicon-killed steel and in present days the Dolo-C bricks are continued to use in metal zone of such application. The advantages of using dolomite lining are:

1. Dolomite-C lining has higher thermal capacity in ladle lining, which allows lower tapping temperature once the ladle has been pre-heated prior to taking it in circulation, and is of a great importance in LD steel making process. This thermal improvement can be utilized for reducing energy consumption.
2. The yield of alloy addition with an affinity for oxygen is improved, and relatively high-quality steels with improved purity can be produced.
3. Desulphurization in dolomite ladle is superior. The low oxygen activity results in a high oxidic purity in the liquid steel compared to high alumina refractory for metal zone application.

MgO enrichment in Dolo-C bricks:

1. MgO enrichment was required to improve high temperature strength and corrosion resistance. In conventional dolomite bricks, the CaO-containing matrix is not saturated and MgO was embedded in CaO matrix. So, to improve corrosion resistance towards basic slag, matrix was made saturated by adding excess MgO.
2. MgO enrichment improves thermal shock resistance by creating micro-cracks at MgO grain boundary and matrix (due to differential thermal expansion between MgO grain and matrix)

(continued)

3. Use of 30% MgO enrichment had shown significant improvement in H MOR (from 25 to 60 kg/cm^2). This may be due to uncontrolled crack formation in MgO grain boundary and CaO-rich matrix.

The main disadvantages of dolomite lining are problem of hydration and poor thermal shock resistance. The tentative properties of different quality dolomite-C bricks are listed below:

Properties	Unit	Dolomite-C	MgO enriched Dolo-C
Bond	Type	Tar/Pitch	Resin
Chemical analysis			
MgO	%	36–39	55–58
CaO	%	54–56	42–46
Al_2O_3	%	0.8	1
Fe_2O_3	%	0.8	0.8–1
TiO_2	%	0.09–0.1	0.01–0.018
SiO_2	%	1	2.5–3.5
Physical properties			
Porosity	%	3.5–4	4–5.5
Bulk density	g/cc	2.95	2.95
Cold crushing strength	kg/cm^2	700–800	700–800
H MOR at 1400 °C	kg/cm^2	21–26	45–55
MOR at RT	kg/cm^2	80–90	100–200

8.4 Ladle Refractory Design for Al-Killed Steel and Ca-Treated Steel

Essentially flat products are manufactured through the route of Al-killing. Strips, rolled coils and plates are manufactured through this route. The slag composition is typically of CaO/SiO_2 ratio > 2.0 and Al_2O_3 is 15–20%. In manufacturing few special grade steel to produce thin slab and cold rolled steel, calcium in added with aluminium, which is called Ca-treated steel. MgO-C bricks are the most compatible refractory to use in slag zone. The corrosion rate of MgO-C bricks becomes slower as the basicity is higher. Increase of Al_2O_3 in slag increases the solubility of CaO and increases fluidity of slag and as a result the corrosion in MgO-C bricks increases. In Fig. 8.10 [4] the hatched (blue colour) area shows the presence of low-melting phases at 1700 °C and with 10% Al_2O_3 and 15% Al_2O_3, which indicates that the corrosion in MgO-C bricks increases with increase of Al_2O_3 content or reduction of CaO/SiO_2 ratio.

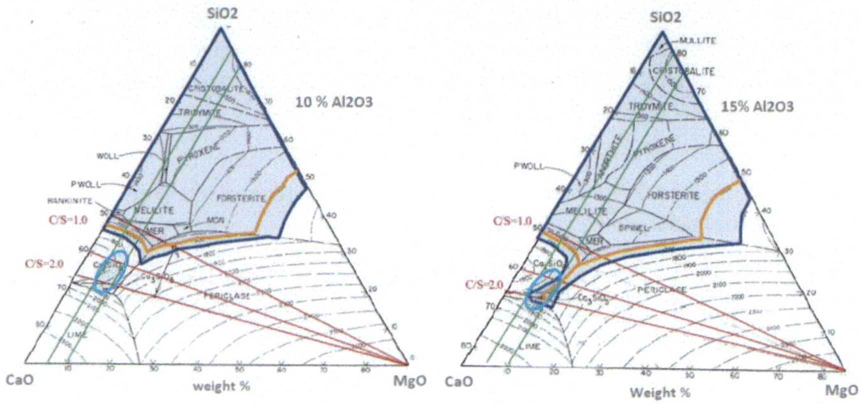

Fig. 8.10 Presence of liquid phases at 1700 °C in CaO-MgO-SiO_2 system with 10% and 15% Al_2O_3

Till the mid-1990s, fired high-alumina refractories with significant amount of silicates, such as andalusite- and bauxite-based bricks, were used in the metal zone of steel ladle lining. They could not perform well in contact with corrosive calcium–aluminate slags because of high wear.

In modern refractory application in steel ladle, metal zone with Al-killed steel and Ca-treated steel, the following types of refractory are in use:

1. MgO-C bricks
2. Alumina spinel bricks
3. Al_2O_3-MgO-C bricks.

8.4.1 MgO-C Refractories

When using MgO-C refractory in metal zone of Al-killed steel ladle, magnesium compound in the form of inclusion in steel is observed. The main reactions in MgO and C in MgO-C refractories are reduction of MgO by carbon and oxidation of aluminium dissolved in steel by carbon monoxide formed at the interface of steel and refractory. Reduction of MgO by C takes place according to the following equation:

$$MgO_{(s)} + C_{(s)} = Mg_{(g)} + CO_{(g)} \tag{8.1}$$

In this case, $Mg_{(g)}$ diffuses towards the free surface at hot face, where it comes in contact with oxygen of higher P_{O2} and is oxidized to MgO, which in turn condenses and forms a MgO enriched protecting layer over MgO-C refractory. This glassy MgO layer protects the further reactions and prevents corrosion.

In the case of Al-killed steel, the CO generated in reducing MgO (Eq. 8.1) diffuses through the interface and comes in contact with metallic aluminium dissolved in liquid steel and oxidized through the following equation.

$$2Al + 3CO_{(g)} = Al_2O_{3(s)} + 3C$$

Depending on the oxygen potential at the interface of liquid steel and MgO-C refractory, the magnesium vapour can be oxidized to MgO or diffused into molten steel where it reacts with Al_2O_3 and form spinel. Spinel thus formed acts as an inclusion in steel, which is a major issue to produce clean steel.

8.4.2 Alumina Spinel Refractory

Alumina spinel fired refractory bricks for steel ladle metal zone lining is one of the recent developments over the last two decades. The purpose of such innovation was primarily to cater the growing need of advanced metallurgy for ultra-low carbon and automobile grade of steel to reduce carbon pick up from refractory to liquid steel. Even as alternative to other carbon-free refractories, the alumina spinel fired bricks are another superior quality fired bricks for achieving high performance in highly corrosive metallurgical environment, such as CaO/SiO_2 ratio 1.5–3.0%, MnO about 2.0%, with Ca addition in Al-killed steel. Such alumina spinel bricks are manufactured by using proper granulometry of very high purity synthetic alumina aggregates, such as white Tabular alumina and white fused alumina and $MgO\text{-}Al_2O_3$ spinel, MgO and calcined alumina in finer fractions. High-alumina bricks with spinel addition has the advantages of lower conductivity and low thermal expansion coefficient compared to MgO-C or Al_2O_3-MgO-C (AMC) bricks, which results is reduced heat loss through the ladle wall, low shell temperature and high thermal shock resistance. The comparative properties are shown in Table 8.2.

The spinel-containing alumina bricks do not contain carbon, and hence dissolution of carbon in ultra-low carbon steel is eliminated. This type of bricks is widely used in producing clean steel. Figure 8.11 [3] shows a microstructure of the hot face of preformed spinel-containing alumina bricks after use. Alumina reacts with CaO of slag to form CA_6 at high temperature. However, in the present case, CA_2 was found to have formed adjacent to the alumina grains and finally forming a mixture of CA

Table 8.2 Comparative properties

Properties	Spinel MgO. Al_2O_3	Periclase MgO	Corundum Al_2O_3
Density (g/cc)	3.56	3.59	3.98
Thermal conductivity (W/mK)	5.7	7.2	6.4
Thermal expansion coefficient [ΔL/L. K.10^{-6}]	7.4	13.5	8.6

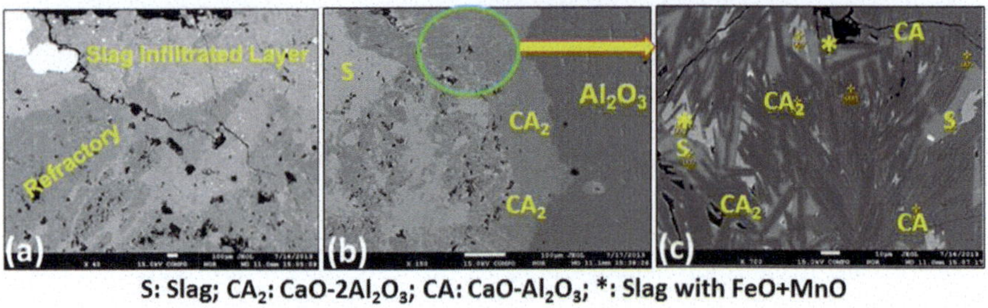

Fig. 8.11 Phases in spinel-containing Al_2O_3 refractory used in Al-killed steel ladles

and CA_2 in a calcium-alumino-silicate slag matrix as the temperature decreased (Fig. 8.11b, c):

$$\underset{}{Al_2O_3} \rightarrow \underset{1850^{\circ}C}{CA_6} \rightarrow \underset{1750^{\circ}C}{CA_2} \rightarrow \underset{1600^{\circ}C}{CA} \rightarrow \underset{1455^{\circ}C}{C_{12}A_7} \rightarrow \underset{1535^{\circ}C}{C_3A}$$

This results in a change in slag composition, thereby making the slag high melting and more viscous and restricting its penetration into the refractory.

$$\underset{}{Al_2O_3} \rightarrow \underset{1850^{0}C}{CA_6} \rightarrow \underset{1750^{0}C}{CA_2} \rightarrow \underset{1600^{0}C}{CA} \rightarrow \underset{1455^{0}C}{C_{12}A_7} \rightarrow \underset{1535^{0}C}{C_3A}$$

Alumina Spinel Bricks for Metal Zone of Steel Ladle Is the Best Practice?

- Pre-reacted spinels have proven their good thermal shock behaviour in practical applications. This behaviour is mainly due to differences in thermal expansion between alumina and spinel. The different expansion leads to micro-cracks in the matrix that act as crack arresters.
- Spinel-containing bricks have lower thermal expansion. Magnesia has the highest thermal expansion of all refractory oxides. Therefore, it improves thermal shock resistance compared to conventional high alumina bricks.
- In steel **ladle bottoms the volumetric stability** under high temperature and pressure is most important. **High erosion resistance** is also important. Bricks containing **pre-reacted spinel are the material of choice** for these applications.
- Impact pads are exposed to high erosion, thus demanding a material with high hot strength. Furthermore, a high thermal stability is required. Spinel-containing bricks are mainly used in this application.

8.4.2.1 Corrosion Resistance of Spinel Refractories

Attack of CaO from Ladle Slag to the Alumina Spinel Brick

$$\begin{array}{ccccccc} CaO + Al_2O_3 & \rightarrow & CaO.6Al_2O_3 & \rightarrow & CaO.2Al_2O_3 & \rightarrow & CaO.Al_2O_3 \\ & & (T_L = 1830^\circ C) & & (T_L = 1762^\circ C) & & (T_L = 1602^\circ C) \end{array}$$

High-melting products CA_6 and CA_2 result in a compressed matrix structure effectively preventing further penetration of ladle slag.

Also, consumption of CaO increases slag viscosity and weakens penetration ability.

This also explains why this spinel performs equally well in Ca-treated Al-killed grades (metallic Ca will reduce FeO and form CaO).

Attack of SiO_2 from Slag to the Alumina Spinel Brick

$$\begin{array}{lll} SiO_2 + CaO.6Al_2O_3 \rightarrow & 2CaO.Al_2O_3.SiO_2 & \text{or } CaO.Al_2O_3.SiO_2 \\ & (\text{gehlenite}/T_L = 1590^\circ C) & (\text{anorthite}/T_L = 1550^\circ C) \end{array}$$

Low-melting compounds (minority phases) provide structural flexibility and infiltration resistance. Although they decrease the refractoriness, its effect is superseded by compression of matrix structure by formation of CA_6 and CA_2 as mentioned above.

A strong feature of all spinels is the tendency to substitutional solid-solutioning, where large percentages of one or both of the spinel components may be substituted by others of the group. For the magnesium-aluminate spinel both magnesium and aluminium cations can be replaced by others with similar size. This is an important feature and a major factor in the advantage of spinel-containing brick for resistance to steel-making slag.

$$FeO_x + MgO.Al_2O_3 \rightarrow (Mg, \mathbf{Fe})O \cdot Al_2O_3$$

$$MnO + MgO.Al_2O_3 \rightarrow (Mg.\mathbf{Mn})O \cdot Al_2O_3$$

- Occupation of interstitial cation vacancies by Fe^{2+} and Mn^{2+}.
- This leads to an expansion of lattice and effectively prevents further penetration of ladle slags.
- The slag after FeO and MnO incorporation in the spinel is more viscous and better positioned to effectively adhere to the refractory surface.

8.4.3 Al_2O_3-MgO-C (AMC) Refractory

Alumina–Magnesia–Carbon (AMC) brick is resin bonded and it contains corundum, brown fused alumina or bauxite, magnesia and graphite as source of carbon. The magnesia content varies from 5 to 30% and 10 to 18% graphite is added. Those bricks are preheated at 280–360 °C, depending on the polymerization characteristics of resin used. During preheating in ladle operation, spinel (MgO.Al2O3) forms and the resulting post-firing expansion can close joints and densify the brick structure. This post-firing residual expansion made it a best choice to use in ladle bottom as well as in side wall. When the ladle is in circulation, the expansion of the bricks takes place gradually due to spinel formation, the expansion begins at about 1200 °C, and is completed at 1500 °C.

AMC bricks after used in producing Al-killed steel had been studied [3] and the microstructure of slag coating after service, over AMC bricks, is shown in Fig. 8.12. A thin but distinct layer of magnesio–wustite with increased porosity was observed along the periphery of the periclase grains. According to the EDS analysis, the wustite consists of 15–20% FeO and 80–85% MgO. The formation of this layer on the surface of MgO grains could impede the dissolution of MgO into the slag. EDS analysis of the bulk slag revealed the presence of undissolved MgO grains along with C_3A phase in slag matrix.

Microstructural analysis of the slag-refractory reveals the presence of large number of MgO islands with smaller sized grains (Fig. 8.13). Rounding of the grain boundary can be attributed to corrosion of the grains by calcium-aluminate slag (Fig. 8.13b).

EDS analysis of the slag indicates that the bulk slag contains mostly CaO, MgO and Al_2O_3. Crystallized laths consisting of 70–72% CaO were also found to be uniformly distributed in the slag matrix (Figure 8.13b). The CaO-Al_2O_3 phase diagram information (Fig. 8.14) suggests the section relating to the solid-solution region of lime and C_3A. Unreacted lime in the bulk slag was also found to be present

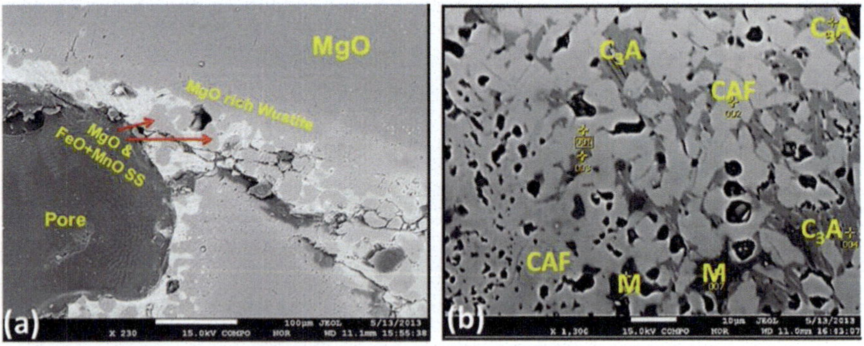

Fig. 8.12 Interface of AMC after service bricks. (**a**) Interface, (**b**) slag of Al-killed steel Al-killed, Ca-treated steel

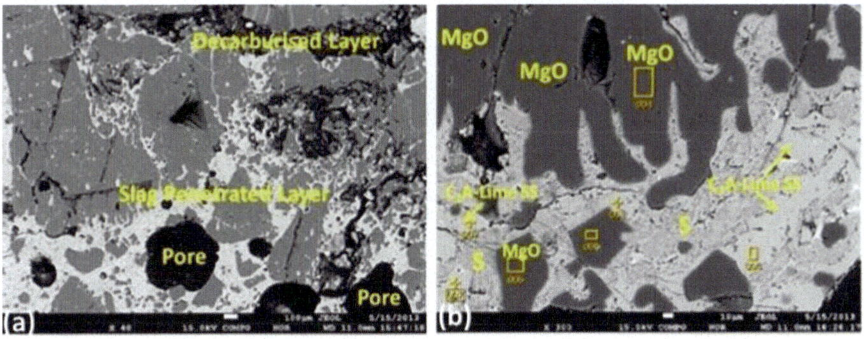

Fig. 8.13 Microstructural analysis of after service bricks

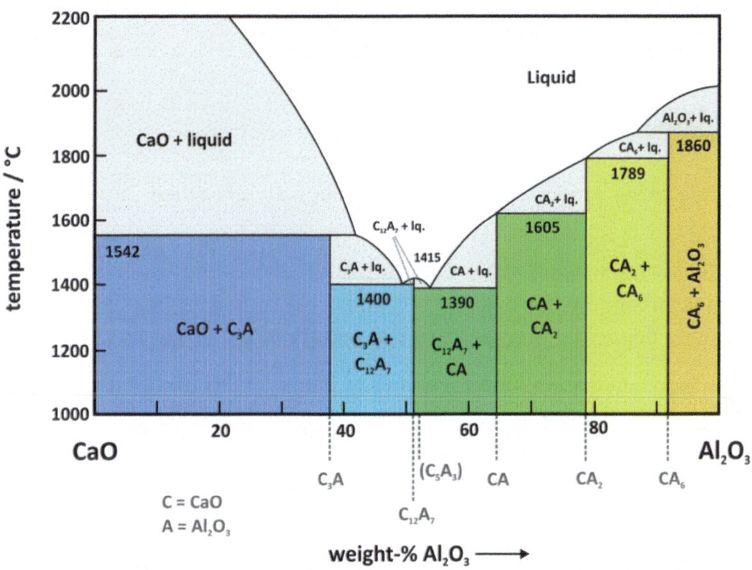

Fig. 8.14 $CaO–Al_2O_3$ system

in few portions which might be due to addition of higher quantity of lime during ladle furnace treatment. Crack between the periclase grain and bulk slag indicates physical adherence of the slag layer to the refractory, which has a chance of getting peeled off during next heat. This in turn increases the corrosion of the refractory lining with successive heats. This increases the chances of refractory wear in subsequent heats.

Formation of Glaze Protection Layer over Refractory

1. Initial slag penetration into the refractory occurs rapidly by slag infiltration at the hot face, but further penetration is slow and occurs by infiltration down the grain boundaries.
2. Formation of C_2S phase near the hot face of dolomite lining prevents further dissolution of the brick in calcium-silicate-based ladle slag.
3. In Al-killed steels, calcium-aluminate phases were found to be the major reaction phases that formed ladle glaze. The formation of these phases shows a positive effect on the corrosion resistance as they inhibit the direct dissolution of Al_2O_3 into the slag. However, addition of Ca during ladle treatment shifted the reaction zone towards lime-rich region in the phase diagram. In addition, the presence of undissolved lime was evidence in case of Ca-treated steels.

8.5 Refractory Used Under Vacuum

Vacuum degassing processes have been traditionally classified into the following categories:

- Ladle degassing process (VD, VOD, VAD)
- Stream degassing process
- Circulation degassing process (DH, RH and RH-OB)

Currently stream degassing process no longer exists. DH is virtually non-existent, while RH process and its variants are the most popular in producing ultra-low carbon and alloy steels. To meet the increasing demand of cold rolled sheets with improved mechanical properties, the demand of ULC (ultra-low carbon) steel with $C < 20$ ppm is increased. The RH process had been modified by oxygen blowing under vacuum and it is known as RH-OB.

Refractory plays a very important role in vacuum degassing process, as the refractories behave differently in high temperature under vacuum than is used under normal pressure. Owing to the recent extensive development of metallurgical processes under vacuum conditions, since last three decades, more attention had been given to develop suitable refractory to use under vacuum and elevated temperature. The vacuum, particularly oxygen partial pressure, leads to a change of valance of some oxides and increases dissociation and vitalization reactions. The different factors for corrosion of refractories under vacuum are namely:

- The presence of iron oxides and attack by slag.
- The influence of vacuum and atmosphere.

The periclase grains are saturated with iron oxide (FeO) associated with swelling effect with dislocation and formation of low-melting compounds under vacuum. The amount of infiltration of slag into refractory is significantly high in vacuum condition and the refractory is worn out by a densification–spalling mechanism. Stability and vaporization of different refractory oxides are explained below:

8.5.1 *MgO and CaO*

Under reduced pressure and depending on partial pressure of oxygen p_{O2} the vaporization takes place as:

(a) $2MgO\ (s) = 2\ Mg\ (g) + O_2$
(b) $2CaO\ (s) = 2Ca\ (g) + O_2$

As per the law of free energy change, under the assumption that the activity of the phase MgO (s) is equal to one, yields:

$$e^{-\Delta Go/RT} = P_{Mg} . P_{O_2}{}^{1/2}$$

ΔGo is the standard free energy change connected with reaction (a), the expression of the element Mg as function of P_{O2}

8.5.2 *Cr–O*

The stability of Cr-oxides depends on partial pressure of oxygen and has been stated as the vaporization of Cr_2O_3 in oxidizing atmosphere and is explained as,

$$Cr_2O_3(s) = CrO(g) + 1/2 O_2$$

The vapour pressure of Cr_2O_3 is high and hence it easily vaporizes to CrO and Cr,

8.5.3 *Al–O*

The vaporization process of Al_2O_3, under reduced partial pressure of oxygen, is as follows:

(a) $Al_2O_3\ (s) = Al_2O\ (g) + O_2\ (g)$
(b) $Al_2O_3\ (s) = 2\ AlO\ (g) + 1/2\ O_2$
(c) $Al_2O_3\ (s) = 2\ Al\ (g) + 3/2\ O_2$

Under vacuum or in low partial pressure of oxygen, the Al_2O_3 refractory is not stable and hence is not suitable to use in vacuum degassing units.

8.5.4 Si–O

The gaseous molecules that develop from the vaporization of SiO_2 (s) are those of monoxide SiO (g) and SiO_2 (g) as shown in the following equations:

(a) SiO_2 (s) = SiO (g) + 1/2 O_2 (g)
(b) SiO_2 (s) = SiO_2 (g)

8.5.5 MgO–C

The advantages of MgO-C bricks in non-vacuum environment are their excellent thermal shock resistance, resistant to slag corrosion due to non-wetting property of carbon component and the in situ formation of a protective dense magnesia layer on the refractory hot face. Under vacuum, the formation of dense magnesia layers do not exist at the hot face owing to the low oxygen partial pressure. It is also reported by Quon and Bell [5] that the laboratory test confirms on MgO-C bricks in vacuum at 1700 °C the loss of refractory components by vaporization, augmented by the reduction of MgO by carbon to magnesium vapour, as shown in Fig. 8.15. Hence MgO-C is not a suitable refractory to use under vacuum.

Today, extremely low final carbon contents of less than 20 ppm can be obtained through RH-OB route, as required for the production of automotive sheets. Dissociation of MgO-C bricks and free carbon increases the carbon pickup in ultra-low carbon steel and makes it unsuitable to use in such operating condition in RH-OB.

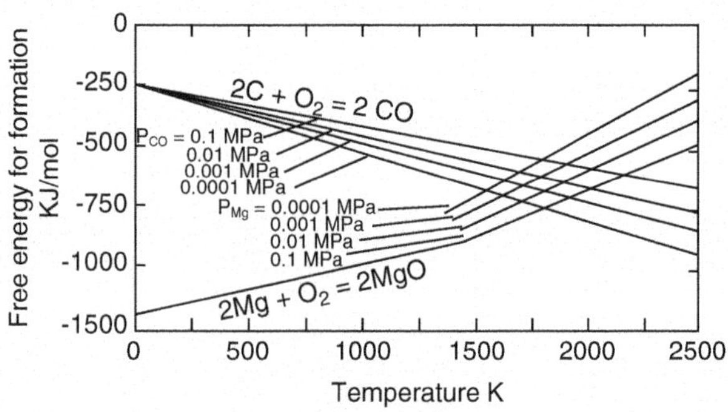

Fig. 8.15 Reduction of MgO by carbon under reduced pressure

8.5.6 RH Degasser

RH process is based on the circulation of molten steel between the ladle and the RH vessel. The rate of steel circulation determines the velocity of the metallurgical reactions and the duration of the process assuming a defined metallurgical target. Melt circulation depends on the geometry of the equipment such as snorkel diameter, the radius of the equipment, and the position and number of lift gas tuyeres [6]. The wear of refractory lining depends on the following operational and design factors:

- Re-circulation velocity of steel from steel ladle to RH degasser
- Diameter of snorkels and diameter of equipment
- Degree of vacuum
- Oxygen purging rate and duration of operation
- Degree of vacuum
- Change of slag chemistry in lower vessel during operation.

8.5.7 Refractory Wear Mechanism

8.5.7.1 Upper Vessel

Upper vessel is exposed to repetitive heating and cooling every time when the lower vessel is exchanged. It is corroded by FeO attack when the oxygen top lance is employed from vessel top to melt the skull.

Lining refractories in upper ducts, alloy addition chute and upper vessel are not in direct contact with molten steel or slag. Longer life is expected by the optimization of an electrode heating method of maintaining proper temperature, when the vessel is not operational. The main problem includes the skull formation over the hot face of refractory lining. The deposited skull causes thermal spalling and mechanical brick peeling by temperature fluctuation accompanied by attachment and removal of the lower vessel.

8.5.7.2 Lower Vessel

Requirement from RH degasser lower vessel bricks is excellent thermal spalling resistance and corrosion resistance to the effects of FeO produced during oxygen blowing through the lance inserted from the furnace top. The bricks lined in the throat portion connected to snorkels are exposed to severe operational condition and high HMOR bearing bricks are recommended. Lower vessel is a very critical part where a large amount of refractory is consumed. The wear of lining bricks is mainly due to structural spalling in which slag is penetrating into bricks to form brittle layers that are subjected to corrosion, and at the same time, cracks parallel to the hot face are formed, resulted in peeling. According to the studies on brick deterioration by

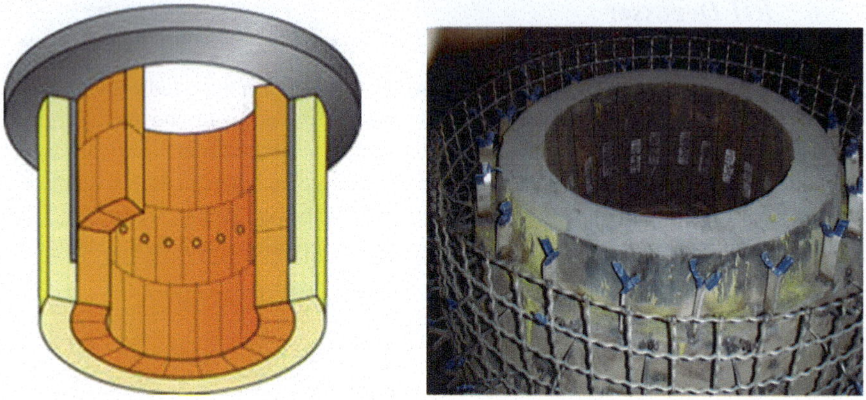

Fig. 8.16 Ring construction in RH snorkels

slag penetration, the penetrated slag is liable to cause thermal spalling; therefore, it is certain that the slag penetration triggers structural spalling. Dissolution of refractory takes place into the penetrated slag and causes severe corrosion. In order to reduce refractory wear, the following counter measures can be taken:

1. Operational aspects: The control of slag chemistry, degree of vacuum, circulation time to prevent slag penetration into lining refractory. The measures of slag removal from the vessel and aluminium addition are effective.
2. Material aspect: Material selection with optimum bond strength to resist slag penetration and corrosion resistance.

8.5.7.3 Snorkels

The snorkel is a complicated refractory assembly, which is inner-lined with the bricks and outer-lined with monolithic refractories on the cylindrical steel core shell. The inner lining bricks of up leg snorkel is abraded by gas blowing and therefore magnesia chrome bricks with excellent resistance against hot abrasion are applied. Deformation of the steel core shell due to long-time operation influences the life of snorkels. The bricks for inner-lining of the snorkel are usually pre-assembled to optimise proper lining in small ring construction as shown in Fig. 8.16. The outside of the snorkel is lined with castable, supported by proper anchoring.

Material Development of Magnesia Chrome Bricks

The magnesia chrome bricks, in particular, direct bonded quality, is most extensively used in RH degasser. In magnesia chrome bricks, corrosion resistance is compatible to spalling resistance. Both the properties are closely related to Cr_2O_3 content in the brick. Chromite has a smaller coefficient of

(continued)

thermal expansion than magnesia and voids are formed around chromite grains after firing, preventing the propagation of micro-cracks, which further prevents in the presence of chromite. Thus, increase of chromite is effective to increase spalling resistance.

To improve corrosion resistance in direct bonded bricks, magnesia or chromite in fines is added to develop spinel in brick matrix. The properties such as spalling resistance and corrosion resistance are associated with the extent of growth of secondary spinel formed at grain boundaries. The growth of secondary spinals is closely related to brick firing process and partial pressure of oxygen. Optimum spinel growth is observed in the case of slow cooling rate or high oxygen partial pressure. The addition of appropriate additives coupled with controlled firing is considered to provide more durable direct bonded magnesite bricks.

The following wear mechanisms of direct bonded bricks are considered:

1. Melting and flow due to lowering refractoriness of bricks by infiltration of CaO, SiO_2, Al_2O_3 and FeO. In the hot face and erosion by flow of hot metal due to liquid circulation under vacuum.
2. Peeling of refractory from crack propagation which occur in the dense zone by penetration of CaO and SiO_2 or at brittle zone behind the dense zone.
3. Brittleness of brick structure by alternative reaction of FeO to Fe_2O_3 due to fluctuation of oxygen partial pressure, vacuum and temperature.
4. FeO produced by oxygen blowing into vessel reacts with direct bonded magnesia chrome bricks forming low melting compound in the presence of CaO and SiO_2.

8.5.7.4 Slag: Refractory Interaction in RH Vessels

An RH slag can be formed in the RH vessel due to entrainment of ladle slag during the immersion of the RH snorkels into the ladle and the dissolution of FeO and Fe_2O_3 attached on the RH vessel wall after the heat taken in previous batch. A thick skull enriched of Fe_2O_3, formed at the lower vessel, is shown in Fig. 8.17. The RH slag contains a large amount of FeO originating from oxidation of the steel skull and metal droplets in the RH vessel and steel oxidation during O_2 blowing stage.

Mun-Kyu Cho et al. [7] had explained the change of slag composition in RH degasser with time of operation, as explained in Fig. 8.18. As the decarburization reaction proceeds, the amount of RH slag and FeO content decreases continuously to supply oxygen towards steel. After decarburization reaction, Al is added as deoxidizer FeO content decreases and Al_2O_3 content increases. According to the model prediction, the slag contains high FeO level during the first 15 min of the RH process, between 40 and 22%, and is known to be aggressive towards the lining refractory. After the Al addition, the slag contains calcium aluminates and MgO.

Fig. 8.17 Thicken Jam in lower vessel

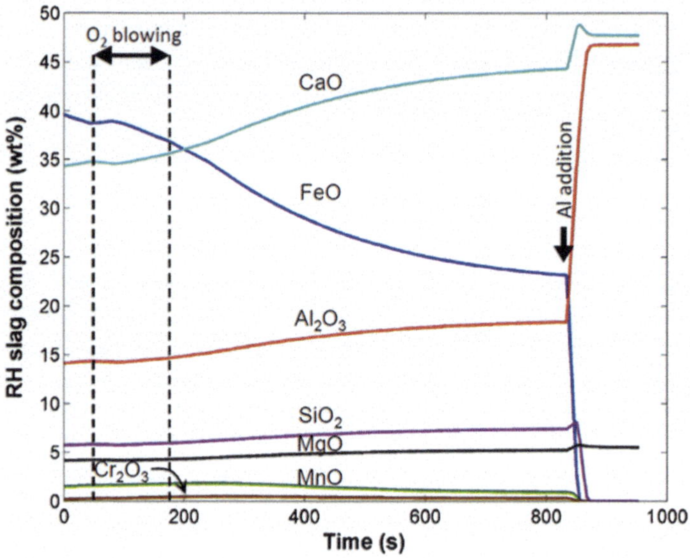

Fig. 8.18 Change of RH slag composition during process, a prediction [7]

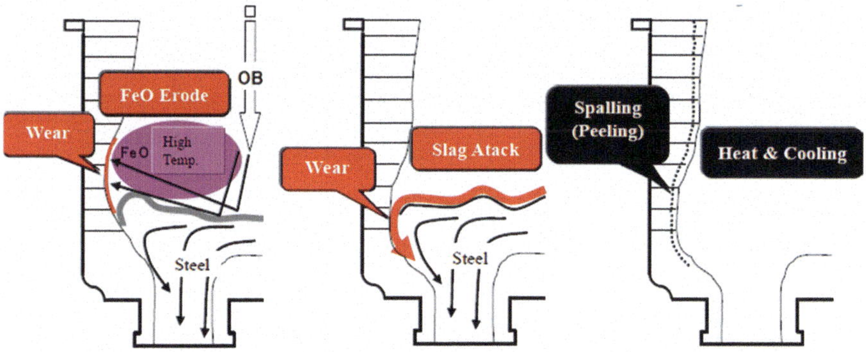

Fig. 8.19 Corrosion mechanism in RH snorkels and lower vessel

Magnesia chrome, rebonded and semi-rebonded bricks have attractive refractory properties, such as high hot strength, corrosion resistance and semi-rebonded bricks have better thermal shock resistance than rebonded bricks. The presence of chromite phase prevents slag infiltration by providing bonding between magnesia grains. Chromite was reported to be more resistant to acidic slag, while periclase is more resistant to high basicity. High alumina content in slag promoted the formation of a spinel ($MgO.Al_2O_3$) layer at the refractory slag interface, which protects the bricks from high slag infiltration. Under low oxygen partial pressure at high temperature, FeO and Cr_2O_3 reduces to metallic Fe and Cr, leading to a severe degradation of refractories. Within the 10–15 min of interaction with high FeO-containing slag of high fluidity, the lining refractory is infiltrated by slag through the open pores of the bricks and periclase grain boundaries and the wear is very severe as shown in Fig. 8.19. In the case of semi-rebonded bricks the effect of slag attack as well as thermal shock resistance had been optimized.

Direct Bonded, Re-Bonded and Semi-Re Bonded Bricks

1. Direct bonded bricks have superior thermal shock resistance but low corrosion resistance towards oxidising slags of vacuum de-gassers.
2. Re-bonded bricks superior corrosion resistance but low thermal shock resistance.
3. To optimise both the properties, semi-re bonded bricks had been developed, which has better thermal shock resistance than Re-bonded bricks as well as better corrosion resistance than direct bonded bricks. Hence the use of Semi-Re bonded bricks have been popular in steel making processes.
4. The comparative corrosion resistance and thermal shock resistance in presence of FeO enriched slag, of Direct bonded and semi re bonded bricks are shown in Fig. 8.20.

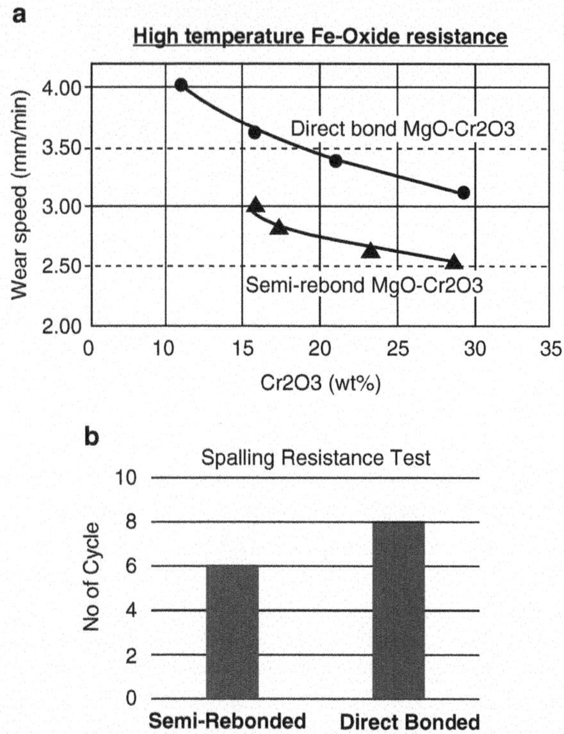

Fig. 8.20 (**a**) Comparative corrosion resistance. (**b**) Comparative spalling resistance at 1000 °C (water quenching)

8.5.8 CAS-OB

Cas-OB is Compositional Adjustment by Sealed argon bubbling with Oxygen Blowing. The CAS-OB process is another secondary steel-making facility where reheating of liquid steel can be done by chemical heating. The process consists of an argon gas bubble ladle fitted with a refractory coated snorkel and bell through which deoxidizing and alloying additions can be made. The schematic diagram of CAS-OB process is shown in Fig. 8.21. The detail of operating practice of CAS-OB is described in the Chap. 2, Sect. 2.2.6.

Good chemical and temperature control of the liquid steel prior to continuous casting at low capital cost. It gives excellent and consistent alloy recoveries because the additions are made to a slag-free area and loss of Ferro alloys due to atmospheric oxidation is significantly reduced and makes use of exothermic reactions between oxygen and Al for reheating of steel. As the open eye is covered by snorkel, the N2 pickup in liquid steel is much lower compared to normal ladle treatment and faster homogenization is achieved.

The refractory is used in the following areas:

- Steel ladle
- Snorkel

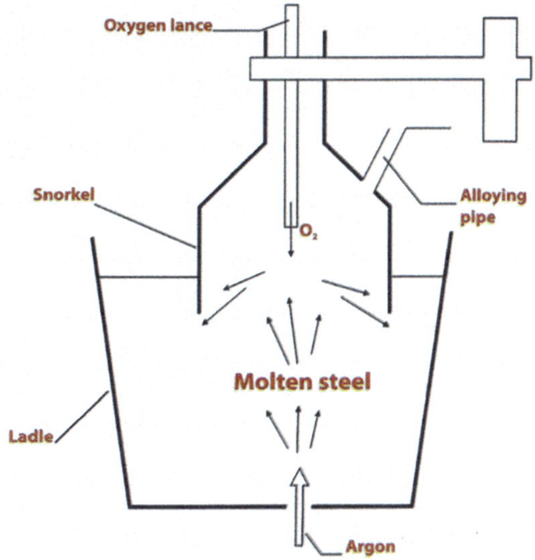

Fig. 8.21 Schematic diagram of CAS-OB

8.5.8.1 Steel Ladle for CAS-OB

The slag during CAS treatment contains higher % of FeO, MnO and SiO_2, which made the slag more fluid at operating temperature. Above 1400 °C, slag became highly fluid. So, the slag is more corrosive at operating temperature, compared to steel ladle for LF operation. As the SiO_2 content is high, the fluidity of slag decreases at lower temperature and results in high jam formation. Further, the slag contains free lime, which is making the slag more reactive towards refractory. Presence of high FeO and MnO in slag, which are making the slag oxidizing and hence it oxidises MgO-C bricks, difficult to achieve higher life by using MgO-C bricks in metal zone. Higher amount of antioxidant needs to be used in MgO-C slag zone bricks to achieve higher life.

The slag corrosion cup test confirms no significant corrosion with spinel bricks at the slag-refractory interface. Higher life can be obtained with the coating formation over the lining bricks. With this type of slag, self-coating is not observed over MgO-C bricks, due to high oxidation and loosening of grains. But by using spinel bricks, alumina in brick composition would react with free lime of slag to form C_2A, CA phases which provides a protection layer over spinel refractory and obtains higher life. The condition of after service ladle lining with spinel bricks is shown in Fig. 8.22.

8.5.8.2 Snorkel

The mechanism of snorkel damage was due to adhesion of oxides and ladle slag and spilling of refractories. When the ladle arrives at the refining station, steel-making slag from the convertor is mixed with the deoxidizing products to form the ladle slag

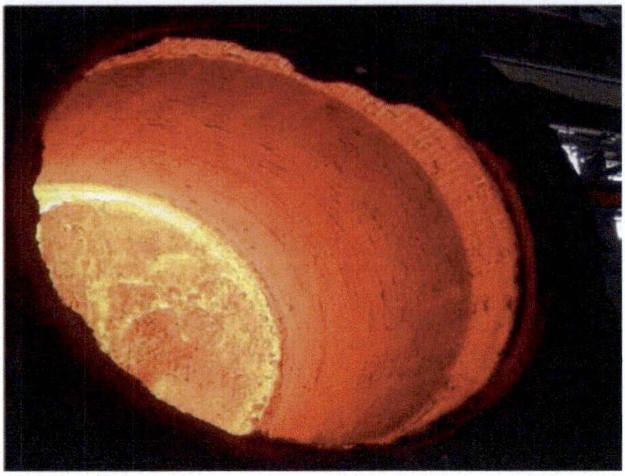

Fig. 8.22 Condition of after service CAS-OB ladle

floating on the liquid steel. Snorkel often thickens with a serious build-up, with the slag sticking on the refractory lining. The snorkel is lined with low-cement castable and the main phases in castable lining are calcium aluminates. In the presence of silica in slag, the silica reacts with calcium aluminates to form the low-melting phases, anorthite ($2CaO.Al_2O_3.SiO_2$) and galenite ($CaO.Al_2O_3.2SiO_2$), at the slag–refractory interface, and causes adherence of slag jam over castable lining. Thus, the weight and volume of the snorkel is increased, which has many undesirable effects on the CAS-OB operation. The increased weight and also the thermal stress result in the cracking of lining refractory and is more often spilling off, so the life of the snorkel is drastically reduced. If the slag adherence around the outer layer became so thick, it would be difficult to lower and dip the snorkel into the steel bath during operation. The alumina-rich slag inherently remains solid at the refining temperature of 1550–1650 °C. The product of Si–Al-killed steel contains a considerable amount of liquid because of oxidation of manganese and silicon. Increased addition of lime during tapping and optimization of production of various steel grades reduce the refractory corrosion. The condition of snorkel after jamming by slag is shown in Fig. 8.23. The issue of slag sticking onto the refractory depends on the following factors:

- Quality of refractory used in snorkel
- Thickness of slag layer in the ladle
- Physico-chemical properties of slag
- Grade of steel produced
- Operating practices
- Operating temperature.

Many steel plants employed $CaO\text{-}CaF_2$ and $CaO\text{-}B_2O_3$ as modifier in CAS-OB slag to avoid slag sticking. The addition of those modifier reduces the melting temperature of slag and increases fluidity and hence reduces jam formation.

Outside of snorkel Inside of snorkel

Fig. 8.23 Jam formation in snorkel (source: Tata Steel, India)

References

1. John Haevey, Samantha Birch, Important factors in the selection of steel ladle lining materials, Tehran International Conference on refractories, 4–6 May 2004, pp 235–242.
2. Kalyani Ravi, Subir Biswas, Sanat Hazra, Use of low carbon content Nano-carbon added Magnesia carbon in steel ladle metal zone at Tata Steel, Jamshedpur, TAIKABUTSU, VOL 71, NO 4, 2019, PP 158–164
3. S Bharati, R Kishore, S Biswas and A R Pal, Effect of refractory quality on glaze formation mechanism in steel ladle metal zone. IREFCON 14
4. Yukinori Matsuo, Yoshiyuki Udoh, Seikichi Higo, Ladle Refractory Cost Reduction, AISTech 2015 Proceedings © 2015 by AIST.
5. D. H. H Quon, and K E Bell, J. Can. Ceramic Soc., 1987, 56, 39–44.
6. Dieter Tembergen, Rainer Teworte and Robert Robey, RH Metallurgy, Millennium Steel, 2008, pp 104–108.
7. Mun-Kyu Cho, Marie-Aline Van Ende et al, Investigation of slag-refractory interactions for RH vacuum degassing process in steel making, Journal of European Ceramic Soc, 32, (2012) 1503–1517.

Chapter 9
Refractory in Ladle Flow Control and Purging System

9.1 Introduction

The flow control mechanism and inert gas purging mechanisms are installed at the bottom of the steel ladles. After completion of secondary refining of steel in steel ladle, the liquid steel is transferred from steel ladle to tundish through the slide gate flow control system. The liquid flow is controlled by opening and closing the slide gate valve as shown in Fig. 9.1 in few steel plants; three plates slide gate valve is used in tundish also to control the liquid flow from tundish to moulds. The slide gate operates on the principle of the parallel displacement of the two perforated plates. When the holes of the upper and lower plates aligned, the system is opened for tapping steel. By sliding the lower plate, the liquid flow would be stopped. The ladle flow control system is consisting of the following items:

- Ladle nozzle
- Slide gate plates
- Collector nozzle
- Well block

The main advantages of slide gate system over stopper rod flow control system are summarized below:

- The slide gate can be operated exclusively from the outside of the ladle bottom.
- Tapping after vacuum degassing became very easy by using slide gate.
- During tapping, only the bore section surface is exposed with the liquid steam.
- Longer holding time and high more nos. of uninterrupted heats can be practiced.
- Ladle availability improved.
- Stopper and stopper rod failure had been eliminated.
- Secondary steel-making processes had been simplified.

Gas purging plugs are popularly used in the steel ladles for secondary metallurgy. Traditional plugs are made of porous refractory structure, blow gas to pass through.

S. Biswas, D. Sarkar, *Introduction to Refractories for Iron- and Steelmaking*,
https://doi.org/10.1007/978-3-030-43807-4_9

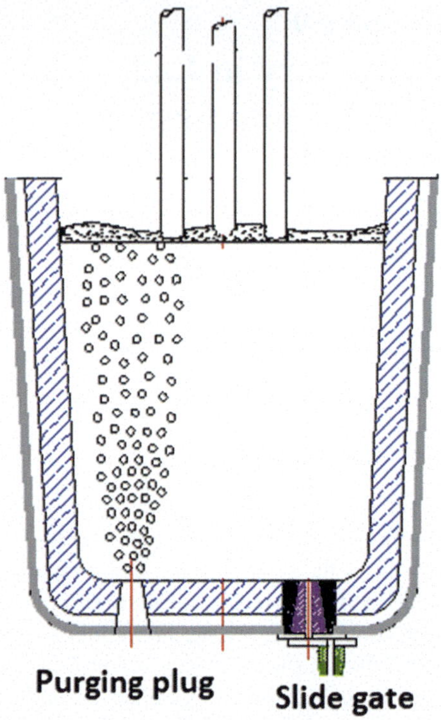

Fig. 9.1 Flow control and gas purging system

Later, different types of plug refractory and design had been developed and used worldwide. The main advantages of using porous plug are summarized below:

- Homogenization of steel chemistry.
- Homogenization of temperature in entire liquid steel bath.
- To enhance the thermodynamic reactions.
- To remove oxide inclusions from steel.
- Very essential to produce clean steel.

9.2 Refractory for Slide Plates

During tapping of liquid steel through slide gate, the steel is passing through the slide plates and hence is subjected to the attack of slag and liquid infiltration in contact surface. Severe abrasion is also experienced during sliding of two plates. Since it is in contact with molten steel at high temperature, high hot strength is required. In closed condition, the ferro-static load is also a concern in lower plate. The refractory requirements for slide gate plates are:

- Resistance to corrosion and erosion towards metal and slag
- Thermal shock resistance

- Lower expansion of plate refractory at operating temperature
- High thermal conductivity
- Crushing strength
- Resistance to severe abrasion.

9.2.1 Alumina Plates

The Al_2O_3-C slide plates has been widely used because of its good thermal conductivity, superior thermal shock resistance and wear resistance. The main raw materials used to produce the plates are tabular alumina, fused white alumina, graphite and addition of Si and Al metal powder to improve oxidation resistance and high temperature strength. Some initiatives had been taken in recent years to improve the physico-chemical properties, as mentioned below:

- Carbon graphitization in phenolic resin by using additives.
- In situ formation of non-metallic oxide reinforced materials such as AlN, SiC and SiAlON.
- Use of nano-size carbon and other oxides to reduce total carbon content, improve oxidation resistance and hot strength.

Addition of carbon in refractory composition of slide plates improves corrosion resistance because of its non-wetting character. It increases thermal conductivity and hence improves thermal shock resistance. Al_2O_3-C refractory with metallic addition (Si and Al metal powder) has superior flexural strength and compressive strength at elevated temperature, which improves erosion and abrasion resistance in sliding surface. Presence of carbon also reduces frictional force within the plates during opening and closing the slide gates. All the above mentioned properties made the Al_2O_3-C refractory suitable to use in slide plates.

However, it is not ideal for the Al_2O_3-C slide plates to be used in producing low carbon, ultra- low carbon clean steel, because of the diffusion and dissolution of carbon into liquid steel. Conversely reduction of carbon content in Al_2O_3-C plates reduces thermal shock resistance and corrosion resistance.

The recent development work [1] indicated the use of Al_4O_4C and Al_4SiC_4 and the advantages are high melting point, excellent mechanical properties, good corrosion resistance, low thermal expansion and hydration resistant. Addition of those compounds reduced the total carbon content. Al_4O_4C can be formed at 1200 °C. Al_4SiC_4 is a congruent melting compound with a melting point around 2080 °C and it can be synthesized at above 1300 °C.

The alumina–mullite plates are fired at high temperature, more than 1650 °C, and then generally impregnated by tar before use. The refractory consists of white tabular alumina, fused alumina, mullite and andalusite.

Phenolic resin is the most efficient binder to use in carbon-containing refractory, because of its many useful properties like high carbon yield, excellent wettability with graphite and other oxides. Though phenolic resin has many advantages, it changes to isotropic glassy carbon during heating and it would become brittle and increases thermal stresses in the material. Gaseous products like CO, CO_2, CH_4, C_2H_6 and H_2O are also evolved during heating, which make the refractory structure porous.

Transition metals like iron and cobalt are used as catalysts to accelerate the formation of graphite structure from various carbon sources. In the case of as-received phenolic resin only amorphous carbon are formed, but in the presence of Ni-catalyst, when the pyrolysis temperature increased from 450 to 1050 °C, crystalline graphite is obtained, specially multi-wall carbon nanotubes are gradually formed. Because of the carbon nanotubes formation, the mechanical properties of the refractory have significantly improved with respect to lowering brittleness in refractory matrix.

Use of microcrystalline graphite by replacing conventional graphite flakes in Al_2O_3-C refractory has improved thermos-mechanical properties. The mechanical properties like modulus of elasticity (MOE) is reduced and hot strength is increased, resulting in improved thermal shock resistance and load-bearing capacity. Formation of SiC whiskers, AlN and Al_4C3 as a result of micronized metal powder addition has been accelerated by adding micronized graphite, compared to addition of conventional graphite flakes. The thermodynamic reactions are accelerated due to high reactivity of micronized powders.

9.2.2 *Al_2O_3-ZrO_2-C*

A major improvement was observed by adding ZrO_2 and ZrO_2-SiC in Al_2O_3-C slide plate refractory composition. Addition of ZrO_2 reduces the thermal expansion under load and hence improves thermal shock resistance. The comparative tentative properties of Al_2O_3-C and Al_2O_3-ZrO_2-C plates are shown in Table 9.1.

The percentage of ZrO_2 in Al_2O_3-C-ZrO_2 refractories is usually 6-9%. Higher the amount of ZrO_2, lower is the corrosion attack of refractories. Addition of silicon carbide (SiC) and boron carbide (B_4C) improves corrosion resistance of Al_2O_3-

Table 9.1 Competitive properties of different grade Al_2O_3-C and Al_2O_3-ZrO_2-C slide plate

Properties		Al_2O_3-C	Al_2O_3-C	Al_2O_3-ZrO_2-C	Al_2O_3-ZrO_2-C
Chemical analysis (%)	Al_2O_3	80–85	76–78	63–68	74–78
	SiO_2	8–10	6–8	8–10	2–4
	ZrO_2	Nil	Nil	8–9	5–7
	C (SiC)	7–10	10–12	10–12	8–10
Porosity (%)		7–10	6–8	10–12	5–7
Bulk density (g/cc)		2.94	2.95	2.98	3.15
Crushing strength (kg/cm^2)		1200	1400	1300	1700
Mod. of rupture (kg/cm^2)		260	350	260	280
Mod of elasticity (kg/mm^2)		6800	6000	5100	5600

ZrO_2-C slide plates. Formation of SiC and Al_4C in Al_2O_3-C refractory has also improved bonding strength and hence improved thermos-mechanical strength. The bond strength in resin-bonded carbon-containing refractory up to 1000 °C is only by carbon derived from polymerization of phenolic resin and there is no much variation in strength up to around 1000 °C. Additional bond derived from formation of SiC, AlN and Al_4C_3 developed at high temperature. SiC phase is formed in the matrix with increase of temperature and is stable in operating temperature of the slide plates.

9.2.3 *Magnesite Plates*

Magnesia plates were developed first in Europe and were used under severe condition of high oxygen-containing steel or high Ca-added steel. Magnesia shows a good resistance to corrosion, but is inferior in spalling resistance because of its high thermal expansion coefficient. Generally, tar-impregnated, 85–95% MgO refractory are used. The material of the plate refractory must be decided considering the bore diameter and the sliding cycles, from the view point of spalling resistance. Alumina and spinel are added in magnesia refractory to improve spalling resistance.

9.2.4 *Slide Plate Refractory for Ca-Treated Steel*

The addition of Ca-alloy in steel ladle becomes popular as a method of removing the non-ferrous inclusions in steel and to control contaminations. Regular alumina carbon slide gate plates show high erosion with such grades of steels which results in unsafe operation and low life. The main reason for the failure of Al_2O3-C plates is the formation of low-melting compounds by the reaction of Ca or CaO with A12O3 and SiO2 of plate composition during use.

The technological needs of slide gate plates for high Ca ppm steel with high dissolved oxygen potential are as follows:

- High resistance to corrosion by high Ca ppm steel.
- High resistance to abrasion at operating temperature.
- High resistance to thermal spalling to obtain multiple heat life.

A steel mill observed a reduction in service life of alumina.

The corrosion in Al_2O_3-C slide plates with Ca-treated steel is shown in Fig. 9.2, considerable corrosion observed at the stroke path as well as bore erosion in both upper and lower plates. The corrosion was mainly caused by the reaction of Al_2O_3 in the plates and Ca or CaO in molten steel and slag. If the plate contains SiO_2, the corrosion is not only caused by the reaction between SiO_2 and CaO but is also accelerated by reduction of SiO_2 by Ca vapour. If the steel contains high oxygen, the corrosion is further accelerated by the reaction between FeO in molten steel and SiO_2 or carbon in plates according to the following equations [2]:

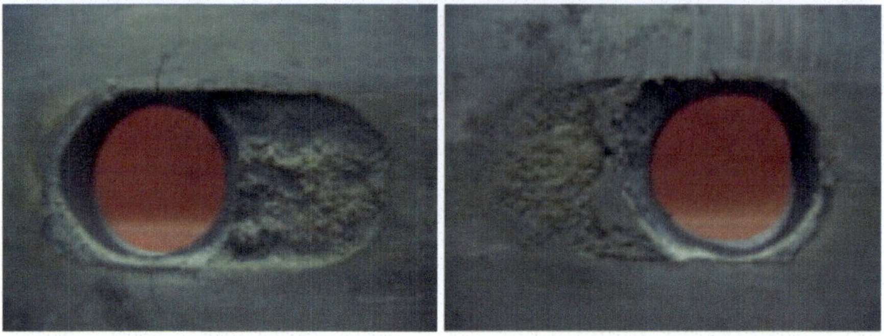

Fig. 9.2 Condition of Al_2O_3-C plates after casting Ca-treated steel

$$SiO_2 + 2Fe = 2FeO + Si$$

$$FeO + C = Fe + CO(g)$$

To reduce the corrosion, silica-free Al_2O_3-C plates are suggested. However, SiO_2-free Al_2O_3-C plates are not working satisfactory because of reaction between Al_2O_3 and CaO to form low-melting compounds.

SiO_2-free Al_2O_3-C plates are used for casting the steel containing relatively low Ca (<25 ppm), spinel-containing plates such as MgO.Al_2O_3-C plates are in use for the casting steel having medium Ca (25–30 ppm) and MgO-C plates are used when Ca is >30 ppm. Due to poor spalling resistance, the use of MgO-C plates is also restricted to use widely. Composite plate with ZrO_2 ring insert in Al_2O_3-ZrO_2-C plate is presently used to cast Ca-treated steel, which results in higher life. Zirconia plates show less wear than alumina and Al_2O_3-C plates with adequate thermal shock resistance. In the presence of Ca or CaO, calcium zirconate ($CaZrO_3$) is formed, which is a high-temperature melting compound. ZrO_2 for refractory use is a partly MgO-stabilized material, which improves the resistance to thermal shock. Properties of different slide gate refractories are summarized in Table 9.2.

9.3 Wear Mechanism of Slide Plate

A typical wear pattern influenced by different factors are summarized in Table 9.3. The wear of refractories in slide gate plates depends on the grade of steel, operating conditions of casting, bore diameter and quality of refractories. The plates are rejected due to the following wear in the plates:

1. Bore diameter increase
2. Corrosion in the sliding surface
3. Decarburization
4. Radial crack formation.

Table 9.2 Properties of slide gate refractories

Properties	Unit	Alumina-C	Al_2O_3-C-ZrO_2	Alumina (fired)	MgO-spinel	Zirconia insert
Chemical						
Al_2O_3	%	90–95	80–85	80–90	8.5–10.5	
SiO_2	%	Tr	Tr	5.5–10.5		
ZrO_2	%		7.5–8.5			97–98
Residual C	%	6.5–10	5.5–6.5			
MgO	%				88–92	
Physical						
Porosity	%	4.5–5.5	6.5–7.5	14–16	14–16	8.5–10
Bulk density	g/cc	3.28	3.26	2.94	2.96	4.54
Cold crushing strength	Mpa	140–160	140–160	98–100	65–75	250–270
Mod. of rupture at RT	Mpa	30–40	35–44	20–22	10.5–15	80–85
Mod. of rupture at 1400 °C	Mpa	15–20	18–22	8.5–9.5	4.5–5.8	38–40
Thermal expansion at 1000°	%	0.78	0.74	0.61	1.15	0.98

Table 9.3 Wear pattern and properties of slide gate refractories

Factors of wear	Wear pattern of slide gates
Erosion and corrosion with molten steel flow	Enlargement of bore diameter
	Erosion of bore edge
	Wear of slide surface
Thermal shock	Radial cracks
	Breaking-off bore edge
	Peeling off sliding surface
Sticking of steel on sliding surface	Peeling off sliding surface
Adhesion of alumina deposits	Clogging of bore

Chemical corrosion and hot abrasion are the main causes of bore enlargement. After significant bore enlargement, there is not enough refractory thickness for a safe operation. For any slide plate, the bore hole diameter should be optimum to ensure the security. The enlargement of bore after use is shown in Fig. 9.3.

The bore of the plate erodes by steel flow and corrodes chemically by reactions. The steel flow attacks the edge of the bore of the slide plates and causes wear. During casting the refractory of the plate reacts with the elements dissolved in liquid steel to create low–melting-point compounds. If the refractory is made of A12O3-C, it becomes corroded due to oxidation of carbon in refractory with dissolved oxygen in liquid steel.

The corrosion of sliding surface is the cause of roughening this part of the plate and often leads to peeling and break out of the grains. Due to this corrosion mechanism, metal tongues are frequently pulled in between the plates, which occurs

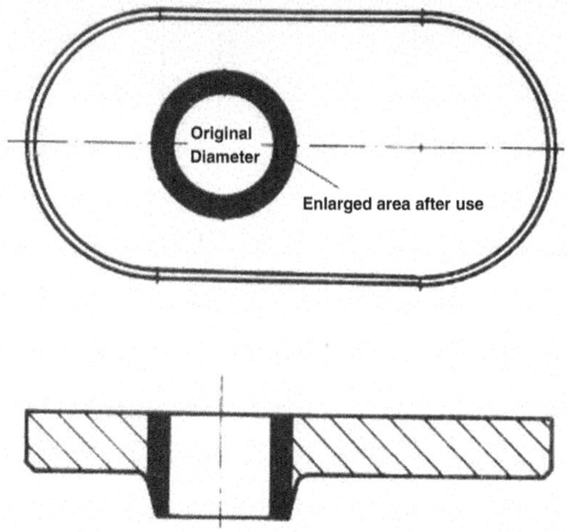

Fig. 9.3 Bore enlargement in slide gate plates after use

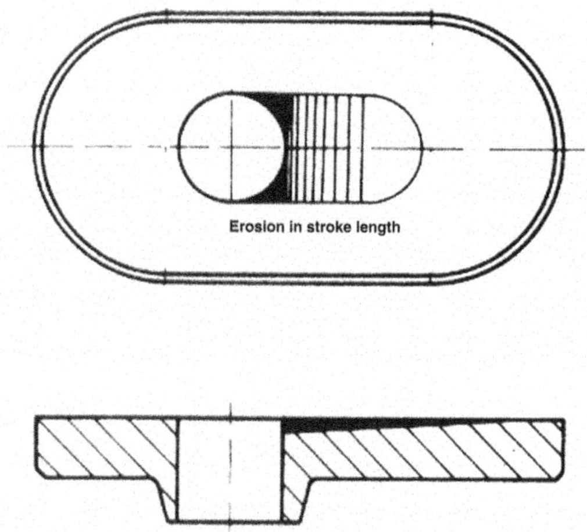

Fig. 9.4 Erosion along stroke length in slide plate after use

due to frequent opening and closing of the plates. Each time of sliding the plates, the steel is lasting on the sliding area with its ferro-static pressure. With continuing corrosion, steel film can remain on the plates, leading to the formation of metal tongues within the plates, which causes the limiting of the slide plates life, as shown in Fig. 9.4.

Decarburization, as it is found in carbon- or rather pitch-containing plate refractory, is caused by the oxidation of carbon in the presence of oxygen in air. The life of the slide plates depends on the damage of the sliding surface. When a plane refractory is exposed to the molten steel flow, carbon on the hot face dissolves into molten steel and a decarburized layer is formed at the surface. The oxidation is reduced by encasing plates into steel casing; however it is not completely avoided. The friction between the plates during sliding increases when the lubricating carbon

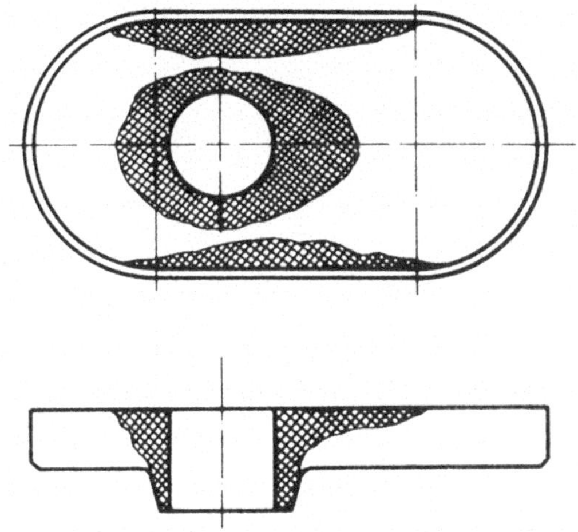

Fig. 9.5 Decarburization of plate refractory after use

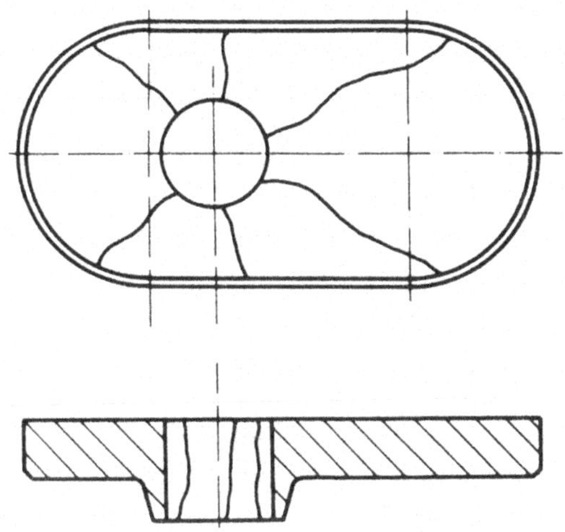

Fig. 9.6 Radial crack formation

constituents are oxidized between the plates, as shown in Fig. 9.5. It also happens during tapping steel containing high oxygen ppm. The plates corrode faster, wear is increased and therefore the service life of tar-impregnated plates become limited.

The formation of radial cracks is very common in all types of plates and it is caused by thermomechanical stress generated during service. The formation of crack and its widening is reduced by encasing the plates in steel casing. The stress is mainly generated due to severe thermal shock. At the first stream of the steel flow through the plate bore, the area around the bore instantly heated up to above 1600 °C while the outside of the plate remains in room temperature. The area around outside of the plate is stressed beyond its structural limits and sustains radial cracks centring from its bore, as shown in Fig. 9.6.

Chemically bonded carbon-added Al_2O_3-C or Al_2O_3-C-ZrO_2 plates show very small crack formation compared to tar-impregnated fired alumina or magnesia plates due to higher thermal conductivity and very high hot strength, which results in superior thermal shock resistance.

9.3.1 *Metal Sticking on Working Surface*

Following multiple use of plates, metal sticking on the sliding surface of the plates is occasionally observed. Once metal sticks to the plate, it further builds up at every stroke of the plates and it causes liquid steel leakage trouble. Al_2O_3-C or Al_2O_3-C-ZrO_2 type of plates are recommended to minimize this metal sticking problem.

9.4 Refractory Design of Purging System

Bottom blow purging plugs experience severe operating conditions of abrasive action of superheated metal, chemical erosion produced by slag of variable composition, and severe thermal shock in the course of pouring operations. The refractory of the plugs must possess the following essential properties in order to maintain stable operations.

- Low open porosity
- Excellent mechanical strength
- Volume stability
- Resistant to erosion against high gas flow rate
- Resistance to corrosion with respect to FeO attack and slag attack
- Resistance to metal infiltration.

Structurally, plug must assure a constant discharge of gas at a given pressure, control over wear of the plug when in use.

Traditional plugs are made of refractories with induced channel pores, through which the purging gas or air were passing through. However, those plugs had several problems and resulted in low life and frequent outage of steel ladle for changing the plugs. The problems are summarized below:

- Cracking due to thermal spalling
- Physical erosion and reduction of plug height
- Chocking of plugs due to liquid steel penetration
- Large fluctuation in flow rate of gas and failure due to back pressure
- Failure of structural integrity.

From the viewpoint of the metallurgical demand and the refractory cost, the design and quality of refractory had been optimized. A schematic diagram of fixing purging plug at the bottom of the ladle is shown in Fig. 9.7.

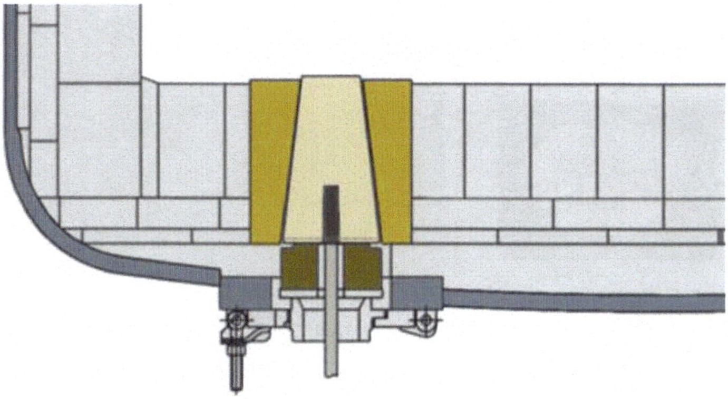

Fig. 9.7 Schematic diagram of fixing purging plug at ladle bottom

9.4.1 Types of Refractory

One of the following two variant approaches is followed in mounting the blow plugs to the bottom of the casting ladle:

1. A system of removable blow plugs. The advantages of such a system include the following: the blow plug may be replaced from the outer side of the bottom of ladle; the plug may be changed in the hot state, i.e., it is not necessary to cool the lining of the ladle prior to replacement of the plug.
2. A system of nonremovable blow plugs. Among the advantages of the system we may note the following: re-equipping of the ladle with a bayonet joint is not required; only a single opening for the gas tube of the blow plug with diameter roughly 60 mm in the sheathing of the bottom of the ladle is needed for feeding of the gas. A drawback of the system is that it is not possible to replace the blow plug in the hot state, i.e., the lining of the ladle must be cooled prior to replacement of the plug.

Major two types of purging plugs are widely used, directional plugs and non-directional plugs. Directional plugs are of two types, one is ceramic plate type, made of ceramic plates at the centre of the plug, and other one is slit type, made of permeable slots in castable body.

Non-directional porous plugs are made of high-alumina refractory with porous structure, permeable to gas flow.

9.4.1.1 Directional Plugs (Segment Type with Ceramic Plate)

Purge plugs made of ceramic plates having gap for purging gas within the ceramic plates. It has a porous wear indicator, slit-type wear indicator at the bottom part of the plug and they are surrounding by castable lining. The complete set of assembly is

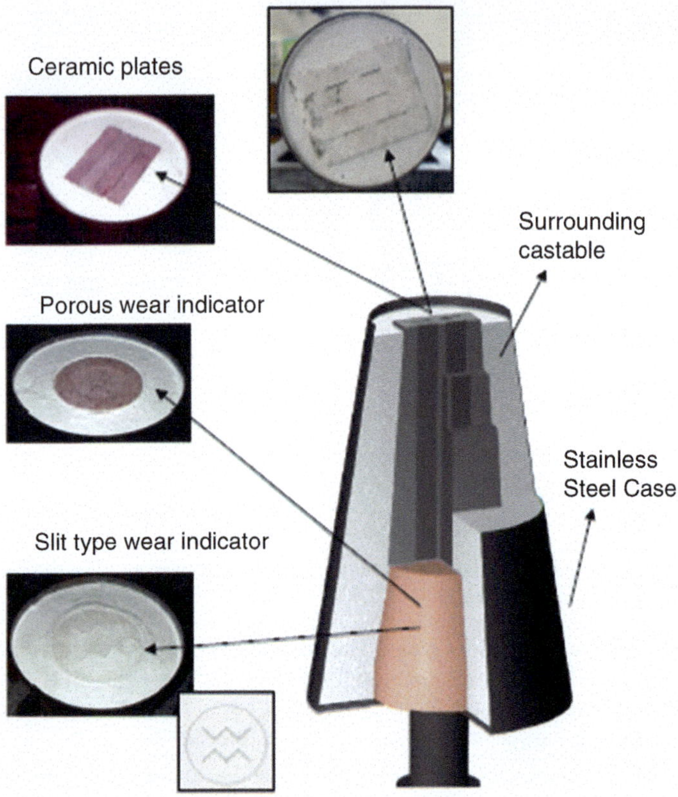

Fig. 9.8 Directional segment-type porous plug assembly

encased in a stainless-steel case. The detail of the different parts is shown in Fig. 9.8. [3].

- Ceramic plate of purge plug is made of high-alumina (Al_2O_3 = 90.0%) or alumina–chrome refractory.
- Surrounding castable is lined with magnesia–alumina spinel-containing castable.
- Optical wear indicator is made of non-directional porous-type high-alumina refractory.

Design of segment-type purging plugs:
Basically, the plug consists of the following parts:

- The segment blocks
- Wear indicator devise at bottom part
- The surrounding castable
- The conical metal jacket
- The gas supply pipe.

The segment block consists of 4–6 numbers ceramic plates, attached loosely to each other. The groove within the blocks is kept within 0.25–0.35 mm, depending on gas flow rate and viscosity of liquid steel and to ensure that there is no penetration of steel due to back air pressure during cleaning of operation.

9.4.1.2 Slit-Type Directional Plugs

Slit-type purging plugs are made of magnesia–alumina spinel-containing refractory; slits are made in the castable body during casting and they are radially oriented as shown in Fig. 9.9. Optimum numbers of slits and the dimension are to be maintained. Optical wear indicator helps to indicate the residual thickness of the plug height, and as soon as this is visible, the plug needs to be changed.

9.4.1.3 Non-directional Plug

Non-directional plugs are made of porous body made of high-alumina refractory ($Al_2O_3 > 95\%$) at the centre part of the plug. The porous body is surrounded by high-alumina spinel-containing castable and optical indicator is provided at the bottom part of the plug. The schematic diagram showing different parts of the plug is shown in Fig. 9.10.

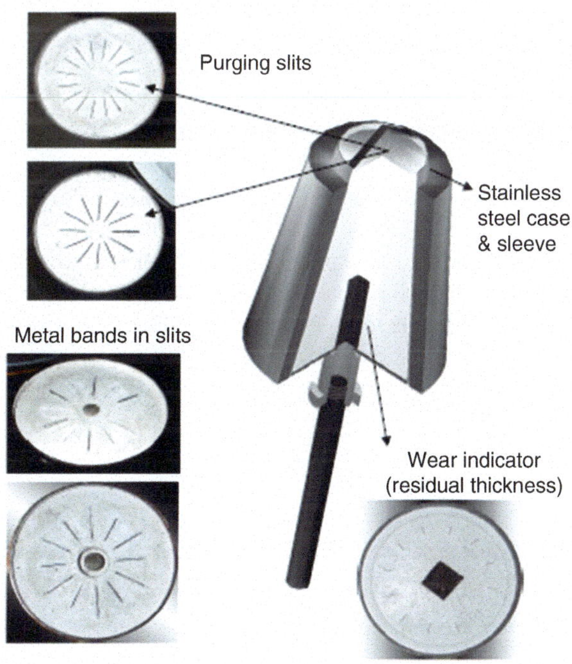

Fig. 9.9 Schematic diagram of slit-type directional plugs

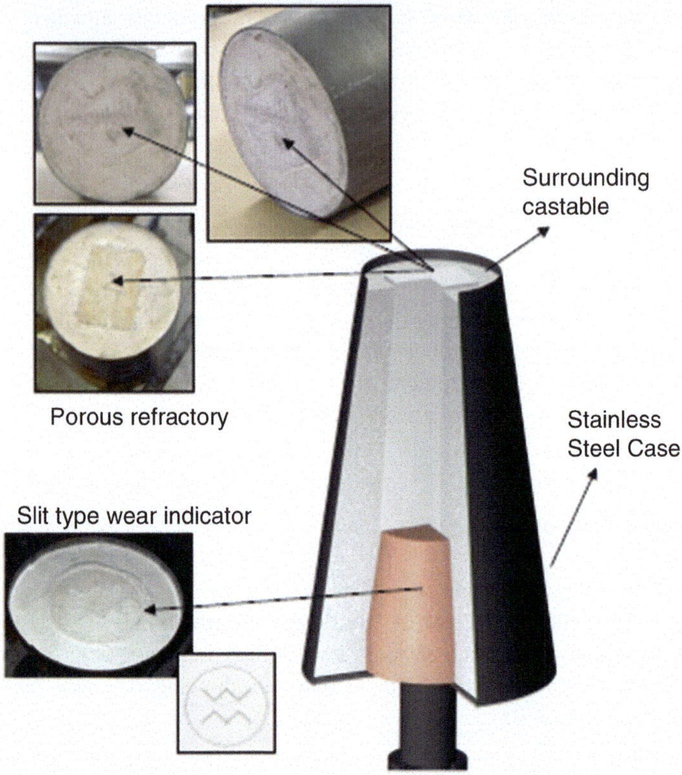

Fig. 9.10 Assembly of non-directional porous plug

9.4.1.4 Function of Optical Wear Indicator

Optical wear indicator is a porous high-alumina refractory with a squire or circular cross-section as per design at a residual thickness of around 120–140 mm of purging plug (depending to the plug design). When the purging plug is at residual thickness, the indicator becomes visible during cleaning of ladle after repeated casting. Due to porous nature, the top section of the indicator in dark colour becomes visible when observed from inside of the ladle in hot condition. As soon as the indicator is visible, the operation is to be stopped and the purge plug has to be replaced by a new one immediately.

9.4.2 Wear Mechanism of Purging Plugs

The main wear mechanisms of a purging plug is summarized below:

1. **Steel infiltration into plug**
   The molten steel is tapped from convertor or EAF to steel ladle, at a temperature of 1600–1700 °C, and the steel comes in contact with the surface of plugs. Due to inadequate positive gas pressure in the plug, the liquid steel penetrates into the plug and thus dense penetrated layer is developed with a different physico-chemical property and causes peeling of the surface.
2. **Hot Abrasion**
   The intensity of the abrasion reaction depends on the gas flow rate and the methods of injecting the inert gas.
3. **Cracking due to thermal shock**
   The hot face of the plug is exposed at about 1600 °C during ladle operation, when at the cold end, near the gas purging channels, the temperature is around 300 °C and this causes very high thermal stress and cracking of plugs. Thermal shock is also experienced after tapping the steel from ladle during continuous casting or the ladle becomes empty.
4. **Corrosion**
   In order to remove slag layer and infiltrated solidified steel from the exposed hot face of the plug, after emptying the ladle, often high-pressure oxygen lance is used. The temperature is increased rapidly, and in the presence of oxygen the remaining molten slag and steel react with plug refractory to form low-melting compounds and the plug is corroded.

Refractory Material Development to Minimize Wear

The wear of purging plugs can be minimized by the following ways:

1. To improve corrosion resistance
2. To reduce metal infiltration
3. To improve thermal shock resistance

Resistance to slag corrosion can be improved by selecting proper refractory materials. The presence of silica (SiO_2) in refractory composition causes low-melting-point phases to develop with calcium–aluminate slags. The formation of anorthite ($2CaO.Al_2O_3.SiO_2$) and gehelanite ($CaO.Al_2O_3.2SiO_2$) are the low-melting phases. So SiO_2 free refractory material is preferable to reduce corrosion.

Addition of spinel improves corrosion resistance. Change of contact angle in refractory and slag interface by adding spinel reduces corrosion. Spinel can be added by adding MgO or by adding pre-formed spinel in the refractory composition. Adding of MgO is not advisable as spinel forms in situ very rapidly with increase of temperature and it involves volume expansion, which increases high stress in plug refractory, causing cracking of plug

(continued)

refractory. Hence, as a usual practice, pre-formed spinel is used in plug refractory composition.

The penetration of molten steel can be explained by the following equation [4]:

$$\cos\theta = \frac{\gamma sv - \gamma sl}{\gamma Lv},$$

where θ = contact angle, γsv = interfacial energy between solid and vapor, γsl = interfacial energy between solid and liquid and γLv = interfacial energy between liquid and vapor.

If the θ is >90°, there is very minimum penetration. Experimentally it was observed that addition of spinel in high-alumina castable has increased contact angle from 85° to 110°, which improves penetration resistance.

Another possibility to reduce liquid penetration is by controlling pore size and pore size distribution in porous refractory structure.

The following conclusions can be drawn to prevent corrosion, improve thermal shock resistance and reduce liquid infiltration in different types of purging plugs:

1. Below a critical ferro-static pressure, there is no penetration in porous type non-directional plugs.
2. Increasing the contact angle would reduce liquid infiltration and increase corrosion resistance in directional plugs. Adding spinel in high-alumina castable increases contact angle.
3. Spinel-containing high-alumina castable is superior in thermal shock resistance compared to conventional high-alumina castable.

9.4.2.1 Corrosion Due to Oxygen Lancing

The cleaning slag skulls over the purging plug by oxygen lancing at the end of every cast is a common practice to ensure optimum plug opening and gas flow rate before taking ladle for next tapping. The lancing operation is continued till the optimum gas flow rate is obtained. On the process of this operation, very high temperature, about 2000 °C temperature, has been reached to melt the slag and clean the plug surface. During this operation, large amount of FeO is generated by oxidation of Fe, which is detrimental to refractory performances of purging plugs. Further, the plug is exposed to severe thermal shock due to oxygen lancing. For those reasons, oxygen lancing is to be minimized as much as possible and should be considered as an emergency, rather than standard operating practice.

Resistance to oxygen lancing and thermal spalling are improved by using silica-free refractory materials and/or addition of spinel. In many cases, high gas flow rates are used to prevent deposit of skulls and liquid infiltration and hence oxygen lancing can be minimized to prolong life of the purging plugs.

9.4.3 Safe Operating Practices

Operation of purging plug for stirring liquid steel is one of the very critical activities with respect to safety of the workmen in shop floor. Any failure of porous plug may cause very serious safety hazards. To minimize the safety hazards, significant development in design had been undertaken. Use of wear indicator is the latest design development. Presently, the most common system for the slit-type and segmented-type directional plugs is based on the use of an indicator made of higher thermal conducting refractory material at the colder side of the plug. With a higher conductivity material at the back side the cooling effect from the gas will cool it earlier than the surrounding material, so that the colour would be darker and it would be easier to the ladle operator to view it from inside of the hot ladle during cleaning. As soon as the operator identifies the dark colour indicator, he will stop using the plug and it needs to be replaced before taking next heat in the ladle.

References

1. Chenhong Ma, Yong Li, Mingwei Yan, Yang Sun, Jialin Sun, Investigation on a post-mortem resin-bonded Al-Si-Al_2O_3 sliding gate with functional gradient feature, Ceramics International, 44 (2018).
2. Tamotsu Wakita, Keiichiro Akamine, Toshihiro Suruga, Jouki Yoshitomi, Keisuke Asano, the basic slide gate plate for casting of ca-alloy treated & high oxygen steel, UNITECR 2005.
3. Nag, M., Agrawal, T., Nag, B., Singh, B., Biswas, S., Study and post mortem analysis of steel ladle porous plug to improve bottom purging efficiency for cleaner steel, Engineering Failure Analysis, Volume 101, Issue undefined, July 2019.
4. Patrick Tassot, Innovative concepts for steel ladle porous plugs, millennium steel 2006.

Chapter 10
Refractory for Casting

10.1 Introduction

In the casting process of steel, a long-time casting is aimed in order to improve productivity and yield of steel. Two types of casting processes are followed in steel plants:

1. Ingot casting.
2. Continuous casting.

Low-grade high-alumina or fire clay refractory was used in Ingot casting. For the demand of high-quality steel and high productivity, the ingot casting is phased out and at present most of the integrated and large size steel plants have adopted hundred percent casting through continuous casting process.

Continuous casting may be defined as teeming of liquid steel in a mould through a rectangular reservoir of steel, called tundish. The tundish regulates the flow rate of liquid steel into the mould. Refractories for continuous casting consist mainly of shrouds, subentry nozzles, slide plates and mono-block stoppers. A schematic diagram of equipment for continuous casting process is shown in Fig. 10.1.

10.2 Ingot Casting

Even though continuous casting of steel is now extensively practiced, ingot casting was the only method of casting for more than 100 years, till the advent of continuous casting. Ingot moulds are made of cast iron, having various cross-sections, such as square, round, and polygonal, depending on the end use of the products. There are various mould designs as follows [1]:

- Narrow end up or wide end up.
- Open bottom, closed bottom or plugged bottom.

S. Biswas, D. Sarkar, *Introduction to Refractories for Iron- and Steelmaking*,
https://doi.org/10.1007/978-3-030-43807-4_10

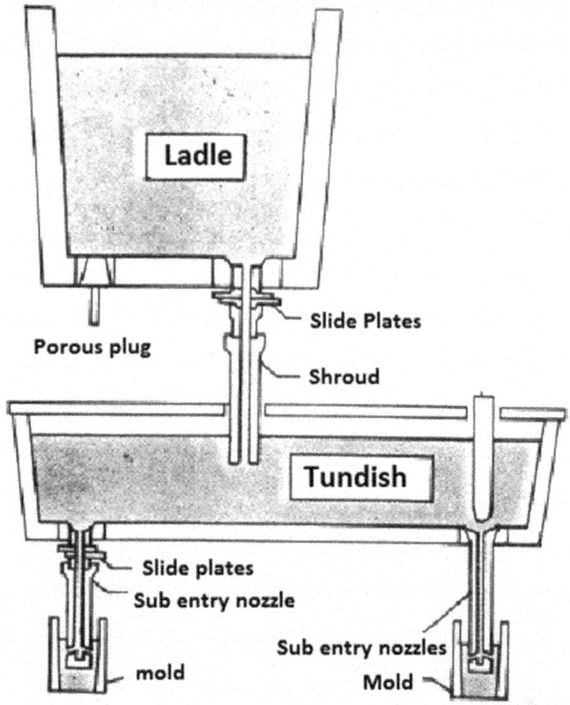

Fig. 10.1 Schematic diagram of continuous casting process

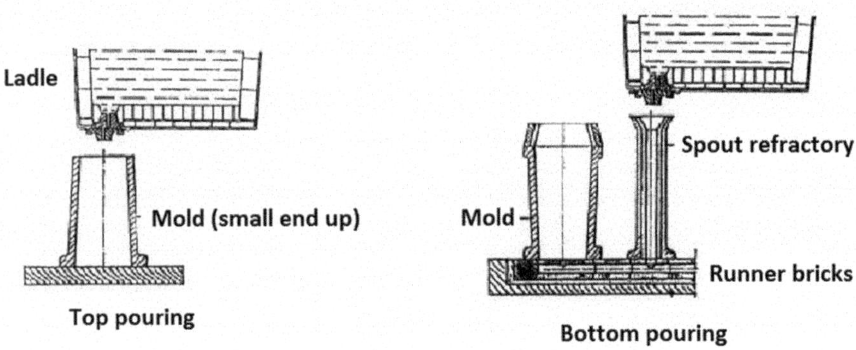

Fig. 10.2 Schematic diagram of ingot casting

– With or without hot top.

Pouring of liquid is usually done by top pouring. However, bottom pouring, i.e. feeding the mould through a bottom opening, is widely practiced for casting ultra-pure low-carbon silicon-killed steel for long products. The processes of ingot casting are shown in Fig. 10.2.

The advantages of bottom pouring casting are [1, 2]:

Table 10.1 Properties of ingot casting refractories

Properties	Unit	Fire clay bricks				High alumina bricks			
		Type 1	Type 2	Type 3	Type 4	Type 1	Type 2	Type 3	Type 4
Al_2O_3 (min)	%	30	35	38	42	55	65	70	80
Fe_2O_3 (max)	%	2.5	2	2	2	2	2	2	2
Refractoriness	SK	30	32	34	35	36	37	38	40
Refractoriness under load, 0.2 MPa (min)	°C	1240	1300	1360	1420	1450	1480	1530	1600
Porosity (max)	%	24	24	22	20	20	20	18	18
Bulk density	Gm/cc	1.9–2.1	1.9–2.1	2.1–2.2	2.1–2.2	2.2–2.3	2.3–2.4	2.5–2.6	2.5–2.6
Cold crushing strength (Min)	Mpa	20	25	30	35	40	50	55	60

Fig. 10.3 Tongue and groove shapes for spout and trumpets

- Marked reduction of ingot cast defects, which may not be dressed out after rolling.
- Improvement in ingot surface finish, largely achieved by the use of mould fluxes and a slow casting rate.
- Use of bottom plate as required by top pouring is avoided.
- Improvement in ingot mould life.

In bottom pouring ingot casting, refractories are used in spout, trumpet and runners. Fire clay and high-alumina bricks are used. The spout and runner are lined with different shaped hand moulded bricks. The properties of bricks are shown in Table 10.1. Bauxite-based bricks of more than 76% Al_2O_3 bricks are commonly used. The spout and trumpet bricks are made of circular shapes with tongue and groove arrangement for assembling nos. of pieces as shown in Fig. 10.3.

Wear of those items are very costly feature. Disadvantages of bottom pouring are the cost of refractories for guide tube, spout and runners. Failure of refractory is a major issue since the liquid steel passes through a complex system of pipes, starting with a trumpet-shaped spout and passing through the tubes, centre block, and various runner bricks, to the base of the mould [3]. The centre brick has a number of off-takes which are fixed with number of moulds, enable to fill up number of ingots simultaneously [4].

Wear of Refractory

Lining erosion generally occurs at area of turbulent flow, especially when combined with reoxidation, high pouring temperature and chemical reactions. The following parameters strongly affect lining erosion:

- Corrosive steel grades, such as high manganese steel, semi-killed grades with high dissolves oxygen, attack trumpet lining, spout and runner bricks.
- Presence of MnO preferentially reacts with SiO_2 of refractory component and corrode the refractory. High-purity Al_2O_3 refractory is suitable for high MnO containing slag.
- Iron-oxide-based inclusions are very reactive and wet the lining materials, leading to erosion in high turbulent areas.
- The dissolved aluminium in steel can reduce SiO_2 in lining and the lining is eroded, hence high alumina and low SiO_2 containing refractory to be used for high Mn containing Al-killed steel.
- Selection of refractory with the grade of steel to be casted has a direct impact on refractory performances.

All the refractory bricks used in trumpet, spout and runners are in high alumina quality of high refractoriness and they must have excellent thermal shock resistance with a fair insulating property, which suggest a relatively high porosity [3]. Numerous attempts are being made to improve insulating property of mould tops by using insulating materials, or of combustibles which actually provide heat to prevent solidification of steel in the pipe and runners.

10.3 Continuous Casting

In present days, continuous casting is the most predominant method of casting molten steel. Bulk of the liquid steel after continuous casting is shaped by rolling either into flat products, i.e. plates and sheets, or into long products, such as rods, angles and rails.

The advantages of continuous casting over ingot casting are:

- It is directly possible to cast blooms, slabs and billets, thus eliminating blooming and slabbing mills and billet mills in large extent.

- Better quality of the cast products, minimizing casting defects to a large extent.
- Higher yield of finished steel (about 10–20% more than ingot casting).
- Improved productivity.
- Higher extent of automation and process control.

Henry Bessemer first propagated the concept of continuous casting of steel in 1846 [5]. However, it took more than hundred years to make continuous casting a reality. It was difficult to apply into steel casting because of higher melting temperature and lower thermal conductivity. The first pilot plant for continuous casting of steel was installed in Germany in 1943. By 1970, many machines were in use throughout the globe. In 1980, about 30% of world steel production was through continuous casting and by 1990 it was reached to more than 50%. In last few decades, significant development had been done in continuous casting technology, and at present, more than 90% of the world steel production is through continuous casting process. The commercial processes of continuous casting can be classified as:

- Slab casting.
- Bloom casting.
- Billet casting.
- Thin slab casting.

The biggest break-through in recent years in the area of steel casting has been the advent of thin slab casting of molten steel.

A schematic showing tundish and other arrangements is shown in Fig. 10.4.

Continuous Casting Process Scenario

The continuous casting process involves the following essential requisites:

- To maintain stable casting rate.
- Prevent reoxidation of steel.
- Maintain steel casting temperature.
- Prevent inclusions in steel.

The essential equipment for continuous casting processes are:

- Tundish.
- Tundish slide gates.
- Mono-block stoppers.
- Dams and weirs, charge pad (Turbo-stop) as tundish furniture.
- Ladle shroud.
- Subentry nozzles.

(continued)

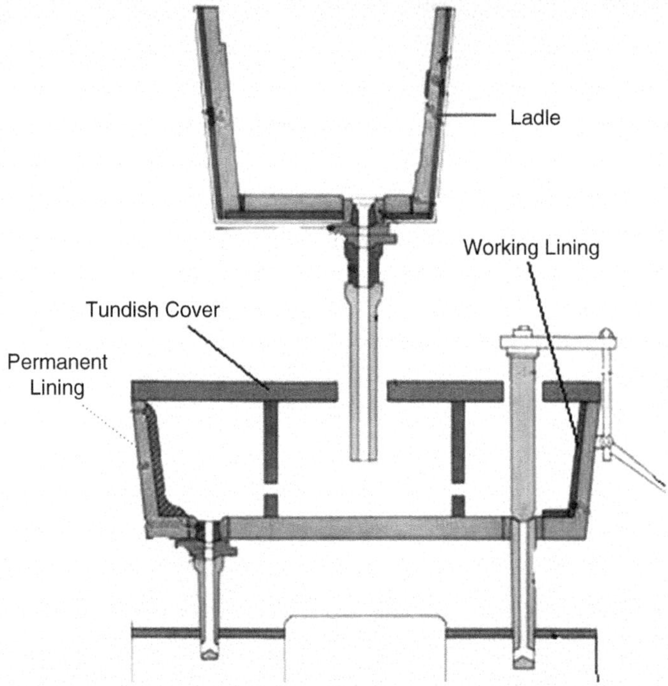

Fig. 10.4 Schematic diagram of continuous casting process of steel making

Tundish is a buffer refractory lined vessel which is located between the steel ladle and continuous casting mould. The tundish serves the purpose of a reservoir and a distribution vessel. Over the years, the function of a tundish is changed from a reservoir to a refining vessel. In present practice the tundish fulfills certain metallurgical functions as:

- Feeding of the liquid steel to the moulds, at a controlled rate.
- Thermal and chemical halogenation.
- Inclusion flotation and removal.
- To control hydrogen pickup in steel.

Shrouds and subentry nozzles are used to prevent reoxidation of steel during casting. Shroud is fixed below the steel ladle and submerged into liquid steel in tundish to prevent deoxidation during transport of steel from ladle to tundish.

The subentry nozzles are fixed below tundish and the liquid steel is transported from tundish to moulds through the nozzles to prevent oxidation of steel during casting.

(continued)

Dams and weirs are fixed in the lower part of tundish and charge pad is used at the bottom. Those tundish furniture help to control the uniform flow rate of steel in all the moulds attached at the bottom of the tundish. The major objectives for using flow control are: maximize residence time of steel in the tundish to allow flotation of non-metallic inclusions, minimize turbulence in the tundish at the pouring area and hence increase of casting sequence length. It allows maximum draining of liquid steel without affecting steel quality, thus improving yield.

10.3.1 Refractory Practice in Tundish

In general, the tundish lining is composed of three layers (Fig. 10.5):

1. The working lining of tundish consists of MgO-based material and is exposed to the liquid steel. Different types of lining materials are presently in use. These include brick, gunning mass, board, spray mass, hot and/or cold setting dry veritable mass (DVM), plasters, etc. [3]. Earlier, people were using silica boards as the working lining. At present, most of the integrated steel plants have switched over to basic lining material.
2. The backup lining is generally made of a high-alumina low-cement castable and of 80–110 mm thick lining. Use of andalusite-based high-alumina-based bricks/castables as backup lining is becoming popular in some of the European plants.
3. The third component of the tundish lining system is the insulating layer. It is installed between steel shell and backup lining and it provides thermal insulation to keep the tundish shell temperature below its critical temperature range throughout the operating time. Some of the commonly used insulating materials are insulation bricks of ASTM23 quality, ceramic fibre paper, ceramic fibre board and magnesium silicate board.

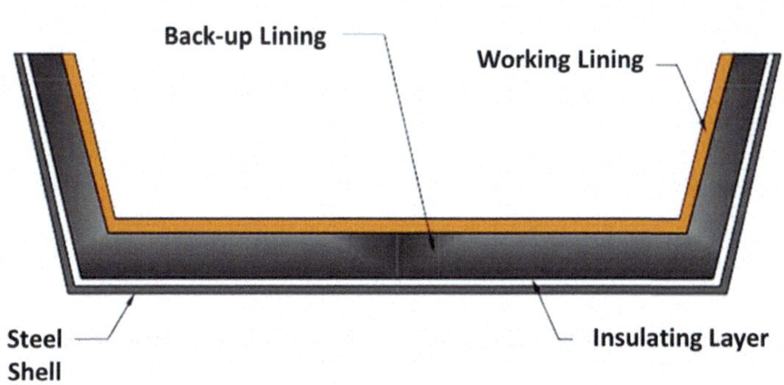

Fig. 10.5 Schematic representation of tundish lining pattern

10.3.1.1 Backup or Permanent Lining

Backup or permanent lining is protecting the shell from overheating and provides thermos-mechanical support to the working lining. This lining needs to be intact during de-skulling and removal of disposable working lining. The demands on tundish permanent lining refractories are:

- Reduced mechanical wear and damage caused during de-skulling and removal of working lining.
- Resistance to thermal shock resistance.
- Resistance to fusion or any chemical reaction between permanent and working lining interface to facilitate easy de-skulling.

In early days, the backup lining was made of high-alumina bricks and the main disadvantages are falling of bricks during changing of working lining, reaction of those bricks with working lining at interface, which caused problem of de-skulling and damage of permanent lining. Thus, it required frequent repair of the permanent lining.

In present practice with the development of high-strength low-cement castable, most of the steel plants have adopted monolithic backup lining, which significantly increases the campaign life of refractory. The advantages of monolithic lining over conventional brick lining are summarized below:

- Joint-less lining, no chances of refractory falling.
- Reduced installation time.
- No adherence with working lining due to low-melting compound formation at interface.
- Easy de-skulling.
- Minimum repair needed during change of working lining.
- Increased thermal shock resistance.
- Increased campaign life of refractory.
- Improved tundish availability.

Andalusite and fused alumina-based 65–70% Al_2O_3 containing low-cement castable is used in tundish permanent lining. Presence of free silica in the form of cristobalite reacts with MgO-based working lining to form forsterite ($2MgO.SiO_2$) and creates problem in de-skulling. Hence Corundum–Mullite-based castable is the most suitable lining refractory to use in tundish permanent lining. The properties of different types of backup lining refractory materials are shown in Table 10.2.

10.3.1.2 Tundish Working Lining and Its Significance

The use of purpose-designed, lightweight expandable boards as tundish working lining revolutionized the tundish refractory practice during the decade 1970–1980. It is a disposable lining, which is easily removed during de-skulling and changed with

Table 10.2 Properties of tundish permanent lining refractories

Properties	Unit	Brick lining			Monolithic lining	
		Insulation	42% Al_2O_3	45% Al_2O_3	LCC 60	LCC 70
Al_2O_3	%	50–55	40–42	43–45	60–65	70–72
CaO	%	0.1–0.5	1.5–2.0	0.2–0.5	1.5–2.5	1.5–2.5
Fe_2O_3	%	0.7–1.0	2.5	1.8–2.0	1–1.5	0.8–1.2
Bulk density	Gm/cc	0.7	2.3	2.4	2.56	2.76
porosity	%		16–18	16–18		
PLC at 1450 °C	%	–0.5	0.5	<0.5	0–1.5	0–1.2
Cold crushing strength at 110 °C, at 1000 °C, at 1450 °C	kg/Cm^2	16–18	300–500	300–550	550 860 1100	650 950 1200
H MOR at 1450 °C	kg/Cm^2				35–40	40–45
Reversible thermal expansion at 1000 °C	%		0.51	0.6	0.58	0.68
Thermal conductivity at 1000 °C	W/mK	0.33	1.08	1.15	1.44	1.68

new lining after completing each sequential heat. The various types of refractories used in working lining are summarized below:

- Expandable boards.
- Spray mass.
- Dry vibratable mass (DVM).

Silica board and basic boards were in use. The properties of boards are shown in Table 10.2. The increased demand on tundish availability for longer sequence casting in large size slabs, blooms and casting of alloy and special steels have prompted a change in tundish working lining. Moreover, tundish is now being more and more considered as a metallurgical reactor than a distributing vessel; the refractory used in working lining must be chosen which can contribute to those purposes.

Use of basic spray mass or gunning monolithic refractory had been developed and those monolithic basic refractories had the following advantages:

- A low bulk density after installation and curing to provide low thermal mass and thermal conductivity.
- Adequate shrinkage at the end of campaign to facilitate de-skulling.
- To reduce hydrogen pickup by steel.
- Good spalling resistance.

Table 10.3 Properties of Tundish working lining refractories

Properties	Unit	Silica board	Basic board	Gunning material	Spray monolithic
MgO	%		75–80	80–85	76–80
SiO_2	%	92–95	15–20	10.5–12	18–20
Bulk density	Gm/cc	1.15–1.25	1.55–1.65	2.2–2.5	1.56–2.5
Permanent linear change at 1500 °C	%			−0.6	−4.1
MOR at 200 °C	kg/cm^2			24–26	28–30
Thermal conductivity at 1400 °C	cal/cm^2/s	0.35	0.45	0.7	0.45
TGA (weight loss)	%	4.0–7.0	4.0–6.0	2.6	3.25

- Suitable chemistry to resist corrosion by tundish slag.
- Adequate adherence to the backup lining.

The properties of the spray mass and gunning mass is shown in Table 10.3 [6].

For proper installation of spray mass, a spray equipment is necessary. The monolithic material is first mixed with water in a special mixer and the wet-mix is pumped by a positive displacement pump at high pressure to a nozzle where it meets with high-speed compressed air to spray onto the permanent lining of the tundish. The spray lining contains 18–20% water, and therefore removal of water and proper drying are necessary before operating the tundish.

The adoption of lightweight spray basic monolithic tundish working lining has gained momentum world over and it is currently in use in many steel plants.

Tundish and its disposable/sacrificial working lining are important parts in the steel-making process. Increased demands upon the performance and longer sequence life of the steel-making tundish have driven a great number of changes in the sacrificial lining. This layer plays a significant role in production of clean steel. High porosity and low thermal conductivity provide insulation to preserve heat in the molten steel. It acts as an inert barrier between the steel and backup lining to avoid any chemical reaction between the corrosive slag and the permanent refractory, and helps in easy de-skulling.

10.3.1.3 Dry Vibro-mass

In practice, basic refractories are widely used as tundish working lining. Basic refractory materials include magnesia, magnesite, dolomite, olivine, or mixtures thereof. In addition to the refractory aggregate, these refractory materials often consist either or all of various organic and inorganic binders, pore formers, plasticizers, etc. to achieve optimal performance during application. In sprayable material, water addition is done prior to use, while DVM is used in dry condition for installation.

Performance of Basic Board and Spray Mass

The mechanism of slag-refractory interaction in tundish working lining is explained as follows:

1. Infiltration of liquid steel into the tundish working lining, mainly in the gunning and spray lining.
2. Steel-oxidized layers formed at the interface of slag and refractory.
3. The preformed MgO board was less prone to steel infiltration than the spray mass and dry vibro-mass. The main difference between them is related to production and application, which influence the porosity of the tundish lining materials. Two aspects contribute to higher porosity in the spray and gunning material:

   - The use of spray material on cold tundish.
   - Water content used in spray mass, which initially fills the pores and further is removed during preheating. On the other hand, MgO boards are preformed materials and there is no concern about water content.
   - De-skulling of the spray lining or gunning is much easier than the preformed MgO boards. Dry-vibro-mass (DVM) has excellent performance in de-skulling.
   - In the case of gunning or spray mass, the formation of spinel layer ($MgO.Al_2O_3$) was observed at steel/refractory interface, which increases inclusion in steel. The spinel formed as per the following equation:

$$3(2MgO.SiO_2) + 4\ Al\ (steel) = 2(MgO.Al_2O_3) + 3Si.$$

Alkali oxides, such as Na_2O and K_2O, are also added in compositions of spray materials. The major disadvantage is, they can react with the permanent lining at elevated temperature, making de-skulling of the disposable lining difficult [7].

Porosity in the refractory lining can be imparted by adding various pore-forming materials, such as inorganic or organic fibres, expanded inorganic or organic materials, or other lightweight filler materials. However, these materials may affect the strength and corrosion resistance of the refractory lining. Polystyrene beads are an improvement over fibrous materials as they decompose during heating leaving residual carbon and also eliminate chances of continuous pore network formation.

Chemical Composition of DVM

The selection of basic raw material is generally dependent on both technical and economic factors. The major components may vary from pure periclase-based dead burnt magnesia (DBM) to olivine. Initially, pure DBM-based DVM was used as the disposable liner to provide a clean tundish. However, it provided no thermal insulation. In last few years, olivine, because of its lower cost and thermal conductivity, is finding a wide application as DVM raw material along with DBM. Now-a-days, use of other raw materials such as dunite has also become popular. In some cases, recycled raw materials are also used without affecting the properties of the lining and can be considered as economically more favourable.

Basic materials, high in MgO with varied amounts of silica, are generally favoured in the industrial applications. Mainly two types of DVMs are used in the plant based on MgO%. One contains 60–75% MgO whereas another contains 70–90% MgO, the base raw materials being periclase and/or olivine. A commercial DVM with two component system generally contains 60–70% MgO, 20–30% silica, 2–10% of iron oxides and minor impurities like CaO, Al_2O_3, Cr_2O_3, TiO_2, etc.

Both periclase and olivine contain SiO_2, FeO/Fe_2O_3 and MnO which can deoxidise the steel in the tundish. The rate of steel reoxidation by these oxides present in the tundish linings has been shown to depend on the amount of fluid silicates present in the disposable lining at molten steel temperatures. Thus, chemical purity of periclase and olivine is important in producing cleaner steels.

In general, phenolic resin such as resol and novolac are used as binders in DVMs. However, its toxicity and carcinogenic vapour release during curing are the major environmental concerns. Caustic sodium silicate and glucose are commonly used as alternative binders in DVMs. Lower refractoriness of the lining due to low strength of sodium silicate and probability of reacting with the backup lining restricts its use as a potential binder. Glucose, on the other hand, is cheap and readily available, with no environmental issues.

Properties of Basic Tundish Lining

Normally basic tundish lining refractories are expected to satisfy the following in continuous casting process:

- Better corrosion resistance to liquid steel and tundish slag.
- Low temperature losses even in long sequential castings.
- Good thermal shock resistance.
- Simple and reliable heating system.
- Less labour requirement.
- Easy de-skulling.
- Better lining integrity required for steel cleanliness.

These requirements can be achieved by controlling the refractory properties of the disposable lining which depend on the chemical composition of the ingredients used. Low thermal conductivity of the lining due to the material used and porosity in the structure prevents chilling of the liquid steel during the start-up of casting period when the tundish is cold. As a result, the heat loss from the molten steel is minimized and a narrow temperature range required to allow the mould to continuously solidify it is maintained.

Fig. 10.6 Refractory lining in Tundish ready for operation

Since glucose is thermoplastic in nature, as opposed to thermosetting in case of phenolic resin, it is possible to install it at a temperature more than 300 °C. One of the disadvantages of glucose binder is its low strength.

Advantages of Dry Vibromass (DVM)

Use of DVM with organic binders (phenolic resin) was first introduced in Europe in 1980s whereas with inorganic binder, mainly silicates, was developed in the mid of 90s. Presently it holds nearly 100% market share. The major advantages of this technology include:

1. Better steel quality (low H pick-up, low oxide inclusions etc.)
2. Better heat balance (lower tapping temperature at vessel and ladle)
3. Higher temp in tundish due to low thermal conductivity of working lining.
4. High energy savings (dry use, no water mixing and heating to remove water)
5. Lower investment, maintenance and labour costs.
6. Better process reliability (less sintering, easy deskulling)

The resinous/silicate bond is activated by relatively low amounts of heat (around 160–300 °C). The hot air is introduced at approximately 300 °C and the heating cycle takes around 45 min with further 30 min for cooling. Once it is cooled, the tundish is ready for use (Fig. 10.6), thus saving a lot of time.

Designing DVM Lining

Performance of the DVM refractory is greatly dependent on its raw material quality, installation and curing process, thickness, slag composition, operating temperature

Table 10.4 Chemical composition of the refractory components used in DVM (wt%)

Raw materials	CaO	SiO_2	MgO	Fe_2O_3
Dead Burnt Magnesia (DBM)	<0.5	1.5–3.0	93–95	<1.0
Olivine	<0.5	35–45	45–55	8.5–10

Table 10.5 Typical chemical composition of DVM (wt%)

	CaO	SiO_2	MgO	Fe_2O_3
DVM	0.5	23.04	61.5	4.13

and duration, etc. The thickness of the lining is being carefully designed to balance the lining wear of different zones in a tundish. In order to ensure proper de-skulling after use, the temperature at the interface between working lining and backup castable should not exceed 1200 °C. Thermal conductivity of the lining material is equally important as it would help to freeze the slag within the working lining. During operation, molten slag penetrates into the porous refractory lining by capillary action. As it moves deeper along the lining thickness, the temperature of the lining gradually decreases due to its lower thermal conductivity. As a result, the slag solidifies inside the working lining, thereby reducing the chance of damaging the backup lining.

Effect of pore size and pore size distribution of the working lining on the performance of tundish have a direct impact on its performance. The penetration of molten steel and slag is more suppressed in a porous lining compared to conventional linings. Thus, high continuous casting sequence was achieved by the application of the highly insulating tundish disposable material.

Materials and Method

In this context, the dry vibratable material (DVM) considered is a commonly used one with Dead Burnt Magnesite (DBM):Olivine in the ratio of approximately 40:60. Chemical compositions of the above refractory components are given in Table 10.4. Phenol formaldehyde-based resin has been used as the green binder system for the mixes. Grain size for mix is in the range of 0–1 mm. Typical composition of the DVM mix is shown in Tables 10.4 and 10.5.

Addition of olivine in DVM improves reactivity as during heating to more than 1450 °C, olivine becomes unstable and forsterite ($2MgO.SiO_2$). This is evident that after heat treatment at 1450 °C. There is an increase of forsterite ($2MgO.SiO_2$) phase and proportional decrease of the Mg-Fe-Si-O-based phase. Further, with the formation of high-temperature forsterite phase, other phases such as magnesio-wustite (MgO-FeO) and fayalite ($FeO.SiO_2$) were formed. Magnesio-wustite is favourable compared to low-melting fayalite phase. Dissolution of FeO in periclase and reaction of FeO with SiO_2 form the new phases, which are low melting. This explains the decrease in periclase content by approximately 8% in DVM after service. The increase of forsterite as well as wustite phases results in the formation of more solid–solid bonding in olivine-based materials.

In tundish working lining, the DVM must be adequately sintered during operation to improve thermos-mechanical strength and corrosion resistance. Formation of

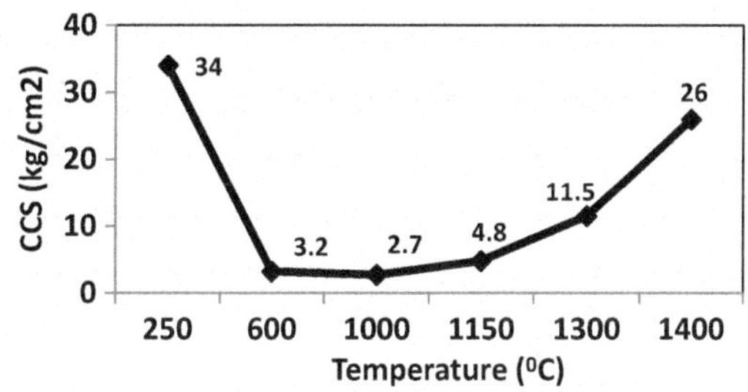

Fig. 10.7 CCS of DVM samples after firing at different temperatures

MgO-FeO solid solution or magnesio-wustite at the periclase-olivine grain boundary is expected to enhance solid state sintering between periclase and olivine. Bygden [6] reported similar inter-diffusion of Fe-Mg ions resulting in periclase-wustite bond formation in a static corrosion study of MgO samples in contact with FeO-CaO-SiO_2 slags. In addition, the extent of grain to grain contact in such types of solid–solid sintered materials is more efficient, compared to pure periclase systems. Hence, the presence of iron oxide as an impurity in forsteritic olivine-based mixes causes an improved bonding in the composition.

Further, at high temperature in the presence of high oxygen partial pressure, olivine became unstable and dissociated into forsterite or ferro-olivine, depending on the composition [8]. Forsterite-rich olivine is used for the DVM mix with low iron content to minimize the formation of fayalite (Fe-end member) to obtain excellent thermal properties. When heated at high temperature, the olivine in the mix got dissociated to the high-temperature phase forsterite (melting temperature about 1900 °C), with no fayalite formation. The absence of the low-temperature-phase fayalite (melting temperature of approximately 1220 °C) is expected to have a significant effect on the high-temperature strength of the mix.

Characteristics of Dry Vibro-mass (DVM) Mix Composition

- Up to 1000 °C it does not have high strength due to the absence of adequate bonding.
- The low strength within 700–1000 °C facilitates easy de-skulling the tundish.
- On further heating, hot strength develops due to ceramic bond formation. The strength develops due to sintering of DVM resulted in formation of periclase–olivine solid solution at above 1450 °C.
- The high temperature strength above 1400 °C is required to withstand the ferro-static pressure of liquid melt in tundish during operation, to eliminate the cause for dislodging the lining material during operation.
- Addition of olivine reduces the thermal conductivity of DVM, which is favorable to keep the shell temperature low.
- The cold crushing strength values of DVM after firing at different temperature is shown in Fig. 10.7.

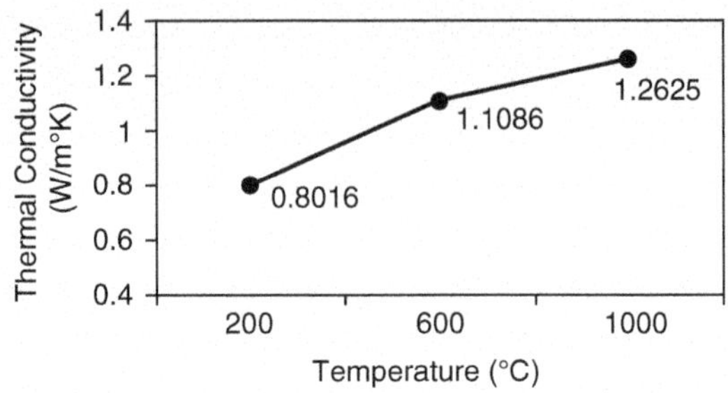

Fig. 10.8 Thermal conductivity of the DVM mix at different temperatures

Addition of olivine in periclase significantly lowers the thermal conductivity of DVM, which favours the lining in tundish to keep the shell temperature low and easy de-skulling. The thermal conductivity of the DVM mix at temperatures 200 °C, 600 °C and 1000 °C, respectively, is shown in Fig. 10.8. The thermal conductivity of refractories depends on mineralogical compositions as well as the glassy phase present in the refractory materials and the application temperature. Due to different coefficient of thermal expansion of different phases present in the mix, disrupts in the heat flow and the thermal conductivity of the mix is reduced [7].

Slag Development in Tundishes

In steel-making process, oxygen is blown into liquid steel bath to reduce the carbon level and oxidize the elements like Si, Fe, Mn and P. These oxides react with lime flux (CaO) and form a slag. During the process, refractory lining of vessel also gets eroded and goes into the slag. The chemical composition of a final vessel slag may vary as listed down in Table 1.1. Depending on steel grade, ferro alloys and other additives are added to the liquid steel during tapping from the vessel. During ladle furnace (LF) treatment, addition of lime, spar and some ferro alloys is a common practice. A variation in ladle slag is observed depending on the steel grade. Covering powders are added in tundish to prevent heat loss from molten steel, its oxidation and also to absorb inclusions present in liquid steel. During casting, some elements in liquid steel get reoxidized by oxygen from the atmosphere or oxides like SiO_2, MnO and FeO present in the tundish slag. Erosion of tundish refractory lining (MgO mass) also helps in slag formation. Some slag from ladle to tundish also may come towards end of teeming due to vortex effect.

Slag compositions of Si-killed steel for vessel, ladle at different stations and tundish are shown in Table 10.6.

10.3.1.4 Refractory Installation in Tundish

In order to realise the benefits of a high-quality low-cement castable lining, it is essential to adopt proper installation technique in refractory installation. The following refractory lining design is used in most of the steel plants:

Table 10.6 Chemical composition of vessel, Online Purging Ladle (OLP), Ladle Furnace (LF) and Tundish Slags (wt%)

Unit	Tot Fe	CaO	SiO_2	MgO	MnO	Al_2O_3
Vessel	16–20	50–55	13–17	0.5–2.0	0.4–0.8	0.7–1.5
OLP	0.7–2.0	47–52	15–27	4–9	1.5–4.0	5.0–12
LF	1.0–2.0	56–62	16–26	4–12	0.2–1.5	4.0–8.0
Tundish	1.0–4.0	4.0–22	35–55	5–20	5–20	5–10

- Ceramic fibre board of 10–20 mm thick over the steel shell.
- Low-cement castable as permanent lining of 80–120 mm thick.
- Disposable working lining of dry-vibro mass (DVM) of 80–120 mm or spray material of 60–80 mm thick.

Cleaning the Tundish The shell of the tundish is required to be cleaned of any residual debris, old anchors should be removed, and any mechanical repair required has to be done. The vent holes of 5–8 mm is to be provided in the shell for steam to escape during curing and preheating the castable.

Fixing of Anchors Metallic anchors are required to be welded to the shell with a proper spacing to provide mechanical stability and supports to the permanent lining. Either "V" or "Y" type stainless steel anchors of 65–75 mm length are to be used. Anchors are coated with bituminous paint or a plastic cap is used at the anchor tip to allow for differential expansion between the metallic anchor and castable lining.

Fixing of Ceramic Fibre Board The ceramic fibre board is fixed all along the tundish inner surface. Holes are to be made in the board to allow the anchors exposed to the castable lining and remaining portions of the holes are to be filled up by ceramic wool. The ceramic fibre board provides adequate insulation to maintain low shell temperature and thermal profile.

Fixing of Former/Shuttering A former fabricated out of about 8–10 mm steel plate and reinforced to resist deformation is generally used. The former is made in several segments and bolted together for easy removal after installation of castable. The warpage in former has to be maintained within 5–10 mm to maintain uniform castable thickness.

Mixing of Castable The water requirement for mixing high-strength low-cement castable is lower than conventional castables. The addition of water is very critical and is restricted within 4.5–6.0%; excess water addition reduces thermos-mechanical strength. Temperature of water should be maintained within 15–22 °C; if the temperature of water is lower than 15 °C, setting of castable would be very slow, and if the temperature is more than 22 °C, it results in very fast setting, which significantly reduces working time. For efficient mixing, a high-intensity mixer is recommended.

Casting of Castable The batch must be placed within 15–20 min of mixing the castable and casting is to be done with poker vibrators to optimize adequate

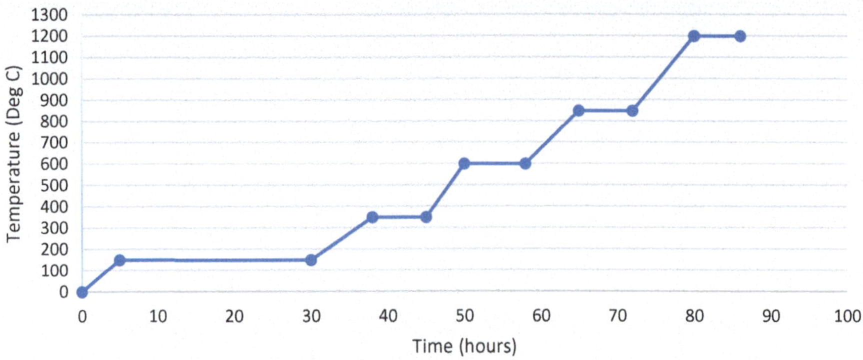

Fig. 10.9 Dry-out and heating schedule of tundish castable lining

flowability of castable enabling to fill-up the cast properly and to eliminate entrapped air. The diameter of the poker can be 25–60 mm, depending on the lining thickness. Vibrations of around 14,000–16,000 vpm and 60–80 mm vertical needle displacement are considered adequate for densification of castable. The bottom of the tundish is to be casted and it is allowed to set, before fixing the former. The casting process is to be continued continuously, till it completed to avoid lamination cracks during preheating. The former is to be removed after proper setting of castable.

Curing, Drying and Preheating The installed castable lining is prevented from drying, to allow air curing for 48 h, after the initial set to avoid cracking due to faster moisture removal. This process can be done by covering the lining with polythene sheet. Subsequently, air drying is followed for about 24 h and efficient drying is possible by using hot air generator, which is used in most of the steel plants.

After complete removal of moisture in the process of drying, preheating is to be done up to 1200 °C. During heating, a slow heat-up schedule to be followed with soaking at various levels are employed for uniform temperature distribution throughout the tundish lining. A typical heating schedule used for large size tundish is shown in Fig. 10.9.

Installation of Working Lining In present days, Dry-vibro mass (DVM) is largely used as working lining. It is a pre-mixed free flow monolithic mass, ready to use. No need to add any additional binder at the time of installation. After preheating the permanent castable lining, it is allowed to cool down and bottom to be rammed by DVM, followed by fixing of former to place DVM in side wall. DVM is to be poured into the tundish with a mild vibration to enable the flow of DVM mass and fill-up the gaps without any air gaps. Use of surface vibrator fixed on the surface of the former is the best method to provide vibration during casting the working lining.

DVM is a non-aqueous resin-bonded material and need not follow slow prolong heating schedule. It is required to heat up the working lining up to 350 °C for polymerization of organic bond and develop strength. Hot air generator is used for this purpose.

10.4 Black Refractory

During the modern steel-making continuous casting process, the molten metal is flowing from the ladle to the tundish and continuously from the tundish to the mould. This metal flow transfer is a key operation of continuous casting which affects strongly the solidified shell profile, floatation of the non-metallic inclusion and mould flux entrainment by large meniscus level fluctuations. In order to improve steel quality and to achieve a better flow control, the ladle shroud, mono-block stopper and submerged entry nozzle (SEN) are used. Carbon is the main refractory component used in manufacturing all the three items; hence they are called as "black refractories". The main functions of the black refractories are:

- Prevent the stream of steel from reoxidation.
- Control flow.
- Flotation of non-metallic inclusions.
- Able to cast the steels of critical grades.
- Prevent N2 pickup by steel during casting.

The shroud is installed at the lower end of ladle collector nozzle, fixed in the slide gate system for the ladle. It is used for supplying liquid steel from the ladle to the tundish. The submerged entry nozzle (SEN) is set to the tundish and mould, and used for non-oxidation casting. Mono-block stopper is fixed with tundish to control flow of steel into the mould. The arrangement of all the three components for continuous casting is shown in Fig. 10.10.

Properties of Black Refractories

The black refractories used in continuous casting process should have the following properties to realize the functions for all the three components [9]:

- High thermal shock resistance.
- Excellent mechanical strength to withstand the thermos-mechanical load of liquid steel stream and vibration.

(continued)

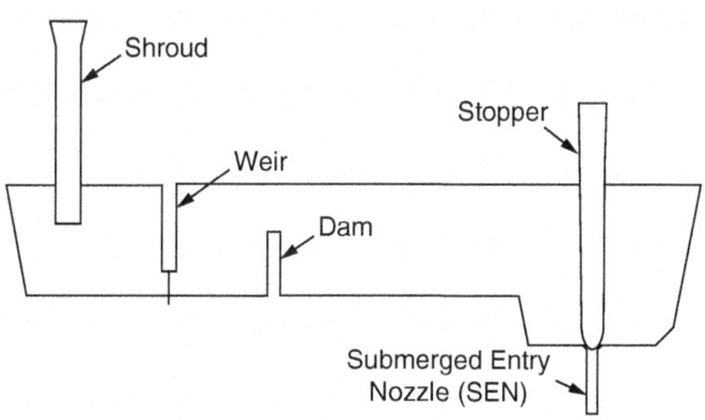

Fig. 10.10 Schematic diagram of con-cast refractory with single strand casting

- High volume stability.
- High resistance to corrosion and erosion from the liquid steel flow.

Resistance to cracking from mechanical shock is particularly essential to the black refractories. The composition of the refractory has changed with the development of the continuous casting process and grades of steel, from fused silica with good thermal spalling resistance to the alumina-graphite or zirconia-graphite with excellent corrosion resistance, which is manufactured by isostatic press.

10.4.1 Mono-block Stopper (MBS)

Mono-block stoppers are used as flow control mechanism for billet caster. The MBS and tundish slide gate valve are sometimes used together in order to prevent molten steel from solidification in the nozzle bore when casting starts or different quality of steel are casted successively. Usually stoppers are made of iso-statically pressed mono-block alumina-carbon materials. In few special cases, zirconia-carbon or magnesia-carbon refractories are used in the slag and molten steel contact zone. Magnesite is used in silicon-killed steel with high MnO and FeO containing slag.

10.4.2 Ladle Shroud

The most important role of the ladle nozzle is to prevent reoxidation of the liquid steel and pickup of nitrogen by protecting the molten steel from air engrossment, while transporting the steel from ladle to tundish. This non-oxidation casting with the shroud has been widely adopted by many steel plants and has contributed to improve quality and yield rate.

The ladle shroud is made of alumina-graphite refractory. The main constituents are alumina, fused silica and graphite. The thermal expansion coefficient and thermal conductivity have a direct impact on thermal shock resistance. Those properties of different raw materials are shown in Table 10.7. As shown in Table 10.7, graphite and fused silica have extremely low thermal expansion coefficient and graphite shows high thermal conductivity; therefore, these constituents are very useful for the good thermal spalling resistance. Alumina shows rather high thermal expansion, but superior corrosion resistance to the liquid steel.

The corrosion resistance has been improved by adjusting the grain size distribution, the Al_2O_3/SiO_2 ratio and the oxides/graphite ratio. The corrosion resistance of ladle shrouds has also been improved by the replacement of pitch binder with phenol resin. Resin-bonded shrouds have higher oxidation resistance due to denser texture and smaller pore size and better hot abrasion resistance, because of higher bonding

Table 10.7 Thermal expansion coefficient and thermal expansion of refractory materials [9]

Refractory materials	Thermal expansion coefficient (X 10-6/°C)	Thermal Conductivity at 1000 °C (W/mK)
Graphite	2.2	63
Alumina	8.8	6.3
Fused Silica	0.5	2.6
Stabilised ZrO_2	10	2.3
Magnesia	13.5	7.1
Spinal	7.6	5.8
Mullite	5.3	3.7

strength at operating conditions, all of which results in excellent corrosion resistance.

The average life of shrouds made of silica refractory was only 2.5–3.5 heats. By introducing the upgraded refractory composition of alumina–carbon composites and change in structural design in holding mechanism as well as introducing anti-oxidant and thermal barrier coating technology, the life has improved to more than 10 heats. Further, use of argon injection and proper sealing and holding mechanism has significantly improved the shrouding effect to minimize the air ingress through joints.

Wear Behavior of Ladle Shroud

1. The ladle shrouds are ordinarily joined to the lower part of the flow control slide gate valve of ladle and molten steel flows down inside the shroud under the condition that its upper part is fixed by the holder below ladle and its lower part is free. Consequently, shrouds are suffered from extreme thermal shock and mechanical stress by vibration caused by molten steel flowing down through the shroud at the beginning of the cast. In particular, ladle shroud is used under severe thermal and mechanical stress, because it is usually used without preheating and it is very long according to the overall layout. Thus, breakage at the joining part with the holder and generation of vertical crack are the main problem in Ladle shrouds.
2. Local erosion is occasionally observed at the upper part of the shroud by the drift flow due to slide gate throttle or during closing of plates at the end of cast. It is necessary to pay attention to the sealing mechanism of the joint, because of sudden loss of refractory observed due to air suction at the upper joint.
3. High wear is caused by different grade of steel. In particular, the wear significantly increases with high oxygen containing steel because high oxygen potential causes oxidation.
4. When Al-killed steel is casted, the inside opening of the shroud becomes narrow and clogging occurs due to alumina deposition at inner surface. The

(continued)

Fig. 10.11 Schematic diagram of ladle shroud

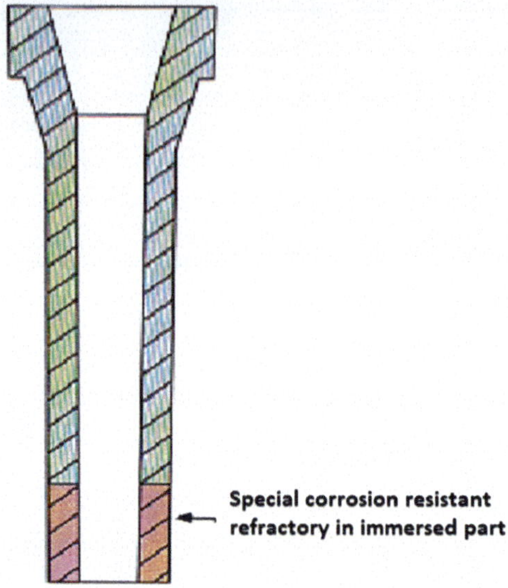

alumina clogging is minimized by changing the refractory composition and providing arrangements of argon purging.

The shrouds are commonly made of alumina-graphite refractory and an anti-corrosion coating is used, such as ZrO_2 and carbon-free inside liner materials. The immersed part of the shroud in side slag and liquid steel is designed with special corrosion-resistant refractory composition, as shown in Fig. 10.11. With the recent development of inner and outer layer lining, additive strengthening, thickness adjustment and cleaning technology, the life of the shrouds has been increased significantly.

Argon Purging in Ladle Shroud

The challenges remain in three driving factors for the advances of ladle shroud: maximizing the shrouding effect from air ingress, exploring multiple functions and solving thermal shock issues.

Air ingress can occur between the ladle collector nozzle and shroud leading to steel reoxidation and nitrogen pickup. In recent design of ladle shroud, this can be eliminated by argon purging arrangement in the ladle shroud. Argon is used to compensate the negative pressure inside the shroud and create an argon seal around the junction. Use of a sealing gasket also prevents air engrossment. Argon injection may be used in conjunction with sealing gaskets. The thermal shock issue is addressed by improving the quality of refractory mix and by applying high emissivity coating on the outer surface of the shrouds.

10.4.3 Submerged Entry Nozzle (SEN)

The submerged entry nozzle (SEN) is used to transport the high-quality liquid steel from tundish to the mould and it can be considered as a vital factor obtaining a stable casting process and a good product quality. It protects the steel from reoxidation and nitrogen pickup during casting. Graphite is used as a major refractory component as the presence of graphite in the SEN-based refractory materials leads to:

- Thermal shock resistance and high thermal conductivity, which make it possible to preheat the SEN at high temperature with higher heating rate.
- No wettability by molten steel.

The continuous casting steel-making process occurs at a temperature more than 1500 °C, so the interaction between the refractory material of SEN and liquid steel is to be considered. The SEN must withstand the corrosion effect with slag and liquid steel, besides its mechanical strength and thermal shock resistance. The interaction between SEN and liquid steel is considered in three different mechanisms [10]:

- The chemical reactions between refractory and impurities in the tundish nozzles and liquid steel.
- External surface corrosion by mould–flux and slag from tundish.
- The attachment of immersed non-metallic inclusions in the molten steel, to the inner surface of the SEN.
- Erosion of the SEN refractory material due to flow of liquid steel.
- SEN also has two other important functions: enables to floating up of non-metallic inclusions and the prevention of entrapping of mould powder in the steel by control of molten steel flow in the mould.

SEN is an iso-statically pressed single-piece refractory tube made of Al_2O_3—C/ZrO2-C refractory and they are preheated prior to actual use and used continuously.

Alumina Graphite Nozzles

The main constituents of alumina–graphite nozzles are alumina, fused silica and graphite. The physical properties of the SEN are also affected by the inner texture of surface to reduce frictional force in liquid steel flow, which is controlled by grain size distribution or by using proper binders and additives.

The corrosion resistance has been improved by adopting proper grain size distribution of the refractory matrix, the Al_2O_3/SiO_2 ratio and the oxide/graphite ratio. In recent years, alumina-graphite nozzles with high density and high strength are used, which has remarkably improved corrosion resistance. Resin-bonded SEN has higher oxidation resistance due to denser texture and smaller pore size and better hot abrasion resistance because of higher bonding strength. Addition of metallic Si and Si3N4 improves high-temperature strength, reduces modulus of elasticity and hence improves thermal shock resistance.

Table 10.8 Tentative properties of Ladle shroud and SEN

Properties		Ladle Nozzle		Submerged Nozzle	
Physical properties	Unit	Body	Holder part	Body	Slag Zone
Apprent Porosity	%	13–14	14–15	14–15	14.5–15
Bulk Density	Gm/cc	2.25–2.35	2.35–2.45	2.4–2.5	3.4–3.7
Modulus of rupture	Mpa	8.5–9.5	12.5–14	8.8–9.5	6.9–7.5
Modulus of elastricity	Mpa	93–110	130–135	110–115	85–88
Thermal expansion at 1000 °C	%	0.25–0.29	0.36–0.4	0.3–0.4	0.44–0.46
Chemical composition	%	40–50	50–55	50–60	nil
Al_2O_3	%				
SiO_2	%	25–30	22–24	5.5–8	nil
ZrO_2	%	nil	nil	nil	70–75
C + SiC	%	25–32	20–25	30–35	20–26

Zirconia–Graphite Nozzles

The zirconia–graphite composition is used to slag line contact of SEN. Zirconia (ZrO2) and graphite are the main constituents and ration of ZrO_2/graphite is adjusted for well-balanced corrosion and thermal shock resistance to meet specific service conditions. The corrosion resistance in slag zone improves with increase of ZrO_2 content; however higher amount of ZrO_2 addition, more than about 80%, causes significant reduction of thermal shock resistance due to the increase of thermal expansion. The tentative properties of Ladle shrouds and SEN are shown in Table 10.8 [11].

Use of Mould Flux in Continuous Casting

In continuous casting process, liquid steel passes through the subentry nozzle and pours into a water-cooled copper mould, where solidification of steel begins. During this process, mould flux, also called mould powder, performs very critical roles as follows:

- It acts as a lubricating agent between copper mould and steel during high-speed casting.
- It absorbs non-metallic inclusions during casting.
- It controls heat flow through wall of highly conducting mould, acts as a thermal insulation.

The major function of the mould flux is to provide lubrication between the solidifying shell and the mould wall. The ability of a flux for lubrication is primarily determined by its viscosity, sintering temperature and the solidification temperature of steel. So, it is important to optimize the viscosity of the liquid at operation temperatures. The lubrication is also influenced by the following process conditions:

- Casting speed.
- Superheat temperature.
- Submerged nozzle (SEN) design.

Table 10.9 Typical basic mould fluxes compositions

Glass formers	SiO_2	15–50%
	Al_2O_3	0–15%
	B_2O_3	0–19%
	Fe_2O_3	0–6%
Basic oxides	CaO	22–45%
	MgO	0–10%
	BaO	0–10%
	SrO	0–5%
Fluxes and fluidizers	Na_2O	0–25%
	K_2O	0–2%
	Florides	2–15%
	MnO	0–5%
	Carbon	2–20%

The mould flux powder is available in many sizes and shapes; it is a synthetic slag and consists of a silica-based complex mix of oxides, minerals and carbonaceous materials. The main oxides are MgO, SiO_2, CaO, Na_2O and Al_2O_3 with a carbon source added to improve lubrication as well as to control melting. The final mould flux can be either a fine powder with 60–90% of −200 mesh, small −35 mesh spheres, or granules in the −10 to +20 mesh size range. The mould fluxes are generally added through the top of the mould on the liquid steel. The additions of mould fluxes on the free liquid steel surface form different layers and each layer in isolation or combined with another one provides the required functions of the powder.

At 1600 °C, viscosity of the steel ranges between 0.002 and 0.006 poise, while that of slag remains above 0.02 poise. The mould fluxes composition is to be selected such that there would be sharp decrease of viscosity around 1450 °C, so that a slag coating forms over the semi-solidified steel in the mould, which acts as a lubricant as well as prevents reoxidation. Mould fluxes of basic composition have this characteristic. Hence, basic mould fluxes are widely used [12]. Few of the oxides used in mould fluxes act as glass former, basic oxides used as modifiers and few oxides are used as flux and carbonaceous materials are generally used as lubricants. Typical basic mould fluxes compositions are shown in Table 10.9 [13].

Slag Line Corrosion of SEN Refractory

Corrosion of the SEN occurs at the contact surfaces of the mold fluxes at outside and contact surface with steel at inner side of the nozzles. Fluxes in the mold powder such as alkalies (Na_2O, K_2O, Li_2O) and fluorides have a direct effect on refractory corrosion in slag line. ZrO_2-C refractory materials are commonly applied to the slag line part of SEN, owing to high corrosion resistance against mold flux attack. In recent few designs of SEN, Zirconia-graphite composition is also used to the inner surface of SEN and the head of

(continued)

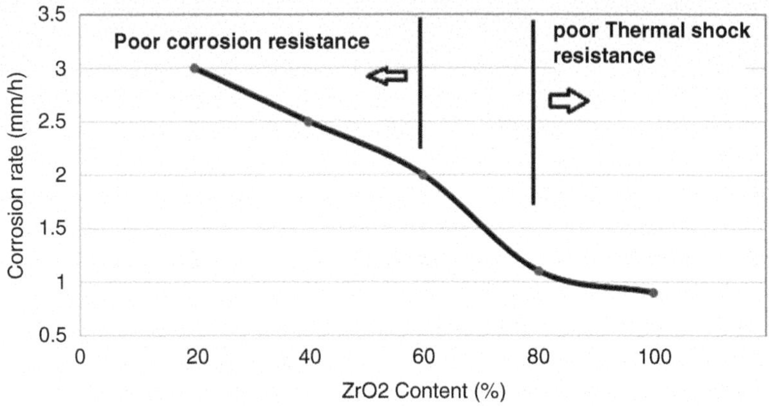

Fig. 10.12 Effect of ZrO_2 content on corrosion rate of Zirconia-graphite nozzle by Mold powder [11]

mono block stoppers. The corrosion of zirconia–graphite nozzles is increased by increasing the Na_2O and K_2O content.

The effect of ZrO_2 content on corrosion resistance of zirconia–graphite nozzles is shown in Fig. 10.12. Below about 60% of ZrO_2 content, corrosion resistance is poor, but above 80% of ZrO_2 addition, poor thermal shock resistance is observed. 70–80% zirconia containing nozzles are generally used for slab casting process, even in the process of high-speed casting of low-carbon steel. Though Y_2O_3-stabilized ZrO_2 has the best corrosion resistance, CaO- or MgO-stabilized ZrO_2 is generally used with baddeleyite, because of high price of Y_2O_3-stabilized ZrO_2. The durability of ZrO_2-graphite material is improved by adding the optimum quantity of un-stabilized ZrO_2, SiC and Boron-nitride. With respect to graphite, the purer graphite is more durable.

Erosion of Inner Surface of the Submerged Entry Nozzle

The erosion of the inner surface of SEN significantly varies with steel quality, and the erosion decreases for Al-killed steel and increases for high manganese steel and high oxygen containing steel. When the O_2 content increases more than 150 ppm, the erosion of nozzle increases significantly [14]. It is caused not only for the oxidation of carbon bond in refractory by free oxygen but also in the presence of FeO and MnO, as shown in the following equations:

$$FeO + C = Fe + CO$$

$$MnO + C = Mn + CO$$

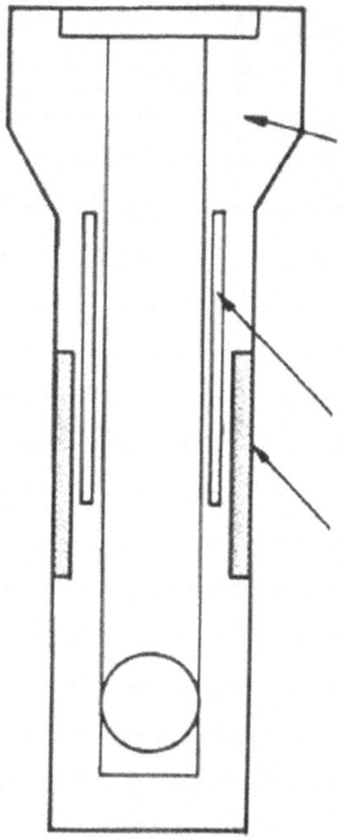

Fig. 10.13 Schematic drawing of SEN

Air suction through the joints is also responsible for oxidation of carbon in refractory. Air suction from joints must be completely prevented to eliminate oxidation. Fused silica commonly used to increase thermal shock resistance is crystallized to cristobalite at the operating temperature, decomposed in the presence of carbon in refractory component, and evaporated, and destruction of refractory bond occurred as shown in the following equation:

$$SiO_2\ (\text{cristobalite}) + C = SiO\ (g) + CO$$

A schematic diagram of SEN showing slag zone and refractory used in rest of the body is shown in Fig. 10.13.

Cracking of the SEN

Crack formation in the SEN is a major problem to achieve higher life. The cracking takes place due to the following reasons:

- Thermal stress.
- Mechanical stress.

Thermal stress can be minimized by optimizing the refractory composition. Increase of thermal conductivity and reducing thermal expansion result in improving thermal shock resistance, which has been discussed, but it is important to discuss the impact of mechanical stress generated, which involves study on design and dimension of SEN. Cracks in materials occur when the mechanical or thermal stress exceeds the strength of the material. It requires high mechanical strength in SEN for its high *L*/*D* ratio and small wall thickness. ($L =$ length, $D =$ diameter) as shown in Fig. 10.13. The crack resistance is expressed in the following equations, considering an endless tube heated from inside [15]:

$$\zeta = K^{`}.a.E.(T1 - T2) = K.a.E.(1/\lambda)$$

$$R_{CR} = (S.\lambda)/(a.E)$$

where,

$\zeta =$ Maximum tensile stress at outer surface of nozzle.
K' & $K =$ Shape factor.
$a =$ Thermal expansion coefficient.
E = Young's modulus.
$T1 - T2 =$ Temperature difference between inner and outer surface.
$S =$ Tensile strength (strength of the material).
$R_{CR} =$ Crack resistance.

On the assumption that $\zeta > S$, cracking occurs, and if the value of RCR is large, the material has excellent crack resistance. Graphite has very high thermal conductivity, low modulus of elasticity and high tensile strength and alumina has high corrosion resistance; hence alumina–graphite composition has a very high crack resistance.

Alumina Clogging at Inner Wall of SEN

The deoxidation products in steel, Al_2O_3 in Al-killed steel, adheres to the inner wall of the SEN, resulting in nozzle clogging which causes discontinue in casting operation and the SEN needs to change. Corrosion by mold fluxes and tundish slag occurs in the slag zone of outer surface, as shown in Fig. 10.14.

In all cases Al_2O_3 is the main constituent of the buildup layer in the SEN. The Al_2O_3 in the reaction layer may originate from:

1. Alumina from the SEN refractory material.
2. Inclusion in steel resulting from deoxidation.
3. In situ alumina formation by reaction of dissolved Al in steel and excess O_2 near the nozzle wall.

Alumina deposits grow up with discharging molten steel in deposit due to high surface tension. It is observed that nozzle clogging suddenly proceeds because alumina which adhered to the inside of nozzle facilitates the generation of clogging and stopping steel flow. Factors affecting the alumina build up and its countermeasures are summarized in Table 10.10.

Gas Injection in the SEN

(continued)

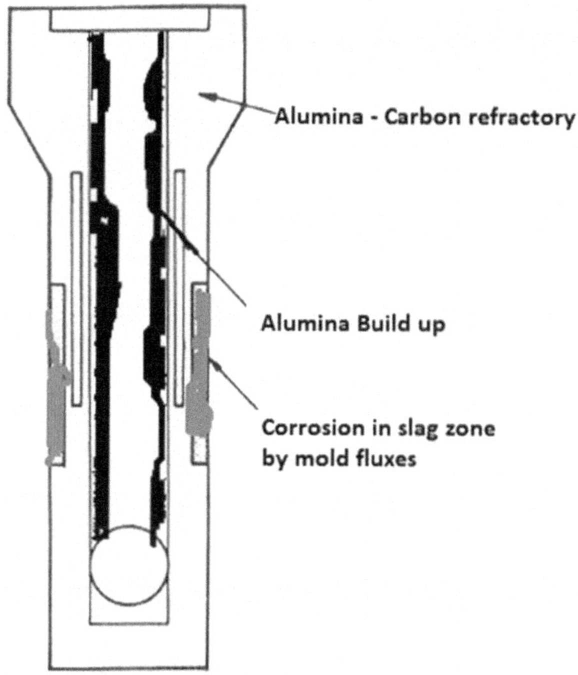

Fig. 10.14 Al_2O_3 clogging and corrosion in SEN

Table 10.10 Factors involved in SEN clogging and countermeasures

Factors for Alumina build up		Counter measures
Deoxidation products	Steel quality	Improving in shrouding system
Secondary oxidation by air	To arrest Air engress	(a) Proper sealing in joints, (b) Argon sealing in slide gate equipment and SEN joints
Reaction of steel and slag with refractories, improves inclusions in steel	Refractory quality improvement	(a) Application of high grade and basic refractory (MgO based) in tundish lining. (b) Application of Al_2O_3 and SiO_3 free, ZrO_2 based refractory
Erosion of SEN refractory		
Increase of melt viscosity and alumina neucleation near inner nozzle surface	Improvement in operating conditions	To maintain high temperature of molten steel
(a) Stagnation at liquid steel and refractory interface (b) Precipitation and build up in SEN inner surface	Improvement in SEN design and materials	(a) Liquid flow control in SEN by improvement in nozzle geomerty and inner surface design (b) Argon injection in slide gate and SEN (c) Improvement in SEN refractory material having resistance to alumina clogging.

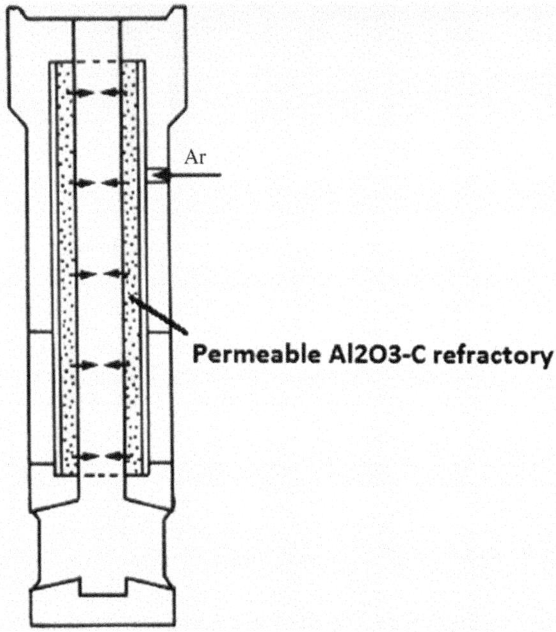

Fig. 10.15 SEN with permeable refractory for argon gas purging

Various slit type gas injection SEN have been developed and put into practice for the purpose of preventing alumina clogging [16]. Standard slit type is shown in Fig. 10.15, which has been most commonly used in steel plants worldwide and is characterized as having the permeable refractory material with a permeability of 8–20 L/min.

References

1. C-A. Dacker, M. Llanos, J. Cordon Velaz, K. Marx, et al, "Innovative process technology for ingot casting by application of simulation and measuring techniques (IPTINGOT)", Technical Report · January 2014, European Commission, B-1049 Brussels.
2. Lifeng Zhang, B.G. Thomas, "State of the Art in the Control of Inclusions during Steel Ingot Casting", Materials & Metallurgical Transactions B on August. 2005
3. M. Wu, J. Li, A. Kharicha, A Ludwig, "Multiphase modelling of the micro-segregation in Ingot casting", 1st International Conference on Ingot Casting, Rolling and Forging, 2012.
4. Chester, "Refractories for Iron and Steel Making"
5. Ahindra Ghosh, Amit Chatterjee, "Iron making and Steel making theory and practice" Prentice-Hall of India privet limited, 2008.
6. N R Khisina, D A Khramov, M V Kolosov, A A Klesehev and Lawrence A Taylor, "Formation of ferriolivine and magnesioferrite from Mg--Fe-Olivine: Reactions and kinetics of oxidation", Phys Chem Minerals, Vol-22; 1995; pp 241–250.
7. J Bygden, T Debroy, S Seetharaman, "Dissolution of MgO in stagnant CaO-FeO-SiO_2 slags", Ironmaking and Steelmaking, Vol-21; 1994; pp 318–323.

8. H B Johnson, "Making the grade for tundish linings: A growth market for DBM", Conference on Industrial Minerals, 'MagMin 2005', Vienna; pp 1–19.
9. Arashk Memarpour, "A study on the submerged entry nozzle (SEN) respecting clogging and de-carburization", Department of Materials Science and Engineering, Royal Institute of Technology, Sweden.
10. Shigeo YOSHINO, "Refractories for continuous casting in Japan", Shinagawa Technical Report, Vol 31. 1988, pp 31–74.
11. B. Mazumder, "A process for developing continuous casting mould-powder for steel industries", Journal of Scientific & Industrial Research, Vol 58, October 1999, pp 773–780.
12. Elena Brandaleze, Gustavo Di Gresia, et al, "Mould Fluxes in the Steel Continuous Casting Process" Chapter 7, Science and Technology Casting process, pp 205–233.
13. Noboru Tsukamoto, "Wear of nozzles for continuous casting", Taikabutsu Overseas, Vol 13, No 4, pp 55 – 61.
14. Hiroshi Shikano, Masahiro Sakai, Toshiaki Kaneko, "Development of alumina graphite nozzle for continuous casting", Taikabutsu Overseas Vol 9, No 2, pp 3 – 19.
15. Hiroshi K KYODEN, Yasutoshi NAMBA, "Prevention of Alumina build up in submerged entry nozzle for continuous casting of steel", Shinagawa Technical report, Vol 31, 1988, pp 85–96.

Chapter 11
Modern Refractory Practice for Clean Steel

11.1 Introduction

During last five decades, the demand of clean steel is escalating and the amount of total impurity elements (C, S, O, N and H) has been reduced from around 600 ppm, in the year 1940, to only <100 ppm in 2000 [1]. Apart from gaseous inclusions, the other inclusions are due to re-oxidation of steel by lining refractories. Hence, development of new refractories and selection are the challenges to reduce inclusions in steel.

Inclusions are categorized in two categories—endogenous and exogenous. The small inclusions that originate from the steel are typically called endogenous or micro-inclusions. Entrapped lining refractory and slag are called exogenous or macro-inclusions. However, agglomerated endogenous inclusions due to a clogging event could also be classified as exogenous inclusions.

Content of Alloying Agents (Few Critical Elements) in Treatments in Secondary Metallurgy [1]

Elements	Content %	Relevant secondary metallurgical operations
C	0.0010–2.5	VOD/VD, RH, RH-OB
Si	0.01–3.70	RH
Mn	0.08–20.0	LF
Cr	0.03–25.0	VD, RH, LF
Ni	0.03–80.0	LF
Al	0.0020–5.5	VD, RH, LF
Ti	0.01–1.5	VD, RH, LF
N	0.0020–0.5	VD, RH, LF

(continued)

S. Biswas, D. Sarkar, *Introduction to Refractories for Iron- and Steelmaking*,
https://doi.org/10.1007/978-3-030-43807-4_11

Ca-treated steel: a metallurgical approach

Ca treatment in steel by adding Ca metal is to modify a solid Al_2O_3 inclusion into liquid inclusions. For thick slab casing by the process of Al-killing, forms Al_2O_3 solid inclusions, typically no attempt is taken to modify the alumina inclusions in steel, but for thin slab casting operation, the steel is treated with calcium to eliminate solid Al_2O_3 inclusions for the production of clean steel, which prevents Al_2O_3 clogging in submerged entry nozzles, results in high life of SEN. The number of heats in these nozzles for casting thin slab is >12 compared to the 3–5 for thick slab caster. Liquid inclusions thus formed result in no clogging of nozzles during casting for long duration of thin slabs. The mechanism for forming liquid inclusion is summarized below:

1. $Ca + O = CaO$
2. $CaO + Al_2O_3$ (solid inclusion) forms C12A7 and CA, which are low melting and remain as liquid inclusion and form slag.
3. Added Ca reduces MgO of spinel (solid inclusion) into the steel and remain liquid. To avoid re-oxidation of Mg, partial pressure of oxygen should be maintained very low.

11.2 Inclusions from Refractories

The development of steel-producing technology and increased demand of clean steel are the main drive for the development of improved and new types of refractories. There are many influences in clean steel production, such as oxygen-blowing practices during melting, slag carry-over and deoxidation methods during treatment, slag type and conditioning, treatment time, temperature, refractory lining in ladle, purging elements and slide gate systems during ladle refining and vacuum degassing. Influence of tundish refractory and shrouding refractories also plays a vital role in producing clean steel. Refractories are one factor that become important when all the other factors are optimized. Adjustment of steel cleanliness is to be taken in the following areas of secondary metallurgy treatments and casting operations:

- Ladle furnace (LF)
- Ladle degassing
- Vacuum arc degassing (VAD)
- Vacuum oxygen decarburization (VOD)
- RH degasser system
- Flow control and purging system
- Tundish refractory
- Use of shrouds, subentry nozzles

Therefore, choice of refractory materials for steel ladle lining and tundish lining is very crucial to produce clean steel. When the steel melt is re-oxidized by lining refractory materials, the oxygen reacts with dissolved Al (aluminium) in steel to form Al_2O_3, which partly floats into slag and partly remains in steel as inclusions.

Refractory materials with lowest oxygen potential result in lowest aluminium losses and hence prevent inclusions by re-oxidation. Lime, corundum and magnesia result in low or very little aluminium loss. In addition, high purity synthetic materials such as alumina, spinel, MgO-C and doloma-C result in very low aluminium loss, and thus refractory lining in steel ladle or tundish made of those materials prevents inclusions in steel. High oxygen potential of SiO_2, Cr_2O_3 and ZrO_2 is responsible for high inclusion by re-oxidation.

Subsequent to steel refining in LF/VD/VOD/CAS-OB or RH degasser, where the liquid steel retains for 20–45 min during teeming, non-metallic inclusions are floated up and prevents direct contact of liquid steel with atmosphere and hence prevent further atmospheric re-oxidation. During teeming, if no precautions are taken, the stream of melt re-oxidized for the following reasons:

1. Atmospheric oxygen absorption by the teeming steel
2. Slag entrainment
3. Erosion of slide gate nozzle refractory of ladles

The oxygen content of steel may vary in the range of 40–400 ppm, depending on the nature and quality of steel. This expedite formation of FeO and MnO enriched inclusions. The atmospheric oxidation can be eliminated using shrouds in steel ladles and submerged entry nozzles in tundish for continuous casting processes. For slag entrainment and to reduce carryover slag, the usual practice is not emptying the ladle completely during teeming.

Monolithic Lining in Steel Ladle

Monolithic refractory lining is one of the latest refractory lining methods for the metal zone of ladles with conventional MgO-C brick lining in slag zone. The castables are alumina-based low-cement castable. This makes them highly suitable for ultra-low carbon steel production. Pre-formed spinel is added in the castable to improve corrosion resistance and they also have excellent oxidic stability. The monolithic ladle lining appears relatively easy and joint-free construction and results in long campaign life of steel ladle. Proper preheating is required before taking the ladle in service. Residual moisture may cause hydrogen pickup in steel, which is the major concern to use monolithic lining. After 6–8 h of air drying, the forma or template is removed and finally the monolithic lining is heated for another more than 50 h to remove the chemical or crystal water.

Stability of Refractory Oxides to Produce Clean Steel

Low oxygen and sulphur can be achieved when re-oxidation of steel by refractory materials is prevented. It requires very low oxygen potential in

(continued)

refractory materials and must be resistant against basic slags during desulphurization. For fully deoxidized steel, the partial pressure of oxygen, Po_2 is around 10^{-15} bar at 1600 °C. The partial pressure of oxygen of refractory oxides at 1600 °C can be estimated from Gibbs free energy and Richardson–Jeffels diagram [2]. To avoid supply of oxygen to steel from refractory lining, the oxygen partial pressure of refractory material must be lower than 10^{-15} bar and it can be concluded as follows:

- MnO, SiO_2, FeO, Cr_2O_3: Po_2: > 10^{-15} bar
- MgO, Al_2O_3, TiO_2, CaO: Po_2: < 10^{-15} bar

As a result, the refractory oxides suitable to produce clean steel are:

- Highly in: MgO, CaO, Al_2O_3,
- Low in: SiO_2, FeO, Cr_2O_3 and MnO

Those types of refractories are MgO, MgO-C, Dolomite, Dolo-C and high purity corundum and spinel.

To achieve good desulphurization, the activity of dissolved CaO in slag must be very high, lime saturated basic slag to be used and the oxygen content in steel to be kept low, therefore reducing condition to be maintained.

11.3 Ladle Refractory Practices for Clean Steel Production

In contrast to primary steel making, the slag in secondary steel making is deoxidized and contains a high proportion of CaO and Al_2O_3. In primary steel making, the design and selection of refractories is synchronized by the life of the refractory lining, resistance to corrosion by slag, and thus overall cost viability and its impact on overall cost. However, in secondary steel making, there are additional concerns about its effect on steel quality, to produce clean steel in particular. The requirements of the refractory lining of the secondary steel-making ladles are:

1. The lining should be stable and inert to liquid steel; otherwise, it will tend to introduce impurities to the steel.
2. Steel cleanliness depends on the interactions between liquid steel and the refractory lining. Erosion of lining increases the exogenous non-metallic inclusions, whereas corrosion causes a change of steel compositions, like increasing its oxygen content.
3. A low thermal conductivity of ladle lining is desirable to prevent heat losses through ladle side walls. However, thermal shock resistance and corrosion must be satisfied.

The lining refractories have great impact on the formation and evolution of inclusions during refining processes. The lining material selection for ladle refining steel is a critical consideration since the oxygen potential in a liquid steel is very low compared to the other steps in steel-making process. Hence, carbon is the most suitable material to use in lining refractory and carbon-containing materials have become popular in last few decades as it can better withstand corrosion by steel melt of low oxygen potential. In ladle metal zone, the three types of refractories are used worldwide, MgO-carbon, alumina and spinel.

11.3.1 MgO-C Refractory in Steel Ladle

When using carbon-containing magnesia refractory in steel ladle metal zone, the producers of clean steel often find inclusions of MgO and MgO-Al_2O_3 spinel in steel. Those inclusions contain different mixtures of alumina and magnesia and they are highly refractory and very hard, which are detrimental to steel properties like fatigue strength. The main reactions that take place between steel and refractory materials are:

- Reduction of MgO by carbon
- Oxidation of aluminium dissolved in steel, by CO formed at interface

MgO Dissolution in Al-Killed Liquid Steel

In MgO-C refractory, magnesia is reduced by carbon in refractory to Mg vapor and it is re-oxidized with the reaction of dissolved oxygen in liquid steel and it leads to formation of spinel inclusion. Some of the dissolved Mg in Al-killed steel is formed by reduction of MgO refractory by dissolved aluminum. The dissolution mechanism of MgO in liquid steel is shown in the following equations:

$$MgO_{(s)} + C_{(s)} = Mg_{(g)} + CO$$

$$MgO_{(s)} = Mg_{(g)} + O_{(g)}$$

$$C\ (\text{Graphite}) + O_{(g)} = CO_{(g)}$$

Reaction of MgO with Al (dissolved in steel)

$$3MgO + 2Al = Al_2O_3 + 3Mg_{(g)}$$

$$4MgO + 2Al = MgO.Al_2O_3\ _{(S)}\ (\text{Spinel}) + 3Mg_{(g)}$$

$$2Al + 3CO(g) = Al_2O_3\ (\text{inclusion}) + 3C$$

The spinel thus formed on the refractory surface. Further, the periclase in MgO-C refractory, in contact with liquid steel melt, continuously supplies Mg

(continued)

to the steel melt owing to reaction with Al. In presence of high carbon in MgO–C refractory, Mg forms by reduction of periclase and it reacts with Al_2O_3 inclusions to form spinel. The effect of dissolution of MgO and inclusion formation during use of MgO-C bricks in steel ladle metal zone is summarized below [3]:

- Reaction of MgO in the refractory with the carbon in refractory to produce Mg gas and carbon monoxide gas.
- Diffusion of the magnesium and CO gases through the refractory to liquid steel.
- Reaction at the steel–refractory interface and spinel formation increases inclusions.
- In every cases, the Al_2O_3 inclusions changed to spinel and the MgO content in the inclusion is greater than that of Al_2O_3 saturated spinel.

It is observed that MgO-C bricks used in steel ladle metal zone to produce ultra-low C steel may not be feasible.

11.3.2 Low Carbon Containing MgO-C Refractory

The transfer of carbon from the refractory lining to the steel can occur through the following mechanisms:

1. Direct dissolution of carbon into the steel

$$\mathrm{C(solid) = C\ (dissolved)}$$

2. Reaction between carbon and refractory components (e.g. MgO) and formation of CO that can act as a source of carbon

$$\mathrm{MgO + C = Mg + CO(gas)}$$

3. Interaction between carbon and reducible oxides penetrated into the refractory body, with subsequent formation of CO that can act as a source of carbon

$$\mathrm{FeO + C = Fe + CO(gas)}$$

$$\mathrm{MnO + C = Mn + CO(gas)}$$

Reducible oxides can be present in the process slag remaining attached to the refractory lining after teeming (glaze) or from the oxidation of steel during ladle filling.

The ultra-low carbon molten steel is produced by two steps decarburization:

1. In the converter, steel produced with carbon concentration of about 350–450 ppm, no further reduction takes place in the ladle furnace. Whereas, in vacuum oxygen decarburization (VOD) process and in RH degasser, reduced partial pressure of CO gas is obtained and thus, carbon can be reduced to 20–40 ppm.
2. Enough oxygen is injected into liquid steel to expedite decarburization process; however, such injected oxygen contributes to the higher concentration of dissolved oxygen in liquid steel after degassing.
3. Consequently, aluminium addition is necessary for complete deoxidation of the ULC steel. This results in the generation of large number of indigenous Al_2O_3 inclusions after the deoxidation step at the ladle.

MgO-C refractories have excellent chemical and thermos-mechanical properties like corrosion resistance, thermal shock resistance, high hot strength, etc., which make them the most suitable refractory for steel ladle lining. Generally, MgO-C refractories contain about 12–18% total carbon and the major source of carbon is graphite. However, with the improved steel-making technology to produce low carbon, ultra-low carbon and clean steel, the disadvantages of using those bricks were found as greater extent of oxidation, high liquid penetration and corrosion as well as carbon pickup by low carbon and ultra-low carbon steel. Higher generation of CO also caused inclusions in steel.

To cope up with these disadvantages, new types of refractory with low carbon containing refractory had been developed. The main advantages of low C containing MgO-C bricks are summarized below:

- Control of carbon pickup in steel
- Very high hot strength, resulted in high thermal shock resistance
- Reduced liquid penetration into refractory bricks
- High thermal shock resistance
- High oxidation resistance

Nanotechnology in Refractory

Use of nanotechnology in refractories is the latest development and nanoparticles are almost in molecular scale, the smallest unit of material commercially available. Due to their very fine size they develop phenomenon improvement in packing, pore size distributions, pore filling, bonding and reduction of addition quantity. Due to the finer size with very large surface area of the nanoparticles, the surface energy is very high, which is helpful to meet all the required properties, even with addition of very small quantity. In magnesia-carbon refractories, use of nano-addition has improved the properties. Recent research trend is inclined to pick up nanoscale materials as it has exceptional advantage on surface-area-to-volume ratio that eventually filled up the very small pores and facilitate formation of secondary phase and microstructure improvement. Use of nano-sized antioxidants, namely metallic aluminium and silicon, greatly improves oxidation resistance and increases the carbide phases during operations, which resulted in increased hot strength. Use of nano-boron carbide as antioxidant in MgO-C refractories showed high hot strength and oxidation resistance at an early temperature and better thermal shock properties. Nano-sized

$MgO.Al_2O_3$ spinel is added in MgO-C bricks to improve corrosion resistance, oxidation, spalling resistance and resistance to liquid penetration.

Low Carbon MgO-C Refractory Containing Nano-carbon
Use of nano-carbon in MgO-C refractories by replacing graphite content showed significant improvement in reducing probability of carbon pickup in low and ultra-low carbon steel and it also reduced inclusions in liquid steel. Nano-type carbon has very high reactivity due to high surface area compared to conventional graphite; hence, it was possible to reduce the total carbon content in new type of MgO-C refractory with nano-carbon addition. The total carbon content is reduced to less than half of the quantity compared to conventional carbon. The addition of nano-carbon had been optimized to 0.9–1.0% [5] to achieve high H MOR and good oxidation resistance and total carbon is reduced to 5.5–6.5%, which reduces the probability of carbon pickup in liquid steel. Increase of nano-carbon content increases porosity as excess amount of nanoparticles do not enter into the inter particle voids and remain as free. H MOR values are found to increase from about 2.5 MPa in conventional MgO-C bricks to 4.5 MPa by adding 0.9% nano-carbon. But further increase of nano-carbon addition did not show any increase of strength. It may associate with the increase of oxidation.

Use of Graphene Oxides
Use of graphene nano-sheets with different thickness, partially replacing graphite in Mag-C refractory [4], showed improved thermos-mechanical properties and reduces total carbon content, which results in reduction of carbon solubility in low carbon containing steel.

Corrosion and thermal shock resistance of the low carbon containing MgO-C bricks were reported to improve by reducing the pore size and to controlling the numbers of pores by optimizing the addition of nanoparticles, which are much smaller than conventional MgO-C bricks.

Catalytic formation of in situ nanostructured high temperature ceramic bonding (e.g. SiC and $MgAl_2O_4$) phases in the matrix of low carbon MgO-C refractories fabricated using new generation graphene as the carbon source to partially replace graphite combined with Al, Si, and B_4C as antioxidant additives is developed and used in few steel plants for low carbon steel production. Nano-carbon particles along with the selection of organic binders and antioxidants control the bond structure and resulted in improvement in thermal shock resistance. These refractory materials offer excellent thermal spalling resistance and good corrosion resistance under service at elevated temperatures. These beneficial properties are derived from the graphite attributed to its high thermal conductivity and low thermal expansion coefficient together with non-wetting nature. However, poor oxidation resistance of graphite can oxidize in air to form both CO and CO_2 gases, leading to increase in porosity of the refractory, corrosion, and poor hot strength. Addition of B_4C with Al metallic reduce oxidation at low and intermediate temperature. However, the demand of developing low carbon containing oxide refractories without comprising the beneficial thermomechanical and thermochemical properties is a challenging task which introduces the use of nanotechnology in the refractory fabrication.

Characterization of Low Carbon Containing MgO-C Refractories

1. Use of nano-size carbon had significantly reduced the total carbon addition in magnesia-carbon refractories, which makes it suitable to use in metal zone of steel ladle for producing low-C steel.
2. The fixed carbon obtained from after heat treatment from graphite and resin in conventional MgO-C composition shows glassy character, but the carbon developing from nano-carbon addition has graphitic character. Being graphitic, the resultant mechanical strength and oxidation resistance have been improved.
3. Aluminium metal powder is an effective antioxidant in graphite-containing conventional MgO-C bricks, but is not effective when nano-carbon is added in the composition. The nano-carbon starts getting oxidized before the oxidation of Al metal powder. Boron carbide forms low-melting dense magnesium borate ($Mg_3B_2O_6$) layer that enhances oxidation resistance, filling up the open pores and thus preventing further interaction with oxygen. The $Mg_3B_2O_6$ forms as per the following reactions:

$$B_4C_{(s)} + 6CO_{(g)} = 2B_2O_{3(l)} + 7C_{(s)}$$

$$B_2O3_{(l)} + 3MgO_{(s)} = 3MgO.B_2O_3$$

Effect of nano-C addition in MgO-C refractories [5, 6]

Properties	Unit	Conventional MgO-C	MgO-C with 0.6% Nano-C	MgO-C with 0.9% Nano-C	MgO-C with 1.2% Nano-C	MgO-C with 1.5% Nano-C
Chemical composition						
MgO (0-6) mm	%	89	95.4	95.1	94.8	94.5
Flake graphite	%	10	3	3	3	3
Nano-carbon	%	0	0.6	0.9	1.2	1.5
Al metal powder	%	0.5	0.5	0.5	0.5	0.5
B4C powder	%	0.5	0.5	0.5	0.5	0.5
Liquid resin	%	2.75	2.75	2.75	2.75	2.75
Pitch powder	%	1	1	1	1	1
Total carbon	%	12.0–14.0	5.5–6.0	6.5–7.0	7.0–7.5	7.5–8.5

(continued)

Physical properties						
Porosity	%	4.5	4.1	4	5	5.8
HMOR	Mpa	2.5	3.8	4.5	4	3.6
Oxidation	%	28	22	20	28	29
Thermal shock resistance	Nos of cycles	6	10	12	9	9

11.3.3 Magnesia–Alumina–Graphite Refractories

The development of magnesia-alumina-graphite refractory is due to the increasing demand of ladle life improvement and increasing demand to produce clean steel. The Al_2O_3-MgO-C bricks are based on Al_2O_3-C bricks that have MgO as additive. When used in ladle and during operation, spinel forms to ensure residual thermal expansion. This expansion of the bricks closes the joints of the bricks and reduces metal penetration. The formation of $MgAl_2O_4$ spinel phase is a direct function of Al_2O_3/MgO ratio, the grain size spinel formation time and temperature. The presence of spinel shows the following advantages [6]:

- High melting point, 2135 °C
- High mechanical strength at elevated temperature
- Better slag resistance than fired alumina bricks
- Low thermal expansion coefficient
- High thermal shock resistance

Higher the magnesia content, better the corrosion resistance due to formation of spinel, but higher the thermal expansion, may cause cracking of bricks and reduce durability. Therefore, the optimum content of magnesia and carbon is determined based on the service conditions and the shape of the ladles. Al_2O_3-MgO-C (AMC) becomes a popular alternative working lining refractory compared to MgO-C in the 1980s, as AMC exhibits controlled residual expansion by in situ spinel $MgAl_2O_4$ formation and results in minimization of joints and wear. The performance of AMC bricks depends on the alumina aggregates used. Bauxite-based bricks are not suitable for higher durability as the presence of silica in bauxite reacts with magnesia in the presence of alumina to form low-melting phases, like cordierite ($2MgO.2Al_2O_3.5SiO_2$). In ladle bottom, high-quality AMC bricks made of tabular alumina or brown fused alumina are used. The magnesia-alumina-graphite bricks contain comparatively less carbon than conventional MgO-C bricks; hence those bricks are suitable to produce low carbon steel.

11.3.4 Spinel Refractory

The most important application of magnesium aluminate spinel in the iron and steel industries is the lining of steel ladles as alumina-rich spinel-containing fired alumina bricks and castables in metal zone of steel ladle are the best suitable carbon-free lining to produce clean steel. Steel ladles require refractories that can withstand high-temperature corrosion, mechanical wear and thermal shock, which are the characteristics of magnesium aluminate spinel. As a result, spinel-containing refractory lining provides excellent performance and service life and is widely used in modern steel-making processes. The main advantage of spinel addition is the increased corrosion resistance, and the betterment is due to the dissolution of spinel itself in penetrating slag, thereby reducing further corrosion. Plenty of work is now available on the in situ spinel-forming alumina magnesia castables as an alternative to the addition of preformed spinel with alumina. The in situ formation of spinel while using alumina magnesia castables is associated with the volume expansion of spinel formation that counterbalances the shrinkage of the castables and presents an outstanding set of properties [7].

Buhr has concluded [1] that for the applications of side walls of steel ladles, where slag resistance, thermodynamic stability and lining flexibility (to cope with the stresses) are more important, spinel-forming castables are preferable. But for ladle bottoms, where high heat strength, creep resistance and very high wear resistance at high temperatures are required, preformed spinel-containing alumina spinel castable-based precast shapes are advantageous over spinel-forming fired bricks or castables [8]. Alumina-spinel refractories are successfully used in ladle side wall for both Al- and Si-killed steel grades. Using fine magnesia powder to reinforce alumina structure by spinel formation was found beneficial. Use of fine magnesia led to a homogeneous distribution of spinel in the matrix and contributed to higher strength. Use of coarse magnesia reduced the thermal stress, and fine magnesia (with a MgO/SiO_2 ratio of 4:8) suppressed the linear change in alumina spinel castables when applied for ladle lining.

Types of Fired Spinel Bricks and Their Characteristics for Production of Clean Steel

Fired alumina spinel bricks are used in many steel plants for ladle applications [9] as an alternative material to different systems, namely bauxite-doloma shapes, magnesia-chrome bricks and conventional MgO-C bricks. Silica content in those bricks is very less, $<1.0\%$, to avoid formation of low-melting phases, and they are successfully used in steel ladles for producing Al-killed steel. Corrosion resistance of spinel bricks depends on quality of slag. Spinel is not stable in contact to slag of Al_2O_3 content less than around 10%, at any temperature, and corrosion resistance increases with increase of Al_2O_3 in slag [10].

(continued)

Table 11.1 Tentative properties of different quality spinel bricks

		Spinel aggregates			Spinel bonded		Spinel aggregate and spinel bond		
Bricks type	Unit	Type 1	Type 2	Type 3	Type 1	Type 2	Type 1	Type 2	Type 3
Chemical properties									
Al_2O_3	%	88–90	75–77	65–68	92–95	88–92	92–96	88–92	84–88
MgO	%	8.0–10.0	18–22	28–32	3–4.5	3–4.5	4.0–5.0	8.5–9.5	10.0–12.5
SiO_2	%		<0.1	<0.1	1	2.5	0.8–1.0	0.8–1.0	0.8–1.0
TiO_2	%				0.2	1.4-1.8	<0.1	<0.1	<0.1
Physical properties									
Bulk density	g/cm^3	2.9	2.91	3.06	3.1	2.85	3.15	3.02	3.01
Porosity	%	16–18	18–22	14–16	18–20	20–24	18–20	18–20	18–20
Cold crushing strength	Mpa	80–85	100–116	140–180	60–70	65–75	85–96	82–89	98–110
P.L.C at 1500 °C/5h	%	0	1	1.1	0.6	2	0.1	0.2	0.1
R.U.L ($t_{1.0}$)	°C	>1500	>1400	>1450	>1550	>1450	>1600	>1650	>1600

1. Spinel bricks: These products are made of sintered spinel and fired at temperature above 1600 °C. Those bricks have high resistance to slag of aggressive quality and are used to produce the clean steel with very stringent properties. Cost of those bricks is very high.
2. Spinel-forming alumina bricks: These bricks are based on corundum or bauxite aggregates with MgO containing raw materials and fired at a temperature above 1200 °C. These bricks are widely used in steel ladle bottom and side wall and they have the advantages of gradual expansion due to in situ spinel formation and closes the joints, and hence reduce the liquid penetration.
3. Alumina spinel bricks with spinel-forming bond: These bricks are made of fused or sintered alumina spinel aggregate with in situ spinel formation as bond. These bricks contain higher amount of spinel, which increases corrosion resistance, and they have expansion characteristics, and those bricks are suitable to use in ladle side wall (Table 11.1).

In the secondary steel-making process also, like RH degasser, few steel plants obtained better corrosion and spalling resistance for magnesia spinel bricks than for magnesia-chrome bricks. Increased spalling resistance of the lining was suggested due to the presence of micro-cracks developed for the thermal mismatch of the two components. Many new applications of spinel-containing refractories are coming day by day in the iron and steel industries. Magnesia spinel slide gate plates are also in use for Ca-treated steel due to its higher resistance against corrosion and thermal spalling.

11.3.5 Corrosion Mechanism of Castable Lining in Steel Ladle

In ladle the secondary refining takes place and the metal chemistry is adjusted by adding alloy elements, blowing inert gases and killing the steel by metallic additions. The steel ladle slag is also basic in nature and it is usually Al_2O_3-rich for alumina-killed steel. On the other hand, when silicon-killed steel is produced, the SiO_2 content in slag is increased and the refractory corrosion mechanism is different.

Besides its main role to facilitate the refining processes, presence of slag is very important as it prevents re-oxidation of steel and heat loss. Though the presence of slag is very important for metallurgical treatment of steel, it has the adverse effect on corrosion of refractory. In general, the corrosion mechanism by slag is summarized below:

- Physical infiltration of molten slag into refractory
- The chemical reaction with slag and refractories
- The peeling off the reacted materials into the melt

The use of Al_2O_3 spinel castable in ladle refractory lining has been started in Japan since late 1980s. To produce clean steel and reduction of penetration of liquid steel was the main objectives. Presently alumina-rich spinel castable is widely used in steel ladle for secondary refining operations. The chemical mechanism by which alumina spinel castables limit penetration of slag is fairly well-understood.

- CaO reacts with alumina to form calcium aluminates which forms a protective layer over refractory surface, $CaO + 6Al_2O_3 = CaO.6Al_2O_3$ (CA6)
- FeO and MnO in slag form solid solution with spinel of the refractory castable lining, $FeO + MnO + MgAl_2O_4$ (spinel) $= (Fe,Mn,Mg)O\ .\ (Al,Fe)2O_3$
- The resultant slag is richer in SiO_2 and viscosity increases.

Zhang and Lee [11] stated that the length of slag penetration may be estimated according to the following equation:

$$L2 = \frac{r.Cos\,(\theta).\gamma.t}{2\mu}$$

where $L =$ penetration depth, $r =$ radius of the pores, $\theta =$ contact angle, $\gamma =$ surface tension of liquid, $t =$ interaction time and $\mu =$ viscosity of liquid slag.

According to the equation, higher the size of pores, voids and cracks in the bricks, higher is the depth of penetration, and molten slag viscosity has a direct effect. Lower the viscosity, lower is the penetration. Hence, besides the reduction of pore size and proper pore size distribution, the attempt had been made to look into the alternatives to increase the slag viscosity in operating steel refining temperature.

The slag viscosity is a function of both temperature and chemical composition of slag. The penetration depth can be decreased by applying a refractory material of low thermal conductivity, which is obtained by creating thermal gradient between hot face and cold face of the lining. As the slag penetrates into refractory, the viscosity increases with depth of penetration increases, and after some penetration through the spinel-containing castable, the further penetration in suppressed due to reduction of temperature. The other factor to reduce the viscosity is the change of chemistry of slag during operation. The spinel structure is able to absorb Fe and Mn ions from slag, and as a result, the slag becomes rich in SiO_2 and viscosity increases, which reduces slag penetration. Hence the use of spinel-based bricks or spinel-added castable is beneficial to reduce corrosion and penetration into refractory lining.

Carbon Pickup

Demand for ultra-low carbon steel is escalating in modern steel-making practices. For such steel grades, carbon is effectively removed in primary steel-making operation (i.e. in the EAF or BOF) by oxygen blowing, and in secondary metallurgy, carbon is further reduced to 20 ppm by vacuum treatment. Since the ladle lining should not be the source of fresh carbon pickup, carbon-containing refractory is not recommended in ladle lining. For low carbon pickup in steel or no carbon pickup in steel low carbon containing MgO-C bricks and carbon free fired spinel containing alumina bricks are widely used and the best compromise is to use conventional MgO-C bricks with 8–12% carbon content in steel ladle slag zone, as the best corrosion-resistant refractory.

11.4 Continuous Casting Refractories for Clean Steel

Exposure of liquid steel to air is a gross source of oxygen and nitrogen and such events can typically be measured by nitrogen pickup in the steel. Most operations attempt to purge argon around the metal transfer points (ladle to tundish and tundish to mould) in order to minimize air ingress, but equally important are the sources of oxygen in reducible oxides. The ladle well-block sand could contain chromite with Fe_3O_4 and Cr_2O_3 but the most important oxide is SiO_2, which could be in the sand, the tundish cover and the refractories. The typical oxidation reactions with SiO_2 are the following:

$$2[Al] + 3(SiO_2) = Al_2O_3(incl) + 3[Si]$$

$$[Mg] + 2[Al] + 2\left(SiO^2\right) = MgAl_2O_4(incl) + 4[Si]$$

$$[Ti] + (SiO_2) = TiO_2(incl) + [Si]$$

It is especially the use of rice hulls as a primary tundish cover or secondary cover over a basic tundish flux that could be significant source of oxygen since the reaction between SiO_2 from the rice hulls and CaO-containing ladle slag carryover can result in a liquid tundish slag with a high silica activity. The kinetics of slag-metal reactions is typically much faster than refractory-metal reactions. Slagging off the tundish might be a required practice in some operations to make clean steel. A basic tundish slag with a low silica activity is desirable for the last-minute absorption of inclusions and to protect the steel from the atmosphere.

Despite extensive efforts, such as argon flushing of the new tundish, tundish covers and basic tundish starting cover, the impact pads are designed to minimize turbulence at the nozzle/cover interface, maximize residence time, promote surface directed flow, and distribute the steel homogeneously in the tundish.

11.4.1 Tundish Refractory

In continuous casting of steel, tundish plays as an intermediate vessel between the steel ladle and mould. In addition to lowering non-metallic oxides inclusions, clean steel requires lowering other residual impurities such as sulphur, phosphorus, hydrogen and nitrogen. The tundish has a very important role with regard to steel cleanliness, thermal homogeneity and providing stable operation.

The presence of slag layer over molten steel during processing in tundish has the following advantages from for the production of clean steel:

- Prevents excess heat loss through the insulating slag layer
- Prevents air entrainment and re-oxidation.
- Improves cleanliness in steel by absorbing inclusions.

A common tundish flux is burnt rice hulls, which is inexpensive, a good insulator and provides good coverage without crusting. However, rice hulls are high in silica (around 80% SiO_2) and they also contain carbon to protect from re-oxidation. Excess carbon (>10%) in rise hulls and presence of other impurities are the source of inclusions in low carbon and ultra-low carbon steel. Hence to cast ultra-low carbon steel, the synthetic basic fluxes based on $CaO\text{-}Al_2O_3\text{-}SiO_2$ with $SiO_2 < 10\%$ are effective used in modern steel-making practice, and it also has been correlated with lower oxygen content in the tundish. For example, if the slag basicity increases from

0.83 to 1.2, the oxygen content is decreased from 25–50 ppm to 19–35 ppm. However, high slag basicity expedites rapid melting rate and high crystallization temperature, and results in the formation of crust on the surface and evolution of an "open slag free eye" around the ladle shroud during teeming. Such prevalence not only exposes the surface for re-oxidation but also allows significant radiative heat loss and discomfort for operators around the ladle-operating zone. In order to avoid such incidence, few steel makers have employed two-layer flux concept, in which low-melting-point basic flux absorbs inclusions at the bottom and rice hulls on top layer provide insulation that significantly reduces the oxygen level from 20–24 ppm to less than 18 ppm [11].

Traditional lining for a tundish is the wet spray which has given very good performances in last few decades and still many steel plants are continuing to use this material. However, for increasing demand for further reduction of hydrogen pickup, non-aqueous, resin-bonded dry-vibrating mass (DVM) is used in most of the steel plants worldwide. Reduction of hydrogen can be obtained only when no hydrogen is increased by the moisture in refractory lining of a tundish. The advantages of using DVM over other tundish lining refractories are summarized below:

- Steel cleanliness by a significant reduction of hydrogen pickup, as there is no source of moisture in DVM.
- As the material is free from carbon, no carbon pickup in the steel
- Very low oxygen level in steel can be maintained in steel
- Long casting duration obtained with no inclusions from lining refractory
- Easy to de-skulling
- With an increased wear resistance, DVM allows a much better stabilization of the steel temperature and chemistry.

11.4.2 Tundish Design and Operation for Clean Steel

The type of flow of liquid steel in the tundish plays a significant role in inclusion flotation, interaction with top slag, and in determining the extent of refractory erosion, all of which are important for the manufacture of clean steel. Flow of liquid steel is affected by the following factors:

- Tundish size and shape
- Use of dams and weirs and turbulence stoppers (Turbos-tops)
- Argon purging at selected locations in tundish
- Use of filters

The inlet stream of steel from the ladle creates turbulence, which make difficult of solid inclusion separation and increases refractory erosion, which further increases the oxide inclusion, hence is not desirable. Refractory weirs and bottom striker pads assist reduction of turbulence. Dams direct the flow upward to facilitate in inclusion flotation.

Large Size Tundish Practice
To meet the demand of high-speed casting, the tundish has been made larger in capacity to maintain optimum residence time. This makes it possible to continuous casting operations at a fixed flow rate, even when ladles are exchanged and to reduce inclusions in steel.

Inert Gas Bubbling
It has been confirmed that the quantity of inclusions can also be reduced by argon gas bubbling in the bottom of tundish. It can be seen that increasing casting speed decreases in the residence time of liquid steel in the tundish and may cause problem of high inclusions. Gas bubbling is also considered effective in solving such problem.

Ceramic Filter
As a recent development on production of cleaner steel, the use of ceramic filters in tundishes has been tried for filtering inclusions. However, this process has not yet been commercially established, although considerable development efforts had been made.

References

1. A. Burh, R. Bruckhausen, R. Fahndrich, "The steel industry in Germany – Trends in clean steel technology and refractory engineering", Refractories world-forum 8 (2016) [1]
2. Gunnar Kunz, "Ladle refractories for clean steel production", RHI Bulletin, No 2, 2010, pp 30–40
3. Chunyang LIU, Xu GAO, et al., "Dissolution behavior of Mg from MgO-C refractory in Al-killed molten steel", ISIJ International, Vol 58, No 3, 2018, pp 488–495
4. Satyananda Behera, Ritwik Sarkar, "Nano-carbon containing low carbon magnesia carbon refractory: an overview", Protection of metals and physical chemistry of surfaces, Vol 52, No 3,(2016) pp 467–474
5. Mousom Bag, Sukumar Adak, Ritwik Sarkar, "Study on low carbon containing MgO-C refractory: Use of nano-carbon", Ceramic International 38 (2012) pp 2339–2346
6. Kalyani Ravi, Subir Biswas, Sanat Hazra, "Use of low carbon content Nano-carbon added Magnesia-carbon bricks in steel ladle metal zone at Tata Steel, Jamshedpur", Taikabutsu, vol 71, no 4, 2019
7. R. Sarkar, "Refractory Applications of Magnesium Aluminate Spinel", Refractories Manual 2010, Interceram Refractories, pp 11–14
8. Hoteiya M., et al.: "Use of an alumina magnesia castable for steel ladle side walls. Taikabutsu Overseas 16 (1996) [4] 104
9. Kimiaki S., et al.: "Magnesia spinel brick containing MgO rich spinel for steel refining ladle". Proceedings of UNITECR (1995), vol. 3, 257–64
10. Axel Eschner, Klaus Santowski, Hans Braun," Wear mechanism of alumina-spinel bricks in steel ladles". UNITECR 1995, pp 250–256
11. E.Y.Sako, M.A.L. Braulio, V.C. Pandolfelli," Microstructural road map to attain high corrosion resistant spinel refractory castables for steel ladles

Chapter 12
Advance Material Design and Installation Practices

12.1 Refractory Material Design

Efforts had been made by refractory suppliers to develop and implement new design and technique; few of them are breakthrough innovation in the area of productivity improvement and reduce downtime by system automation. In practice, the biggest challenge is the handling of thermal shock performance and adverse chemical reaction of refractory during metallurgical applications. Around the globe, the refractory material design stringently targeted to enhance the number of heats or throughput, and some of the classic attentions are listed below:

(a) Existence of certain degree of impurities in natural and synthetic raw materials are common incidence; however, ceramic engineer has to select and design the raw materials to obtain the desired properties of refractories for a particular zone of interest. This practice can enhance the RUL and PCE of refractories; however, costly raw materials should not hamper business interest.
(b) Exhaustive crushing, grinding, sieving and milling are common practice to obtain wide range of few millimetres to micron particle that eventually produce different grade of densification; thus, laboratory-scale tap density of mixed particles can predict the resultant density of compact refractories. However, inter-particulate interaction reduces the particles flowability, resulting in a difference between tapped and bulk densities. Thus, compressibility index and Hausner ratio predict the particle compressibility efficiency.
(c) Elastic modulus reduction is an important criterion, where material is subjected to damage by thermal shock. For example, the alumina–zirconia–carbon refractory has less elastic modulus than alumina–carbon system; thus, former is applied in exhaustive thermal shock resistance and crack arresting zone. It also improves corrosion resistance.
(d) Refractory raw materials with large crystals and less grain porosity improves resistance to slag and metal penetration and corrosion. Significant improvement

S. Biswas, D. Sarkar, *Introduction to Refractories for Iron- and Steelmaking*,
https://doi.org/10.1007/978-3-030-43807-4_12

in corrosion resistance in steel ladle was observed by using dead burnt magnesia of crystal size of few mm, compared to sintered magnesia.

(e) Same volume percentage pore content may have different degree of thermal conductivity of two bricks, because of pore size, distribution and connectivity (dendritic or entrapped). Thus, starting raw material and subsequent fabrication protocol can ensure the reliability and consistency of the product.

(f) Cement content in castable has a direct impact on high-temperature strength and other thermal properties. Cement free (no-cement castable) is widely used in most critical area of applications [1]. Gel-bonded castable is introduced to produce large size precast shapes and has additional advantages of faster drying and heating cycle.

(g) Silica bricks of dense quality and the recent development of fused silica-based bricks are used in coke oven for permanent oven lining and repaired zones for their high thermal conductivity, volume stability and thermal shock resistance.

(h) High-alumina refractories are widely used in backup lining of steel ladle and tundishes. Addition of spinel in high-alumina refractory made it feasible to use in steel ladle metal zone hot face lining due to improved corrosion resistance and thermal shock resistance. Al_2O_3-SiC-C bricks have lower thermal expansion and are more resistant to spalling than MgO-C bricks. This is the important reason for their use in torpedo and hot metal ladles.

(i) Tap hole clay performances in large size blast furnace have been significantly improved by addition of SiC and Si_3N_4, which improved corrosion resistance and reduced clay consumption.

(j) Dolomite bricks with MgO addition and resin bond system replacing tar/pitch improves corrosion resistance in steel ladles.

(k) Mag-chrome bricks made of rebonded and semi-rebonded compositions improve both thermal shock resistance and corrosion resistance and the physico-thermal properties are not affected when used under vacuum in RH degassers.

(l) SiC refractory with Si_3N_4 and SIALON bond is widely used in blast furnace stack lining due to its improved high-temperature strength, high thermal conductivity and excellent corrosion resistance with hot metal slag.

(m) Refractory lining design in BF trough is upgraded by using SiC-containing high-alumina PCPF blocks as backup lining. The hot metal throughput has been increased from few thousands to more than 10 lakhs in modern large size blast furnaces.

(n) Zircon brick has a critical application in steel plant for its good wear resistance, poor wettability by molten steel and slag. In service, zircon is dissociated into ZrO_2 and silica at 1540 °C and the presence of SiO_2 densifies the refractory by reducing the porosity to less than 1%, which is used in tundish nozzles and it is also widely used to produce zirconia-mullite refractory. Zirconia-mullite refractory has a wide application in slide gate plate refractory.

(o) Use of nano-carbon in carbon-containing refractory is introduced to reduce the total carbon content in MgO-C and Al_2O_3-C refractory to produce clean steel and to improve oxidation resistance. Due to high reactivity of nano-size carbon it was possible to reduce total carbon content from 12–14% to 6–8%, which significantly improved the high-temperature strength, reduced MOE and as a result it improved thermal shock resistance. This type of refractory is widely used in steel ladle slag zone to produce ultra-low-C steel.

(p) Graphene addition in Al_2O_3-C and MgO-C systems has improved the thermo-mechanical properties. Graphene forms an interlocking structure by reacting with Si metal and other metallic to form SiC whiskers. It showed significant improvement in hot strength, oxidation resistance and corrosion resistance. Due to its large surface area, the reactivity is very high to form high-temperature bond. Carbide formation with metal addition is possible in much lower temperature to protect carbon from oxidation at intermediate temperature.

(q) Use of ceramic filter during steel casting improves cleanliness. The ceramic filters have the capability to remove oxide and nitride inclusions from liquid steel. Contamination of castings by mold flux agents, which have not dissolved, can also be prevented by filtration. The quality of steel is generally improved if using ceramic filters in the casting system. Several types of ceramic filters are in use, fabric filter, honeycomb filter, deep bed filter and ceramic foam filter. Ceramic foam filter is the most common filter used in the foundry industry and it is made of Al_2O_3, SiC or ZrO_2. For using ceramic foam filter, the flow improvement and purification of liquid steel can be optimized by adjusting the pores. Inclusion up to a size of 0.5 μm can be filtered out.

(r) In modern refractory design, micro-porous insulation is used in steel ladle and tundishes as an excellent thermal insulation to protect shell from over-heating. Micro-porous insulating materials have an extremely low thermal conductivity, with a pore size <0.1 μm. This means that the pore spaces are the size of the path length of the gas molecules, which results in heat transfer by convection and radiation. The infrared permeability is further reduced by adding opacifiers. It is observed in many steel plants that the shell temperature is reduced by 60–80%, compared to using conventional ceramic insulation boards.

In recent, special attention has been given on monolithics and chrome-free refractories and their details are summarized as below.

12.1.1 Colloidal Silica-Bonded Cement-Free Castable

Calcium aluminate cement is the most used hydraulic binder in refractory castable composition. The addition of cement in high-alumina castable includes liquid formation at more than 1500 °C, so for high-temperature application cement-free castable has been developed and the other important concerns for the high-alumina cement-containing castables are the curing and dewatering steps, which is to be carefully followed to avoid explosive spalling [2]. The main advantages of using cement free castable, also known as no cement castable, are:

– No liquid phase formation at high temperature

- Faster curing and drying compared to cement-containing LCC or ULCC
- Increase of strength with temperature
- Higher strength above 1450 °C due to presence of mullite and corundum phases
- Faster heating reduces installation or repair times

Colloidal suspension of silica, also called colloidal silica, is the binder usually used to produce cement-free castable. Colloidal silica sols are stable water-based binder, containing up to 50% of nanometric spherical amorphous silica particles of 8–15 μm diameter.

When colloidal silica sol combines with other solid micronized particles, it can be linked together and gelation takes place by water removal. During the drying stage, the hydroxyl group (Si–OH) on the surface of the particle generates siloxane (Si–O–Si) bonds, which help to develop strength [2, 3].

Gelation can also be influenced by pH changes by addition of water soluble organic solvent, named gelling agent [1]. With the proper selection of gelling agent and maintaining pH value, the colloid is gelled around the solid aggregates and develops strength. The resulting structure is permeable enough so that the composition can be easily dried in faster rate without showing cracking or explosive spalling.

Characterization of Gel-Bonded No-Cement Castable

At low temperatures, the un-dries cement containing castable are very dense and have very low gas permeability. During the fast drying and heat up, initially the physically bonded water evaporates at a temperature of 110–120 °C and subsequently the chemical-bonded water in the hydrated phases is lost at 350–600 °C [2]. Since a variety of hydrated phases are present and the de-hydration characteristics of those phases are different, the entire de-hydration process takes within a wide temperature range and continues up to 600 °C. The entire drying and heating process must be very slow to allow gradual moisture removal; otherwise, the concrete would spall due to high steam pressure. In comparison to cement-bonded castable, the gel-bonded castable has a different bond system and casted concrete made of gel-bonded castable has high gas permeability and can be dried and heated in faster rate. The other advantages of using gel bonded castable are summarized below:

- The setting mechanism is based on condensation reaction which involves release of small amount of water and formation of Si–O–Si bonds. Like cement-bonded castable no hydrated phases are formed.
- There is no loss of bonding strength during the de-hydration process and formation of more Si–O–Si bond increases the strength with temperature increase. High permeability of casted concrete of gel-bonded castable facilitates de-hydration.
- Reduced impact of ambient temperature on the properties of final products.

(continued)

- Easy adjusted to the setting time.
- Higher refractoriness compared to cement-bonded castable due to no melting CaO containing phases are forming.
- Improved resistance to corrosion and erosion with hot metal and slag.
- Longer shelf life compared to low-cement and ultra-low-cement castables.
- Carbon-containing gel-bended castable shows higher refractoriness under load by 200–400 °C, compared to cement-bonded castable of same composition, despite the presence of significant amount of SiO_2, which is introduced via liquid binders.

During sintering at high temperature, its high surface area may induce the formation of mullite and other high-melting alumina-containing phases, which improves corrosion resistance, thermal shock resistance and hot strength. The no-cement gel-bonded castable is successfully used to produce precast and pre-fired shapes, and those shapes are needed to be pre-fired at high temperature, above 1450 °C, so that ceramic bond, mullite would be formed to provide high-temperature strength.

12.1.2 Alumina-Oxi-Carbide-Bonded Castable

Carbon-containing high-alumina castable is widely used in modern steel plants. Few critical applications of those castables are in blast furnace trough lining and snorkel of CAS-OB. One of the major disadvantages of carbon-containing refractories is the oxidation at operating conditions. In MgO-C or Al_2O_3-C refractories bricks with non-aqueous, resin- or tar-bonded bricks, metal powders, such as Al-metal and Si-metal, are used as antioxidant, which protect carbon from oxidation by forming carbides (Al_4C_3 or SiC) at high temperature. But in carbon-containing hydraulic cement-bonded castables, such as low-cement and ultra-low-cement castable, addition of metallic antioxidant may not be feasible due to oxidation of carbide phases and formation of Al_2O_3 and SiO_2 and also cause for explosive spalling in the presence of hydraulic bonds.

An idea behind the development of the new type of aluminium-oxi-carbide bond was to replace the traditional calcium-aluminate cement in carbon-containing castables. All oxi-carbide-bonded castables are completely cement free and due to the absence of Ca-hydrate phases there is no chemically bonded water in the cured products. Therefore, a safe rapid heating is possible without and explosive spalling. The other advantages of using oxi-carbide-bonded castable include:

- Excellent thermal shock resistance
- High corrosion resistance towards acidic and basic slag
- Hot erosion and corrosion resistance
- Increase of strength with increase of temperature.

As an aluminium-oxi-carbide the two types are formed, Al_2OC and Al_4O_4C, and they are produced synthetically. The Al_4O_4C is stable at very high temperature and excellent in oxidation resistance, corrosion resistance and thermal shock resistance. Especially, Al_4O_4C is used in carbon-containing hydraulic-bonded cement-free castable, which acts as an excellent anti-oxidant. The oxi-carbide is formed by heat treatment of fine powder of alumina and graphite in an argon atmosphere. Along with Al_4O_4C formation, other compounds like Al_4C_3 and Al_2OC are also formed and Al_4O_4C is the most stable phase at high temperature and becomes inert in presence of hydraulic bond. To maximize this stable phase, the $C/Al_2O_3 = 1.5$ is to be maintained without presence of Al_2OC or Al_4C_3. Formation of Al_4O_4C from carbon and high-alumina raw material, with a particle diameter of <100 μm, is progressed according to the chemical reactions:

$$2Al_2O_3(s) + 3C(s) = Al_4O_4C(s) + CO(g)$$
$$2Al_2O_3(s) + CO(g) = Al_4O_4C(s) + CO_2(g)$$
$$CO(g) + C(s) = 2CO(g)$$

Aluminium oxi-carbide is produced either by sintering at more than 1700 °C or a melting process in an arc furnace by melting carbon and alumina raw materials. The formation of Al_4C_3 must be as minimum as possible, because it reacts with bond water of castable to form methane gas.

Corrosion Resistance

The oxi-carbide-containing castable showed very high resistance to hot metal and slag compared to conventional LCC and ULCC. The other observations are summarized below:

- In oxidation, a very thin decarburized zone of a few millimetres was formed below the surface.
- Wettability by slag, hot metal and steel was significantly reduced due to the presence of carbon, compared to conventional LCC and ULCC

Application of Oxi-Carbide-Bonded Castable

The new bond system had been successfully implemented in lining the snorkel of CAS-OB and is in regular use worldwide. This bond can be used in carbon-free no-cement-castables (NCC), and the precast shapes are produced, such as well blocks and seating blocks for steel ladles.

CAS-OB snorkel, also called bell, is lined with castable in both inner and outer wall. The two types of castables are used, one in lining the portion, dipped into slag and liquid steel, and the other is in rest of the portion, above slag line. Since only the lower part is dipped into the steel bath during the CAS-OB process, this part is the most stressed by extremely high temperature, thermal shock and corrosion. This castable lining in lower portion is experiencing severe thermal shock due to thermal cycling between each heat, which limits the life of the bell. The condition of the after service bell due to thermal shock and high corrosion is shown in Fig. 12.1. To

Fig. 12.1 After service condition of the CAS-OB snorkel

address the thermal stress, the bell is lined with oxi-carbode-bonded castable and significant improvement of the lining condition with life improvement was observed.

12.1.3 Magnesia-Containing Castable

Basic castables made of magnesia composition are usually limited because of the expansion followed by hydration of MgO. This expansion is related to the difference in density between magnesium oxide (MgO) of 3.5 g/cm^3 and its hydroxide, Mg (OH)2 of 2.4 g/cm^3, which leads to the rupture of castable shapes. Use of MgO in hydraulic-bonded castable is not widespread due to the fundamental problems. The disadvantages of using MgO as a refractory aggregate in castable are summarized below:

- Excessive expansion due to hydration, when used with cement-containing hydraulic bond.
- Very fast setting
- High thermal expansion of MgO at steel-making temperature
- Low strength at high temperature.

Basic refractory has high demand to protect basic slag in steel-making processes and hence lot of efforts have been made to develop magnesia anti-hydration techniques. Magnesia is one of the most refractory oxide known, with a melting point at more than 2900 °C. However, at a temperature 827 °C lower than the melting point, an intense vaporization of the oxide is observed; hence, its use is up to the maximum temperature of 1927 °C, still high above most of the refractory oxides used in steel plant applications.

The development of additives and binders, which can provide both hydration resistance and adequate physico-chemical properties at high temperature of a castable, has been the most important topic of the magnesia-containing castables.

Binders should provide suitable strength after curing as well as after firing at high temperatures. The binders studied for magnesia castable were either hydraulic cement, sodium silicates or phosphates. However, due to the chemical nature of magnesia, its reactions with some of the components of this systems such as silica, phosphates and the combination of lime and alumina lead to decrease of refractoriness and high volumetric change after firing at a temperature lower than steel-making process, which made them unsuitable to use as basic castable with adequate flow properties. The corrosion resistance with basic slag is also negatively affected, when compared to magnesia-containing shaped refractory. However, use of micro-silica was found a suitable bond for basic castables.

Silica Bond to Magnesia-Based Castable
A proposed binder is micro-silica and fine magnesia that can react with water for casting MgO-based self-flow castable. This gives high strength at elevated temperature. The use of micro-silica prevents hydration, resulting in preventing castables to crack and disintegrate.

The characteristics of using micro-silica as a useful bond is summarized below:

- Setting mechanism is reaction of MgO fines with micro-silica, no calcium aluminate cement is required.
- The setting reaction is the formation of magnesia–silica–hydrate phase.
- Improved flowability is achieved by adding around 4% micro-silica, together with appropriate dispersing additives, in the cement-free castable.
- Above around 1000 °C, forsterite ($2MgO.SiO_2$) formed, which has a melting point of 1890 °C, eutectic temperature of the system, $2MgO.SiO_2$–MgO is 1860 °C, which is suitable to use in steel-making process.
- Addition of alumina substituting micro-silica usually lowers the flowability, fast setting and volume expansion due to spinel formation. This can counteract the shrinkage from forsterite formation when used in the presence of silica.
- The combination of SiO_2 with Al_2O_3 and MgO forms low-melting phases at high temperature. However, use of pre-formed spinel in silica-bonded MgO containing castable may improve corrosion resistance, thermal shock resistance as well as improve hydration resistance

12.1.4 Chrome-Free Refractory

Cr_2O_3 is the most corrosion-resistant refractory oxide; its addition to the other refractory oxides such as addition in MgO-containing refractory produce magnesia-chrome, addition in Al_2O_3 refractory produce Al_2O_3–chrome refractory and Cr_2O_3 is also added in wide range of castables. Chrome-containing refractory has a wide application in glass industry and cement and steel plants. It is also extensively used in gasifiers and waste melting furnaces. However, the formation of toxic, carcinogenic and water-soluble hexavalent chromium [Cr(VI)] compounds restricts the application of Cr_2O_3(III) in refractory composition. Related to the execution of ISO 14001, the use of magnesia–chrome or aluminium–chrome bricks, which develop hexavalent chromium during use, has been restricted. Hence, it is felt to develop Cr_2O_3 free and free of hexavalent chrome refractory.

In nature, chromium exists as chromite, consisting of complex spinel (Fe^{+2}, Mg^{+2}). (Cr^{+3}, $Fe^{+3}Al^{+3}$)$_2$. O_4, in which chromium is present as Cr(III) and it is not toxic as well as not soluble in water. However, as mentioned earlier, Cr(VI) is toxic and highly soluble in water, and can easily contaminate soil and ground water. The safe disposal of Cr(VI)-containing waste refractory is a great concern. Here, development of new products without addition of Cr_2O_3 or chromite is described.

New systems were developed, such as magnesia-zirconia, magnesia-spinel-zirconia, magnesia calcia-zirconia, magnesia-hercynite and magnesia galaxite. Great efforts are concentrated in $MgOAl_2O_3$-ZrO_2 and CaO-MgO-ZrO_2 systems. Despite its hydration problem, they have good corrosion resistance. Magnesia-spinel-zirconia is quite a recent new development and its results have been promising.

One of the methods for obtaining magnesia spinel brick composition is the addition of alumina and titania (TiO_2) to a magnesia composition [4] and spinel is formed during firing of bricks by an in situ spinel formation process. However, spinel is unstable under the corrosive effect of few steel-making slag and bricks are corroded. To overcome this problem, by latest development, the use of TiO_2-containing magnesia refractories has been undertaken. It is observed that magnesium-titanate phases develop a strong bond between in situ magnesia-spinel as well as magnesia aggregates and this also causes the increase in strength and corrosion resistance. The solid solutions are formed according to the following equations:

$$2MgO + TiO_2 = 2MgO.TiO_2(ss)$$

$$MgO + Al_2O_3 = MgO.Al_2O_3\ (spinel)$$

$$2MgO.TiO_2 + MgO.Al_2O_3 = [2MgO.TiO_2 - MgO.Al_2O_3]\ (ss)$$

The refractories made of MgO-TiO_2-Al_2O_3 system would show satisfactory performance in lining RH degasser lower vessel [5].

Chrome-Free Castable

Chrome is added in high-alumina castables to improve thermal shock resistance and corrosion resistance and Cr(VI) is developed during use of those chrome-containing castables. It is known that Cr_2O_3 oxidizes in the presence of CaO, and at around 800 °C, it forms $CaCrO_4$, where Cr(VI) is generated according to the following reaction:

$$4CaO + 2Cr_2O_3 + 3O_2 = 4CaCrO_4$$

High-alumina cement is used as hydraulic bond and in the presence of different calcium–aluminate phases, Cr_2O_3 added in castable forms chrome–hauyne compound, such as

$Ca_4[(Al, Cr^{+3})_6O_{12}]$ $(Cr^{+6}.O_4)$, and co-exists with $CaCrO_4$ and other calcium–aluminate phases. Hence, castables that contain Cr_2O_3 when comes in contact to oxidizing atmosphere or with oxidizing slag, Cr(VI)-containing phase, $Ca_4Al_6CrO_{16}$ is formed and is soluble in water, which restricts the use of chrome-containing castables. The formation of $Ca_4Al_6CrO_{16}$ depends on the ratio of CaO/Al_2O_3; this indicates that the presence of different types of hydraulic phases in castable influences the formation of Cr(VI). The formation of $Ca_4Al_6CrO_{16}$ phase in 50%, 60% 70% and 80% Al_2O_3 composition with around 4% Cr_2O_3 are shown in Table 12.1 [6].

During heat treatment, Cr(VI) compound like $CaCrO_4$ formed within 500–1100 °C and hauyne ($Ca_4Al_6CrO_{16}$) formed at 900–1300 °C. In case of low-cement castable (LCC), a solid-state oxidation of chromium in the order Cr (III) to Cr(VI) to Cr(III) follows from room temperature to more than 1500 °C.

From the above analysis according to CaO-Al_2O_3 and CaO-Al_2O_3-Cr_2O_3 system, it is observed that the water leachable Cr(VI) is the main secondary phase in composition of 50%, 60% and 70% Al_2O_3 and very small amount is present in 80% Al_2O_3 composition. Hence, Cr_2O_3-added castable is not suitable to use due to Cr(VI) due to health hazards, and it is required to develop Cr_2O_3-free castable. To apply Al_2O_3–CaO–Cr_2O_3 castable system safely without generation of Cr(VI) it is necessary to reduce the cement content and mitigate the Cr(VI) below the standard limit of 5 mg/L leachable Cr(VI) according to ISO 140001 and Environment Protection Agency (EPA), USA. Several studies [6] confirmed that Cr (VI) converted to Cr(III), once it formed by using clay minerals, glass ceramics, silica and phosphates. Silica has found most effective source in inhibition of Cr (VI) and so low-cement castable (LCC) and ultra-low-cement castable (ULCC) are doped with proper silica source in such a way that there is minimum generation of Cr (VI) and at the same time castable properties remain unchanged. Addition of micronized silica by substituting high-alumina cement with a CaO/SiO_2 ratio to around 1.4 shows promising results [6] which promotes formation of Cr(III) at 1500 °C and almost eliminates the formation of $CaCrO_4$ and $Ca_4Al_6CrO_{16}$ phases within the temperature range from 900 to 1300 °C.

Use of Cr_2O_3-Containing Refractories: Injurious to Health

Chrome-containing refractories such as magnesia–chrome and alumina–chrome has a wide application in steel plants worldwide due to their superior corrosion resistance and thermal shock resistance. Since long time, those refractories have been used mainly in steel ladle, RH degasser and other areas particularly in extremely harsh operating conditions. However, the formation of toxic, carcinogenic and water-soluble hexavalent chromium, Cr

(continued)

Table 12.1 Formation of Cr(VI) with different CaO/Al_2O_3 ratio

Compositions	Unit	Al_2O_3	CaO	Cr_2O_3	C_3A	$C_{12}A_7$	CA	CA_2	CA_6	$Ca_4Al_6CrO_{16}$
50% Al_2O_3	%	50	46	4	10.96	59.53				29.51
60% Al_2O_3	%	60	36	4		0.56	62.92			30.52
70% Al_2O_3	%	70	26	4				73.39		26.61
80% Al_2O_3	%	80	16	4				58.17	41.42	0.41

(VI) compounds restrict the application of Cr_2O_3 in refractories. Due to water solubility of Cr(VI) in magnesia–chrome or alumina–chrome after use bricks, disposal is becoming very critical and a matter of great concern.

Use of Cr_2O_3 in high-alumina castable is also restricted due to formation of Cr(VI)-containing compounds at high temperature. The formation of Cr (VI) containing in Al_2O_3-CaO-Cr_2O_3 castables is explained below:

- Cr_2O_3 used in refractories as chromite, which consists of a complex spinel (Fe^{+2}Mg) (Cr, Al, Fe^{+3})$2O_4$, in which chromium is in Cr(III) form and is not toxic and hardly water soluble.
- In the presence of CaO, Cr_2O_3 oxidizes at a temperature of 800 °C to $CaCrO_4$, where Cr(III) is oxidized to Cr(VI).
- Most of the high-alumina castables contain cement and in the presence of Al_2O_3 and CaO with Cr_2O_3 oxidizes at a temperature of more than 900 °C to form $Ca_4Al_6CrO_{16}$, where chromium is hexavalent, Cr(VI). Hence in Cr_2O_3 or chromite-added castable, both hexavalent compounds, $CaCrO_4$ and $Ca_4Al_6CrO_{16}$ co-exist within a temperature range of 900–1300 °C; use of Cr_2O_3 in high-alumina castable containing cement is restricted.

Use of Alternate Refractory—Today's Need

- Use of TiO_2-added spinel-containing magnesia bricks in place of magnesia–chrome bricks shows encouraging performances.
- Castable containing very low cement and addition of silica and other additives found suitable to use as a Cr_2O_3 free refractory.

12.2 Best Installation Practices

12.2.1 Endless Lining in Steel Ladle

Modern steel mills are nowadays under huge pressure to achieve highest quality products while minimizing time and respecting demanding budgets. Optimization is the leading factor, both in time and in space—especially for steel ladle bottoms.

Endless lining is a systematized lining method developed from casting monolithic refractory in steel ladle, blast furnace troughs and tundishes, first introduced in Japan in 1974. The monolithic castable would be preheated within short time. The advantages of the endless lining are:

- *Smooth, joint-free lining*: Joint-free refractories substantially improve the safety during the steel ladle operation as they decrease the risk of leakage and infiltration into the lining through joints (in the case of brick lining).
- *"Endless lining"*: In contrast to traditional bricks which need to be completely exchanged at the end of the ladle campaign, an "endless" refractory layer can

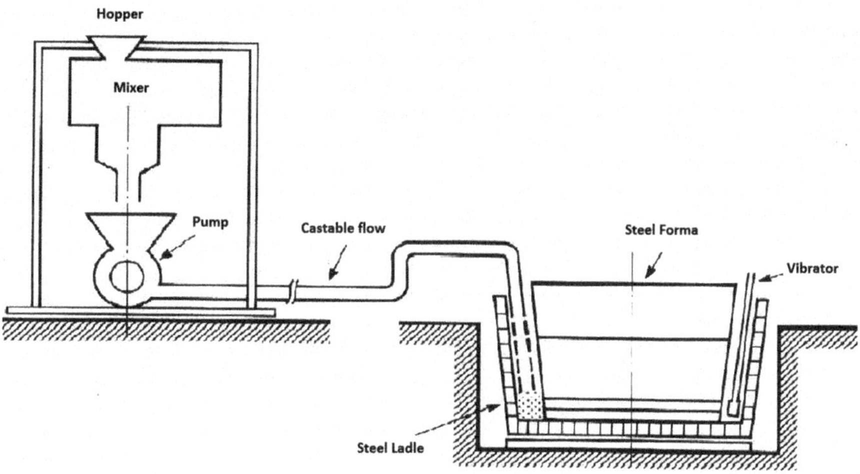

Fig. 12.2 Arrangement of castable casting of steel ladle

simply be replaced when needed. These smaller maintenance repairs only substitute which is necessary and can therefore result in substantial cost savings.

- *Potential increase of ladle capacity*: When the walls of a ladle are installed with self-flow castable instead of bricks, a thinner refractory layer is the result, thus increasing the inner volume of the ladle. And cast after cast, day after day, the gained space means increased tonnage, lowered costs.
- *Improved steel production and product quality*: With a thinner refractory layer, steel preserves the temperature better, contains less non-metallic inclusions and avoids carbon pickup.
- *Wide range of applications*: From the safety lining of all steel mill vessels (ladles, tundish…) to their working lining, including covers, sealings elements, runners, etc. can be achieved with the help of self-flow castable.
- *Chemical and physical strength*: High resistance to steel and slag contact, compatibility with all the used covering powders, abrasion resistance, high mechanical strength, all coupled together with an enhanced ease of installation
- *Easy repair maintenance*: In addition to the overall quick lining process, it can restore partly damaged lining to their original lining thickness without major replacement of old lining. The lining life can be extended indefinitely by repeating this quick repair processes. There have been cases, when the ladle life is extended to more than 1000 heats before complete relining of steel ladles.

The process of lining castable in steel ladle metal zone is shown in Fig. 12.2 and restoring the original lining thickness during intermediate repair is shown in Fig. 12.3. Reduction in refractory consumption is substantial up to 6–10 campaign.

Regarding the quality of castable to be used in endless casting of ladle, the quality had been changed from alumino-silicate to zircon, when the endless casting was introduced. However, with the increased demand of productivity and to produce

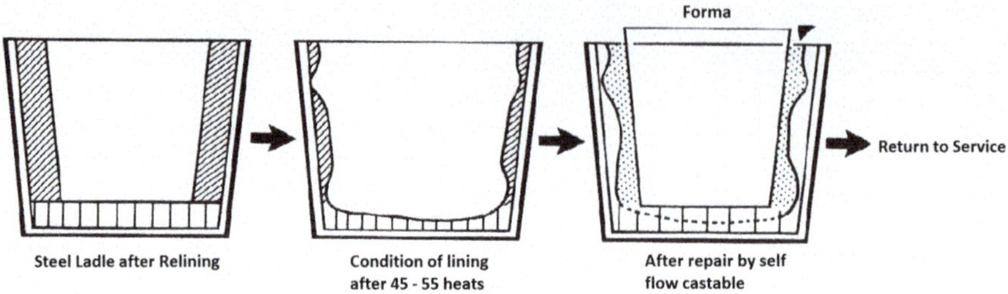

Fig. 12.3 Typical repair methodology for endless casting

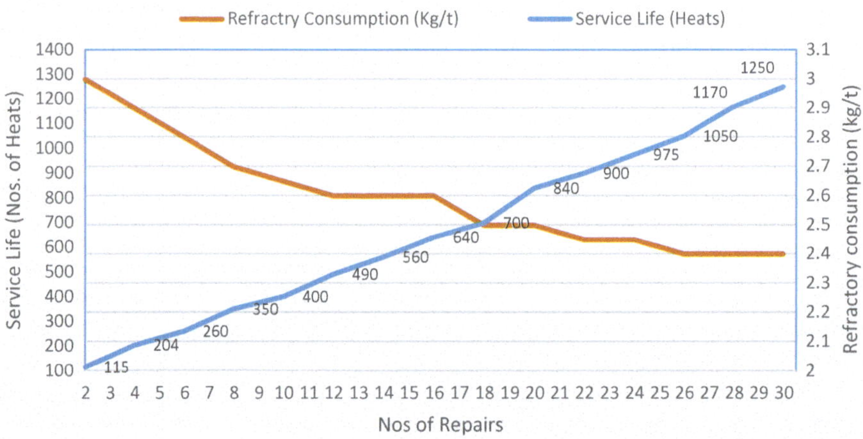

Fig. 12.4 Performance results of ladle life improvement

clean steel, the high-quality silica-free castable was enhanced during the last few decades, high-alumina-based silica-free castable with spinel addition was used in metal zone and bottom of steel ladle due to high durability, and simultaneously it contributed to produce quality steel. The present practice is to use dense quality precast shapes in ladle bottom, spinel-containing castable in side wall of metal zone and MgO-C bricks in slag zone.

After obtaining the straight life of 45–55 heats the slag zone is changed by new bricks, bottom and metal zone side wall are repaired by casting or by shot-creting and another 45–55 life is obtained and then the repair activity is being repeated and thus the benefit of high life of more than 1000 heats are obtained before undertaking the next complete relining. A typical ladle life performance trend in the process of endless lining is shown in Fig. 12.4. The consumption is high up to around 350–400 heats (10th repair) and then the consumption is reduced from 2.7 to 2.4 kg/t of crude steel production with the increase of life to more than 1000 heats.

12.2.2 *Micro-wave Heating of Castables*

The refractory castables are hydrated materials that require special precaution during dry out physical water and chemically bonded water. Depending on the heating schedule applied, steam that is generated may become pressurized due to faster heating, occasionally causing cracking or even explosive disintegration of the product. Hence, processing of castable refractory is generally long time taking and high energy consuming including curing for around 24 h and complete drying and heating for another 2–3 days with gas or electrical energies, which resulted in low productivity as well as high energy consumption.

An environment friendly process innovation for shortening the dry-out process time of castable, a microwave heating system has been introduced. Dry-out of ladle lining castable and processing of precast shapes would be faster by microwave heating process. In this heating system, the refractories are warm-up quite efficiently with very short dry-out and heating process, compared to conventional processes. In this system, heat is generated from inside of the material due to rotation of di-pole of water molecules without depending on the thermal conditions of the casted products.

Ceramics such as refractories are conventionally heated by the combustion of fuel or electric resistance heating. In this conventional heating process, the refractory material is heated from outside by radiation and convection. Temperature gradient in the material causes stress generation with the increase of heating rate. High heat-up rate causes damage of the material due to crack formation; hence the rate of heating must be slow enough to control the internal stress formation. Development of microwave heating solved this problem by introducing fast and uniform heating by heating from inside of the material. Most important advantage is uniform heating of the material which minimizes the development of internal stresses and eliminates the causes of cracking of cast material during fast heating. In Nippon Steel Corporation, Japan, the microwave heating system has been used for heat-up of monolithic lined steel ladle and snorkels of RH degasser [7]. Microwave heating is also introduced in many refractory manufacturing plants to produce precast shapes, such as well blocks and seating blocks.

Microwave Generation and Heating Principle

Microwave

A generic term of microwave is electromagnetic wave and it has the frequency between 300 MHz and 300 GHz, as shown in Fig. 12.5 [7].

Microwave heating

Microwave can heat dielectrics such as water, food and wood and magnetic substances such as ferrites. Castable with hydraulic bond can be dried and heated. Some ceramics can be heated to the higher temperature of more than 1000 °C by irradiating microwave. The mechanism of drying and preheating of hydraulic bonded castable is the water molecules with di-pole interact with

(continued)

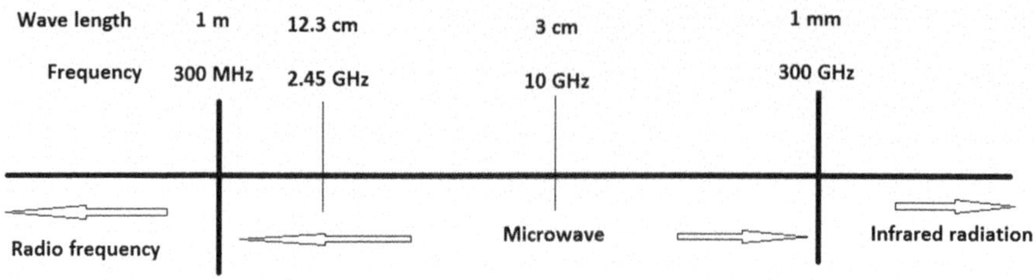

Fig. 12.5 Microwave frequency

microwave, and it starts rotating following the change of electro-magnetic field. The rotating water molecules produce friction with the molecules of the circumference, and the friction generates heat. The advantages of microwave heating of castable refractories are summarized below:

- Rapid heating reduces production time, and thus increases productivity
- Uniform heating of the product
- Heating from inside of the material
- No cracking due to faster heating rate during curing and dry-out
- Significant energy saving
- Environment friendly heating system due to no emission of gases

Microwave Drying System in Steel Ladles
It is reported by Hatsuo et al. [7] that in Nippon Steel Corporation microwave heating system was developed in drying 100t ladles in 1979. Then the system had introduced for drying 250t ladles in 1985. Kimitsu works had adopted this system for 300t ladles in 1998, and for 250t ladles in 1999. A schematic diagram of the heating arrangement for microwave heating in steel ladle is shown in Fig. 12.6.

Uniformity in Microwave Heating
Microwave is generated by a microwave generator and introduced through a waveguide to the application area such as castable lining in steel ladle, monolithic lining in RH degasser and precast shapes inside the micro-oven or furnaces. Most important factor for efficient microwave heating is the uniformity of microwave irradiation. It depends on arrangements of waveguide, mechanical stirrers and reflectors.

Function of Hot Air
In conventional drying and heating system hot air is applied as a source of heat, but in microwave drying and heating hot air is not a source of heating and it has two important functions, it prevents the condensation of moisture over the surface of casted material (applicators) and exhaust moisture from the applicators.

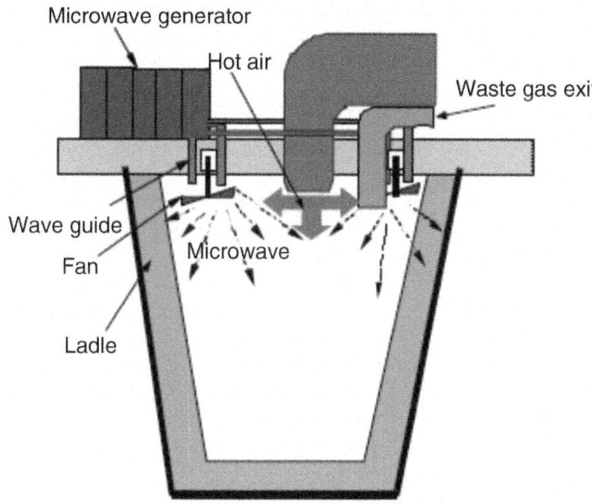

Fig. 12.6 A schematic diagram on arrangement of microwave heating in steel ladle

12.2.3 Flame Gunning and Ceramic Welding

It has been about more than 50 years since the flame gunning application started for furnace repair. Basic gunning material such as MgO-based material can be used in flame gunning to repair LD vessel to extend the campaign. This technique can also be used in other equipment in steel-making processes. In flame gunning, a mixer of dead burnt natural magnesite or sea water magnesia powder and coke powder is used as the gunning mass. The gunning mass is conveyed by N_2 gas, and is discharged together with oxygen into the hot furnace from the lance tip introduced into the furnace. Coke discharged through the nozzle is burnt in the presence of oxygen to produce very high temperature which melt partially the gunned powders. The refractory powder containing impurities are fused and adhere to the furnace wall. For a satisfactory flame gunning operation, the equipment factor as well as the material characteristics, coke ratio and grain size distribution play very important role. For a dense flame gunning in LD vessel the following parameters are to be maintained [8].

- Furnace wall temperature before gunning operation to be more than 1200 °C
- Discharge quantity of gunning material is around 220 kg/min,
- Coke ratio in gunning mass to maintain around 35%
- Optimum gunning distance is 2.1–2.5 m

A schematic diagram of flame gunning arrangement is shown in Fig. 12.7 [8].

The Advantages of Flame Gunning Over Conventional Wet-Gunning

- As no water containing bond is used, the thermal shock due to lowering furnace temperature at the interface of gunned surface and refractory is eliminated.
- Short gunning time
- No need for curing the gunned material before operation starts

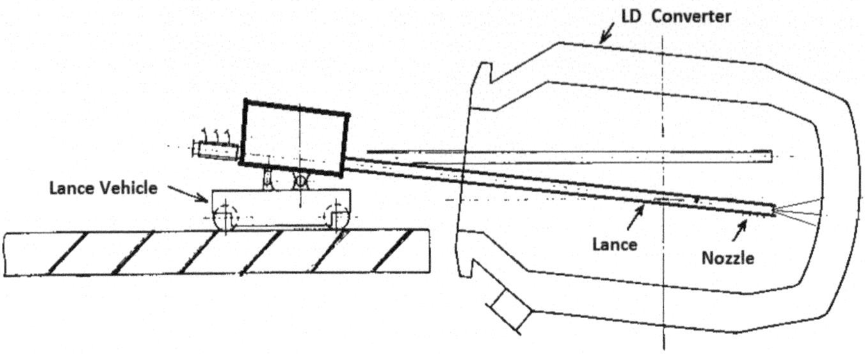

Fig. 12.7 Schematic illustration of flame gunning operation

- High corrosion resistance of the gunned layer
- Low heat source cost
- Excellent adherence with furnace refractory.

Characteristics of Flame Gunning Materials

1. The flame gunning material must have good flowability to convey by N_2 gas from hopper to the burner, where it is mixed with oxygen. The grain size is to be maintained above 13 μm and maximum grain size is about 210 μm, based on the theoretical fact that the grains to melt completely in about few fractions of a second. Hence, the grain size is controlled from between an upper limit of 210 μm for proper melting, to above 13 μm for adequate fluidity.
2. The melting characteristics of the powders relate closely to the density and strength of the gunning materials and the melting rate of lower limit is around 70%; if the melting rate is over 70%, the gunned material will have good strength and density [9]. Hence, if the gunned material is melted more than 70% in the flame, it will stick to the target well and must have good strength and density.
3. The gunned material is not durable for long time under rapid heating and cooling, because with its high density, the thermal shock resistance is generally poor. One method is tried in recent material development to use course grain to improve thermal shock resistance [9]. The coarse grain of 1–3 mm is added to the fine grain of less than 210 μm, which completely melts. The maximum quantity of coarse grain to be added is 30 wt% so that 70 wt% of fine grain (<210 μm) would completely melt to develop strength and good adherence with the furnace wall.

Ceramic Welding

Ceramic welding is a similar process like flame gunning. In both the processes gunning or injection of refractory material is performed in hot condition to develop new refractory lining on the affected target area. Ceramic welding uses a mixer of refractory raw materials fines, such as fused grain of silica/alumina and metallic particles as solid fuel, like aluminium metal powder, which are conveyed in an oxygen or oxygen/air mixer through a system of hoses and specialized cooled pipes to the repair area.

(Before) (After)

Fig. 12.8 Ceramic welding in coke oven (Tata Steel Ltd, India)

The heat available from the application area initiates an exothermic reaction between the metal powder and oxygen to attend temperature more than 2000 °C. Ceramic weld powder melts and forms a ceramic bond with the original base refractories. Upon impact, material fuses to brickwork and cools forming a restoring brick to their original state. Ceramic weld results from fusion of chemically and physically matched materials. Deposition layer can range from 2 cm to >30 cm to restore damaged areas or provide a protective layer.

Ceramic welding repairs are always carried out in hot ovens. Heat is retained by fibre door panels. Areas to be repaired are thoroughly cleaned using specially designed heat-resistant cleaning tools before ceramic welding hot repair process is applied. The weld mass fuses to the base refractory and is built up to reconstruct the refractory and re-establish stability. Gas tight welding repairs reduce energy loss and assist in environmental issues. Any area of the oven can be welded.

Pressurizing flues allow leaks to be pinpointed. Once detected, open joints and cracks are cleaned and welded, sealing hard-to-see voids which may generate leaks. Stack emissions are eliminated and the refractory life is extended. Repaired jambs are restored with a high-density solid weld mass that is fused to the base refractory.

The ceramic welding process is extensively used in coke oven repair. An example of ceramic welding process in coke oven is shown in Fig. 12.8, where the condition of refractory lining before and after welding process is shown.

Features of Ceramic Welding Process

1. Ceramic welding to be done on hot surface of base refractory, temperature to maintain more than 1200 °C

(continued)

2. In flame gunning process, coke is used as fuel, but metal powder is used in ceramic welding.
3. Silica- and alumina-based refractory lining can be repaired by ceramic welding. The Al_2O_3/SiO_2 ratio in refractory powder can be adjusted according to the composition and properties of base refractory lining. A new refractory layer of 30–300 mm can be developed over the base refractory lining.
4. Excellent thermal shock resistance of the new welded layers of refractory.
5. During welding process, highly reactive metal powder reacts with oxygen in available high temperature from base refractory to generate heat through exothermic reaction, which is utilized to melt the refractory oxide powders and form a permanent bond with base refractory lining materials.
6. It is a fast repair process with environment friendly operation.
7. Eliminates heat-up time and risks.
8. Cost-effective long-life high-quality repair.

References

1. Milos Blajs, Rene von der Heyda, et al., "COMPAC SOL – The new generation of easy, safe, and fast heat up No cement castables", RHI Bulletin, No 1, 2010, pp 13–17.
2. Milos Blajs, Bernhard Pribil, Garald Reif, "Sol bonded Alumina silica monolithics – New developments, recent experience, and future possibilities", UNITECR 17, pp 202–204.
3. Jurgen Schutz, Alexander Maranitsch, Milos Blajs, "New Oxy-carbide refractory products demonstrate outstanding properties – First practical results", RHI Bulletin, No 1, 2012, pp 16–19.
4. S. Ghanbarnezhad, A. Nemati et al. "New development of Spinel bonded chrome free basic brick", Journal of Chemical engineering and Material Science, Vol 4(1) January 2013, pp 7–12.
5. SEyoung Um, Sangwon Lee, Kangyone Lee, "Fired Chrome- Free basic bricks development for 2nd Refining furnace", UNITECR 2007.
6. Mithun Nath, Tengteng Xu et al. "Utilization of Al2O3-CaO-Cr2O3 refractory castable system without Cr(VI) generation", UNITECR 2019.
7. Hatsue Taira, Yoshitoshi Saito et al "Microwave drying of monolithic refractories in Nippon Steel Corporation", 21B-19, UNITECR 2003.
8. Tadashi Morimoto, Shigeru Ogura, et al, "Application of Flame Gunning to Hot Fettling of BOF", Kawasaki steel technical report, No.7 (March 1983).
9. Refractories Hand Book, The technical Association of Refractories, Japan, 1998, pp 256–261.

Index

S. Biswas, D. Sarkar, *Introduction to Refractories for Iron- and Steelmaking*,
https://doi.org/10.1007/978-3-030-43807-4

The manufacturer's authorised representative in the EU is Springer Nature Customer Service Centre GmbH, Europaplatz 3, 69115 Heidelberg, Germany. If you have any concerns regarding our products, please contact ProductSafety@springernature.com

Printed and bound by CPI Group (UK) Ltd, Croydon, CR0 4YY
23/09/2026
02221471-0004